Subcellular Biochemistry

Volume 71

The book series SUBCELLULAR BIOCHEMISTRY is a renowned and well recognized forum for disseminating advances of emerging topics in Cell Biology and related subjects. All volumes are edited by established scientists and the individual chapters are written by experts on the relevant topic. The individual chapters of each volume are fully citable and indexed in Medline/Pubmed to ensure maximum visibility of the work.

More information about this series at http://www.springer.com/series/6515

E. Anibal Disalvo
Editor

Membrane Hydration

The Role of Water in the Structure and Function of Biological Membranes

Springer

Editor
E. Anibal Disalvo
Laboratorio de Biointerfases y Sistemas Biomimeticos
Centro de Investigacion y Transferencia de Santiago del Estero
Universidad Nacional de Santiago del Estero-Consejo Nacional de Investigaciones Científicas y Técnicas
Santiago del Estero, Argentina

ISSN 0306-0225
Subcellular Biochemistry
ISBN 978-3-319-19059-4 ISBN 978-3-319-19060-0 (eBook)
DOI 10.1007/978-3-319-19060-0

Library of Congress Control Number: 2015951200

Springer Cham Heidelberg New York Dordrecht London

Printed on acid-free paper

Springer International Publishing AG Switzerland is part of Springer Science+Business Media (www.springer.com)

SANTIAGO DEL
ESTERO,
ARGENTINA

Water, water, everywhere,
And all the boards did shrink;
Water, water, everywhere,
Nor any drop to drink.

The Rime of the Ancient Mariner
Samuel Taylor Coleridge (1797)

Preface

Biological membranes are unique material in terms of surface and mechanical properties due to its contact with water, and nowadays important attempts to mimic their properties in the search of biotechnological inputs in human health, food industry, crop, and energy have been developed. Thus, hydration in membranes gets new insights from the prospect of nanosystems.

Hydration is an emerging subject in the field of material sciences. In particular, in biological systems water is organized in proteins and membranes. In this last case, the amount of water average is no more than 20–25 molecules per lipid. If it is considered that it may be distributed in discrete sites of different chemical features, water environments are restricted to less than ten water molecules. With this criterion, studies of water in biological systems in general, and in membranes in particular, are within the scope of nanosciences.

This book is an effort to enlighten the importance of this subject in relation to biology and biophysics. This project has been possible due to the enthusiasm of all the authors of the chapters to which I want to particularly thank for their work.

I also like to recognize those who for different reasons could not contribute to this edition and hope that they may enrich future ones.

Among the authors I am particularly grateful to Zoran Arsov, Stephanie Tristram-Nagle, Helge Pfeiffer, and Gustavo Appignanessi for their help, advice, and comments along the preparation of the manuscripts.

The ideas about membranes and water have been built along years, and therefore, it is the product of what I have been able to collect from excellent teachers, colleagues, and friends.

For this reason I want to specially express my gratitude and recognition to Dr. Jorge Arvia and Dr. Hector Videla from INIFTA (Universidad Nacional de La Plata) with whom I began my feeling for research in biophysical chemistry and bioelectrochemistry during my PhD thesis, to Prof Raul Grigera who showed me the importance of water, and to Prof Hans de Gier from Utrecht University who introduced me in the world of lipids as a postdoc.

Also, I learned thermodynamics with Ernesto Timmermann and lipid monolayers with Bruno Maggio. With all of them I had exciting and vigorous discussions.

A special place is reserved for Sid Simon and Tom McIntosh with whom I spend my sabbatical enjoying science, tennis, and drinks.

Finally I want also to thank all the students who went through my laboratory in the Universidad Nacional de La Plata, University of Buenos Aires, University of Tucuman, and, in these last years, the University of Santiago del Estero, because along their works, their doubts, their achievements, their challenges, and their irreverences I, found new routes to pursue in this research.

I hope that this book will encourage them and the next ones in the fascinating field of biophysics of biological membranes.

Santiago del Estero, Argentina E. Anibal Disalvo

Acknowledgments

The editor is grateful to

Prof. Gerardo Fidelio (Universidad Nacional de Cordoba),
Prof Laura Bakas (Universidad Nacional de La Plata),
Dr. Maria de los Angeles Frías (CITSE, Universidad Nacional de Santiago del Estero-CONICET (RA)), and
Prof. Daniel Rodrigues (Universidad Nacional del Litoral)

for their contributions in the revision of the chapters.

E. Anibal Disalvo
(Editor)

Contents

Contributors

S.A. Accordino Departamento de Química and INQUISUR-UNS-CONICET, Universidad Nacional del Sur, Bahía Blanca, Argentina

Laureano M. Alarcón Departamento de Química and INQUISUR-UNS-CONICET, Universidad Nacional del Sur, Bahía Blanca, Argentina

Gustavo A. Appignanesi Departamento de Química and INQUISUR-UNS-CONICET, Universidad Nacional del Sur, Bahía Blanca, Argentina

Zoran Arsov Laboratory of Biophysics, Department of Solid State Physics, "Jozef Stefan" Institute, Ljubljana, Slovenia

Luis A. Bagatolli Membrane Biophysics and Biophotonics Group/MEMPHYS-Center for Biomembrane Physics, Department of Biochemistry and Molecular Biology, University of Southern Denmark, Odense M, Denmark

John H. Crowe Department of Molecular and Cellular Biology, University of California, Davis, CA, USA

Srinivasan Damodaran Department of Food Science, University of Wisconsin-Madison, Madison, WI, USA

E. Anibal Disalvo Laboratorio de Biointerfases y Sistemas Biomimeticos, Centro de Investigacion y Transferencia de Santiago del Estero, Universidad Nacional de Santiago del Estero-Consejo Nacional de Investigaciones Científicas y Técnicas, Santiago del Estero, Argentina

Axel Hollmann Laboratorio de Biointerfases y Sistemas Biomimeticos, Centro de Investigacion y Transferencia de Santiago del Estero, Universidad Nacional de Santiago del Estero-Consejo Nacional de Investigaciones Científicas y Técnicas, Santiago del Estero, Argentina

John Katsaras Biology & Soft Matter and Biosciences Division, Oak Ridge National Laboratory, Oak Ridge, TN, USA

Zoltan Kovacs Biomeasurement Technology Laboratory, Kobe University, Kobe, Japan

Department of Physics and Control, Corvinus University of Budapest, Budapest, Hungary

Yosuke Kubota Biomeasurement Technology Laboratory, Kobe University, Kobe, Japan

M. Florencia Martini Instituto de Química y Metabolismo del Fármaco, IQUIMEFA UBA-CONICET, Facultad de Farmacia y Bioquímica, Universidad de Buenos Aires, Buenos Aires, Argentina

J.M. Montes de Oca Departamento de Química and INQUISUR-UNS-CONICET, Universidad Nacional del Sur, Bahía Blanca, Argentina

Marcela A. Morini Departamento de Química and INQUISUR-UNS-CONICET, Universidad Nacional del Sur, Bahía Blanca, Argentina

Jonathan D. Nickels Joint Institute for Neutron Sciences, Oak Ridge National Laboratory, Oak Ridge, TN, USA

Department of Physics, University of Tennessee, Knoxville, Knoxville, TN, USA

Viviana I. Pedroni Departamento de Química and INQUISUR-UNS-CONICET, Universidad Nacional del Sur, Bahía Blanca, Argentina

Helge Pfeiffer Department of Metallurgy and Materials Engineering (MTM), University of Leuven (KU Leuven), Leuven, Belgium

J.A. Rodríguez Fris Departamento de Química and INQUISUR-UNS-CONICET, Universidad Nacional del Sur, Bahía Blanca, Argentina

M. Belén Sierra Departamento de Química and INQUISUR-UNS-CONICET, Universidad Nacional del Sur, Bahía Blanca, Argentina

Stephanie Tristram-Nagle Biological Physics Group, Physics Department, Carnegie Mellon University, Pittsburgh, PA, USA

Roumiana Tsenkova Biomeasurement Technology Laboratory, Kobe University, Kobe, Japan

Abbreviations

$\|F(q_z)\|$	Form factor
18:0:22:5PC	Stearoyldocosapentaenoyl-phosphatidylcholine
18:0-22:6PC	Stearoyldocosahexaenoylphosphatidylcholine
$2D_C$	Hydrocarbon thickness
AFM	Atomic force microscopy
A_L	Area/lipid
B	Bulk modulus
D, D-space	X-ray lamellar D-spacing
d, d-space	X-ray wide-angle chain spacing
D_B	Bilayer thickness
D_H'	Headgroup thickness
DHPC	Dihexadecanoyl-phosphatidylcholine
diC22:1PC	Dierucoylphosphatidylcholine
diphytanoylPC	Diphytanoyl-phosphatidylcholine
DLPC	Dilauroylphosphatidylcholine
DLPE	Dilauroylphosphatidylethanolamine
DLPG	Dilauroylphosphatidylglycerol
DMPC	Dimyristoylphosphatidylcholine
DMPE	Dimyristoylphosphatidylethanolamine
DMPG	Dimyristoylphosphatidylglycerol
DMPS	Dimyristoylphosphatidylserine
DOPC	Dioleoylphosphatidylcholine
DOPG	Dioleoylphosphatidylglycerol
DOPS	Dioleoylphosphatidylserine
DPhPC	Diphytanoylphosphatidylcholine
DPPC	Dipalmitoylphosphatidylcholine
DSC	Differential scanning calorimetry
DSPC	Distearoylphosphatidylcholine
EggPC	Egg phosphatidylcholine
EPR	Electron spin resonance
FTIR	Fourier transform infrared resonance

$I(q_z)$	X-ray intensity
interdig.	Interdigitated
K_C	Bending modulus
MD simulation	Molecular dynamics simulation
MLVs	Multilamellar vesicles
NIR	Near-infrared
NMR	Nuclear magnetic resonance
n_W	Number of waters/lipid
n_W'	Steric number of waters/lipid
OSM	Osmotic stress method
PCA	Principal component analysis
PLS	Partial least squares
POPC	Palmitoyloleoylphosphatidylcholine
POPG	Palmitoyloleoylphosphatidylglycerol
PPM	Piezotropic phase transitions method
PrP	Prion protein
SFA	Surface force apparatus
SIMCA	Soft independent modeling of class analogy
SOPC	Stearoyloleoyl-phosphatidylcholine
SOPG	Stearoyloleoylphosphatidylglycerol
Tm	Main transition melting temperature
TMCL	Tetramyristoylcardiolipin
TPM	Thermotropic phase transition method
ULVs	Unilamellar vesicles
u_n	Vertical displacement
V_L	Molecular volume/lipid
V_W	Molecular volume/water
WAMACS	Water matrix coordinates
WASP	Water spectral pattern
η	Fluctuation parameter

Chapter 1
Membrane Hydration: A Hint to a New Model for Biomembranes

E. Anibal Disalvo

Abstract The classical view of a biological membrane is based on the Singer-Nicholson mosaic fluid model in which the lipid bilayer is the structural backbone. Under this paradigm, many studies of biological processes such as, permeability, active transport, enzyme activity and adhesion and fusion processes have been rationalized considering the lipid membrane as a low dielectric slab of hydrocarbon chains with polar head groups exposed to water at each side in which oil/water partition prevails. In spite of several analyses and evidence available in relation to membrane hydration, water is not taken into account as a functional component. For this purpose, new insights in the water organization in restricted environments and the thermodynamical and mechanical properties emerging from them are specifically analysed and correlated.

This chapter summarizes the progress of the studies of water in membranes along the book in order to give a more realistic structural and dynamical picture accounting for the membrane functional properties.

Keywords Water penetration • Interphases • Hydration water • Confined water • Complex systems • Membrane models

1.1 Introduction

Cell membranes are the physical limit between the living and the non-living world. If the interior of the cell is considered the actual place of reproduction, compartmentalization gives the appropriate environment to biological metabolic reactions and the highly selective kinetic barrier properties of the membrane regulate the exchange of matter and energy with the surroundings (Yeagle 2004).

Biomembranes are self organized assemblies of lipids and proteins, The classical view of a biological membrane is based on the Singer-Nicholson mosaic fluid model

E.A. Disalvo (✉)
Laboratorio de Biointerfases y Sistemas Biomimeticos, Centro de Investigacion y Transferencia de Santiago del Estero, Universidad Nacional de Santiago del Estero-Consejo Nacional de Investigaciones Científicas y Técnicas, 4200 Santiago del Estero, Argentina
e-mail: disalvoanibal@yahoo.com.ar

E.A. Disalvo (ed.), *Membrane Hydration*, Subcellular Biochemistry 71,
DOI 10.1007/978-3-319-19060-0_1

in which the lipid bilayer is the structural backbone (Singer and Nicolson 1972). In this context, the lipid membrane was usually described by a low dielectric slab of hydrocarbon chains with polar head groups exposed to water at each side in which oil/water partition prevailed (Overton 1889; Al-Awqati 1999).

Many studies of biological processes such as, permeability, active transport of ions, enzyme activity and adhesion and fusion have been rationalized under this paradigmatic model of sticks and balls in which different proteinaceous particles were inserted to explain biological activity that the single non polar slab could not explain.

In the last years, several critical reviews have introduced changes in the proposal of Singer and Nicholson by including lipid mixtures, lipid heterogeneities, rafts, local curvatures and protein-lipid interactions (Israelachvili 1977; Ti Tien and Ottova 2001; Bagatolli et al. 2010; Goñi 2014). However, in spite of several analyses and evidence available in relation to membrane hydration (Jendrasiak and Hasty 1974; Jendradiak et al. 1996; Israelachvili and Wennerström 1996) the principal feature of the classical view still remained: water is not taken into account as a functional component. To put emphasis in the importance of the structural/thermodynamic properties of water relevant to membrane response, a thorough revision is imperative.

As said elsewhere, water has been for biologists as the canvas for the painters. All is stabilized by water but consideration of water incorporated in the final structure and the dynamical (thermodynamic and mechanical) properties it imposes to the ensemble is far from being a routine.

As pointed out by Damodaran in Chap. 10, "although it is well recognized that structural evolution of proteins and formation of lipid vesicles and cell membranes are simple manifestations of the hydrophobic effect, i.e., a consequence of energetics of interaction of water with the apolar moieties of these molecules, the possibility of water playing a vital role in the very functioning of these biological systems is often overlooked".

Moreover, a non-negligible number of works deals with biological mechanisms as they would occur in an anhydrous state or ignoring water in structure and kinetics processes.

This book is an attempt to organize in a rationale way the progress of the studies of water in membranes that give place to different biological phenomena.

For this purpose, new insights in the water organization in restricted environments and the thermodynamical and mechanical properties emerging from them are specifically analysed and correlated, in order to give a more realistic structural and dynamical picture accounting for the biological functional properties.

The main points that deserve special attention are:

- water penetration and distribution along the lipid molecules
- water mediation in the interaction between lipid membranes (adhesion, fusion) or a lipid membrane and a protein (lipid-protein interaction) through water structure (hydration force, hydration layers)
- surface pressure and its implications in surface water activity.
- water mediation in enzyme activity

1.2 Water Penetration and Distribution Along the Lipid Molecules

In terms of Tristram-Nagle (Chap. 2); “why should we be concerned about the number of waters per lipid?”. The reason given is that if lipids become too dehydrated, their physical and structural properties change.

Lipid membranes are not formed in the absence of water. With this premise, life is not possible without water (Crowe, Chap. 11). This view justifies the role of water for the thermodynamic stability and assembly of lipids in membranes. However, this is just a beginning and there are several additional points to consider in terms of the bilayer as a complex material system. Some compounds may stabilize biological structures in the absence of water mimicking its hydrogen bonding patterns (Chap. 11). However, for a biological system to work (grow and reproduce) water as a liquid should be present. Thus, the peculiar physical chemical properties of water as a liquid should be considered (Chaplin 1999).

In the seventies, some results in permeability gave place to think about an active role of water in membrane processes although in a naïve and indirect way. Permeation of polar solutes through lipid bilayers, such as glycerol and erythritol among others, was explained by the rupture of H bonds of the solute with the water phase before partitioning into the membrane. These papers (de Gier et al. 1971; McElhaney et al. 1973) explained the process considering that the lipid membrane was composed by a hydrocarbon phase in which only non polar solutes could dissolve. However, at this point no consideration of the role of water in the membrane structure was done, i.e. the permeability barrier of the lipid membrane was imagined as a low dielectric slab, based on Overton rule (Overton 1889; White 1976) which rejected polar and charged compounds and readily dissolved non polar solutes. That is why the authors proposed a dehydration and a particular conformation of the permeant polyols to explain partitioning into the non polar phase.

Further analysis of the permeation of erythritol and glycerol (Disalvo and de Gier 1983) demonstrated that the thickness of the lipid barrier was not only composed of the bimolecular lipid leaflet but, also of a slab of 10 Å thick at each side due to non-solvent water. Equilibrium dialysis experiments used to measure excluded volumes for the non-electrolyte permeant [U-^{14}C] erythritol in lipid bilayer systems indicated a measurable amount of water associated with the lipid membrane similar to that found by calorimetric measurements (22–24 water molecules per lipid in liquid crystalline membranes of phosphatidylcholine) (Chapmann 1971; Chapmann et al. 1974).

Further studies located this region, named as the interphase, between two ideal planes: one at the carbonyl level that defines the interface between the hydrocarbon region and the polar head group region and the other at the external plane along the hydrated phosphates of the phospholipids (McIntosh et al. 1989).

The membrane system was then described as a composite element consisting of the lipid bilayer and adjacent water layers on both sides from which permeant is excluded showing that the water layers contribute to the permeability barrier.

The notion of excluded volume of lipids and water for solutes, implied that membrane matrix was able to offer empty spaces into which molecules of different sizes were able to insert in order to cross it. Although in principle this was treated in terms of geometry, the thermodynamic counterpart was appearing. In this regard, analysis of the permeation of non electrolytes showed that permeability activation energies were consistent with the hydration state of the liposomes. When liposomes were previously shrunken in a hypertonic medium, the activation energy value was considerably lower in comparison with that obtained when liposomes were maintained in an isotonic state (Disalvo 1986 and references therein). This indicated that the main barrier for permeation kinetics was affected by the water/lipid ratio n_W.

The inclusion of water in the lipid bilayer does not only affect the thickness, influencing the permeability coefficient derived from the first Fick law, but also the phase transition temperature. As pointed out in Chap. 2, when the hydration level in phosphatidylcholines drops below 20 wt.% water ($n_W = 12$), the measured T_m (main chain melting) increased dramatically.

In addition to thickness, increase in n_W generally increases the area per lipid, another parameter affecting permeability, both in the gel and fluid phases. Small Angle X-Ray Scattering (SAXS) indicated that the phase transition of lipid membranes leads to a membrane thickness decrease and an area per lipid increase (Ipsen et al. 1990). Concomitantly, in this condition, the increase in the trans-gauche isomers in the lipid acyl chains produces a higher disorder state identified as "kinks", where water molecules can fit (Träuble 1971).

The larger lipid areas require more water, according to Tristram-Nagle (Chap. 2). On the other way round, water exchange between lipids in the membrane and the bulk can occur with changes in the area per lipid. This point is essential to the approach developed in Chaps. 9 and 10 where changes in monolayer surface pressures are discussed in terms of water activity.

The headgroup structural water, shown in Fig. 2.7 of Chap. 2, is considered to be a lower limit to measurements of tightly bound water. This water includes the first six water molecules bound to the phosphate and additional six in carbonyl groups (Goñi and Arrondo 1986; Disalvo et al. 2008; Nagle and Tristram-Nagle 2000). Nowadays, water beyond this limit appears to play a role in the process of several peptide and amino acid insertion and enzymatic reactions (Chaps. 9 and 10).

The interpretation of the permeability processes described above maintained the paradigm that the bilayer was totally impermeable to water. But this could not be sustained longer after the finding of Van Zoelen et al. (1976) based on the Kedem and Katchalsky Thermodynamic of Irreversible Process (TIP) formalism. Kedem and Katchalsky (1958) demonstrated the simultaneous non-independent permeation of non-electrolytes and water, i.e. water permeation affects the permeation of the non-electrolytes and vice versa. According to de Gier (1989), the exchange of water molecules across the membrane under equilibrium conditions is disturbed by the solute, producing a net additional flux of water molecules coupled with it. This assumes that non-electrolyte is acting as a carrier or inducing an additional flow of water into the membrane. In thermodynamic language, this can be rephrased by saying that the diffusion of the non electrolyte into the membrane phase produces a change in the water activity driving a water flux to compensate it. This can be explained considering a chemical equilibrium when lipids are dispersed in an excess of water. Membrane structure is stabilized with a given amount of water forming part of its structure. When a solute enters the lipid matrix, probably dissolving in water between the acyl chains occupying the kinks, a difference in the chemical potential of water is produced and a water influx occurs to reach again the equilibrium. This process is similar to osmosis and the swelling of the membrane matrix may produce water paths expanding the membrane. This is congruent with the leak of ions contained in the liposome interior during hypotonic swelling. These arguments not only introduce water as a structural component in the membrane phase but as an unavoidable element in the thermodynamic response.

The formalism of Thermodynamics of Irreversible processes (TIP) denotes that water not only associates with lipid head groups (Griffith et al. 1974), but also with many of the other membrane's functional groups – in addition to occupying free volume – and considers the possibility that water may be found also in the hydrocarbon region in a limited amount. According to Träuble (1971) and Haines and Liebovitch (1995) the movement of water molecules across membranes can be produced as a consequence of the thermal fluctuations of the conformational isomers in the hydrocarbon chains of the membrane lipids resulting in the formation of so-called "kinks" (Trauble 1971 and, Flory 1969).

Details of the correlation between chain conformation and water states have been published recently (Disalvo et al. 2013).

The need to introduce the interphase region in a model to account many of the molecular details regarding ions, lipids, and more importantly, the solvent permeation is anticipated by Nickels and Katsaras in Chap. 3. The order parameter defined by the geometrical and dynamical restrictions imposed to water molecules by the bilayer, bring the presence of hydrogen bond partners, hydrogen bond lifetimes, and dynamical retardation factors that are analysed later on Chap. 7 by Alarcon et al.

According to Nickels and Katsaras, the rate controlling step is the head group region, which is being modulated by the area of the lipid head group relative to the area per lipid (Mathai et al. 2008). This important insight arises from a deep structural understanding of lipid hydration mentioned in this chapter. From those studies, it is clear that the area per lipid scales with the number of water molecules associated with the head group region. It is natural then, to take into account this information to model water permittivity following an improved criterion of the solubility-diffusion but considering a more complex structure than that corresponding to a single non polar slab as previously considered.

In this regard it is worthwhile to recall that some permeation processes has been interpreted with an approach considering pore formation in contraposition to the solubility/diffusion one. This approach describes water transport considering the line tension, γ, which stabilizes the bilayer, and the surface tension, Γ, which stabilizes a pore of radius r. These quantities, line tension and surface tension as parameters relating to bilayer stability, imply that pores appear more frequently in unstable bilayers. The classical view of a pore comes from the idea that water molecules pave paths across the membrane as a stationary phase and mobile waters displace as in bulk water in the center of the pore. This view can only be sustained assuming large amounts of water in the pore. The radius of the pore is in terms of number of water molecules of only two to four water molecules. The meaning in this context of instability is a matter of discussion that is analyzed by Bagatolli (Chap. 5), Disalvo (Chap. 9) and Damodaran (Chap. 10).

1.3 Water in Membrane Structure and the Interfacial Properties

To relate surface tension and contraction/expansion phenomena exclusively to pore formation is questionable. A rigorous analysis of the thermodynamics of lipid monolayers and bilayers demonstrates that solubility/diffusion criterion also involves membrane area and surface pressure changes (Chaps. 9 and 10), which shows that classical partition of solutes is not correct in lipid phases constituted by a two molecular thick layer apposed to each other by the hydrocarbon chains. Moreover, the heterogeneous distribution of water in the lipid matrix in which different water species, in terms of the number and type of hydrogen bonds of water molecules between them and with the lipid residues, can be found, is against the notion of partition in a single hydrocarbon slab.

The expansion of the membrane by water penetration involves other properties of the lipid membrane. In this regard, the chapter by Pfeiffer (Chap. 4) brings about several important considerations of lipid membranes as a unique material.

In this Chapter, an overview on the theory of hydration forces, ranging from polarisation theory to protrusion forces, based on a selection of appropriate experimental techniques, such as X-ray diffraction, atomic force microscopy and calorimetry, is presented.

It mainly calls the attention to three important features: polarization of the lipid interphase, protrusion of the lipid molecules to the aqueous phase and water as a plasticizer element. All of them should be rigorously analyzed to have a realistic model for lipid membranes. The first one is related to water dipoles at the lipid interphase, the second with topology, in relation to head groups exposed to the water phase, the third is an important component in mechanical response.

The polarization brings about the idea that organization, mainly given by hydrogen bonding between water molecules themselves and with the different membrane groups, confers special mechanical and dynamical properties (see also Chap. 3). In this regard, when water is absorbed in solid, dehydrated phospholipids, it will act as a so-called "external plasticizer", i.e. after gradual ingress of the solvent, the mobility and deformability of the membrane strongly enhances.

However, the limiting factor that ensures global integrity of the bilayer is its mechanical properties, which prevents its dissolution in an aqueous environment. What is essential for is that the mechanical properties, may also modulate permeants and water permeation. This limiting factor arises from the entropic effect ensuring lamellar stability of lipid bilayers in water (Ben-Shaul 1995). This is therefore why solubility diffusion phenomena cannot be sustained with the classical assumption that the bilayer is an autonomous phase acting as a homogenous solvent for non polar solutes, only.

Water acts like a spacer when entering the headgroup region of lipids forcing free volume in the hydrocarbon chains leading to liquid phases. This criterion is used to explain why some molecules such as trehalose preserve membrane structure upon dehydration (Chap. 11).

Hydration enables an enhancement of the configurational space for additional degrees of freedom in water populations and most probably also in lipids (Ge and Freed 2003).

The debate on the validity of polarisation or protrusion models is not yet decided (Gordeliy 1996) but it seems that they are not independent, since protrusion changes polarity of the interphase by showing different groups to water. Moreover, the arguments discussed by Bagatolli in Chap. 5 suggest a distribution of different sites (or cavities), in the bilayer surface in which the fluorophore molecules can reside (Parasassi et al. 1997). Fluorophores are able to detect the local dielectric properties in which they are inserted. Thus, providing that the location in the membrane is known, they may be used to infer polar and non polar environments and to relate them to the presence of water. With this methodology, sites are characterized by a different number of dynamically restricted water molecules (Parasassi and Gratton 1995).

The average number of water molecules at the location of the LAURDAN fluorescent moiety 10 Å from the center of the bilayer (Antollini and Barrantes 1998) was estimated to be no more than two or three (Parasassi et al. 1997).

This few number of water molecules probably relates with those involved in the pore forming approach (Chap. 3). It is obvious to see that this number cannot physically define a pore, in the classical way described above, but rather defects of packing of nanomolecular dimensions, either in the acyl chain region or in the polar head groups, into which water may be organized. Water present in defects (Nagle and Tristram-Nagle 2000), illustrated by gaps in lipid film layers (see Fig. 2.4 in Chap. 2), form as irregularities in lamellar structure and because of imperfect bilayer alignment. Also protrusion may be considered a defect in which water polarization is changed.

The heterogeneity reported by these probes inserted in natural membrane systems could be interpreted in terms of distinct "structured water domains" (Almaleck et al. 2013; Heimburg 2010), not exclusively related to lipid packing, but to the combined ability of different membrane constituents to generate areas of different water content and dynamics. The dimension of these areas are of at most two to five water molecules as discussed in Chap. 9.

These considerations open the question formulated by Arsov in Chap. 6: *what about water molecules farther away from the membrane?*. This would be water beyond the water hydration hard core shell immobilized by polar head groups described Tristram Nagle,

How far does the influence of the membrane surface propagates into the water and what are the manifestations of such propagation? (Berkowitz and Vácha 2012)

These questions suggest that water beyond the hydration shell of the headgroups or near groups less abide for water (non polar residues) of the lipid structure is labile and could be perturbed easily by solutes in the aqueous phase. The properties of this kind of water could be relevant for membrane response in terms of the concept of "responding structure" coined by Sparr and Wennerström (2001). In this regard, the impact of hydration beyond the hydration water on the function of biomembranes should be discussed in terms of the lability of the solvent structure facing membrane surfaces of different polarities (i.e. polar or non polar groups). In thermodynamic terms, the lability is related to excess free energy, (surface tension) that is the driving force for insertion of compounds from the water phase into the membrane.

Results obtained by Arsov in Chap. 6 by attenuated total reflection Fourier transform infrared (ATR-FTIR) spectroscopy show that perturbation extends through several hydration layers beyond the first hydration shell. Consequently, water in confined membrane environments can be different than in the bulk.

In this chapter, Arsov defines different water species such as free water (bulk-like, far water), perturbed water (intermediate, freezable interlamellar water), and interfacial water (bound, buried, neighboring, non-freezable interlamellar water) (Kiselev et al. 1999; Kodama et al. 2001; Murzyn et al. 2006; Pinnick et al. 2010; Debnath et al. 2010). Such classification has been introduced very early also for hydration of proteins and peptides (Kuntz and Kauzmann 1974) and can be based on different attributes, such as water distance from the bilayer, water hydrogen bonding (HB) characteristics, as well as dynamic and thermodynamic properties.

The interfacial water is represented by water molecules that directly interact with lipids or reside in the interface region, while the free water denotes putative bulk-like molecules. The perturbed water corresponds to water in the transition region whose properties are still influenced by the presence of lipid membranes. It is concluded that the water structure is perturbed throughout the whole interlamellar space with thickness of about 1.8 nm in DMPC (Nagle and Tristram-Nagle 2000).

With the venue of computational calculations, water has been considered as a component in terms of stability, and different models for water structure (specifically the intermolecular interactions by hydrogen bondings) has been used (Berkowitz et al. 2006; Berkowitz and Vácha 2012; Bhide and Berkowitz 2005).

In this context, recently, the model visualizing the membrane as a low dielectric slab has been challenged in order to explain polar amino acid's penetration. Contrary to prediction of partition in non polar phases (White 1976; Wimley and White 1996; Preston Moon and Fleming 2011) charged aminoacids such as arginine are stabilized in the bilayer, which has been explained by means of molecular dynamics in terms of water pockets (Mac Callum et al. 2008; Herrera et al. 2012). In this regard, it has been recognized that water may penetrate the lipid bilayer reaching the region of the carbonyl groups (Simon and McIntosh 1986; Sovago et al. 2009; Disalvo and Frias 2013).

Alarcon et al. in Chap. 7 makes an interesting and novel evaluation of the water properties near different materials in comparison to lipid membranes. They investigate the properties of water buried in phospholipid membranes facing the lipid head groups and apolar alkyl chains (Malaspina et al. 2009).

Preliminary studies on membrane hydration consider that the effects of these interactions depend on the local environment. Results on the hydration and water penetration in phospholipid membranes indicate spatial distribution of water molecules around the different groups of the lipid molecules of the bilayer. The tendency of the water molecules to sacrifice the lowest hydrogen bond (HB) coordination as possible at extended interfaces reveals that the first hydration layers are highly oriented, in some situations even resembling the structure of hexagonal ice. A similar trend to maximize the number of HBs is shown to hold in cavity filling, with small subnanometric hydrophobic cavities remaining empty while larger cavities display an alternation of filled and dry states with a significant inner HB network.

An important conclusion relevant to biological phenomena is that the combination of different non-covalent interactions can produce clearly non-additive effects.

1.4 Water Beyond the Interphase: Water in Complex Systems

A more ambitious picture is proposed by Tsenkova in Chap. 8. She provides a framework for understanding changes in water giving a holistic description related to system functionality. Aquaphotomics, in addition to the other –omics disciplines, is based on the presumption that all the components of the system shape up the water matrix and would allow to describe the coupling between bio-chemical-physical perturbations and related changes in hydration structures. Water molecular conformations, for example dimers, trimers, solvation shells, etc., are known to contribute very specifically to its Near Infrared (NIR) spectrum. Direct evidence by means of FTIR spectroscopy shows that water band profiles are changed whether lipids are in the solid state, in the gel state after heating and cooling across the phase transition, or in the fluid state (Disalvo and Frias 2013). The different bands found in each case were assigned to different H-bonded water populations in agreement with the exposure of carbonyl groups.

1.5 Water and Surface Pressure: Its Implications in Membrane-Membrane and Membrane-Peptide Interactions

As pointed out in Chap. 2, the headgroup structural water, is considered to be a lower limit to measurements of tightly bound water. In monolayers, (Chap. 9) this limit is found at pressures corresponding to 40–45 $mN{\bullet}m^{-1}$, depending of the lipid head group. This system allows to investigate further the hypothesis put forward by van Zoelen in liposome and to link permeability processes in bilayers with properties in the interphase region. The limit of packing given by the contact of the head group hydration water is defined as the cut off (critical) surface pressure, at which no effect of aminoacids, peptides or proteins on monolayer surface pressure is found (Chap. 9).

The perturbation of the surface pressure is produced when the area per lipid is just above 4 % larger than that corresponding to the hydration shell of the phospholipid head groups found in the cut –off, and therefore at lower surface pressures with respect to the cut off. The possibility that this area increase is related with the increase in water beyond the hydration shell, as anticipated by Bagatolli (Chap. 5) and studied by Arsov (Chap. 6) is discussed. The change in interfacial water activity is related with the surface pressure according to the Defay-Prigogine interphase model, which considers the interphase region as a bidimensional solution of head groups in water. As predicted by solution chemistry, the increase of surface pressure is independent of the protein nature but depends on the water surface state determined by the lipid composition.

With this background it is possible to analyse again the dynamics of different membrane processes such as solute and peptides penetration. Van Zoelen et al. (1976) claimed that in the isosmotic condition outflux and influx of water are counterbalanced by the outer solute concentration.

Liposomes and lipid vesicles swell when water penetrates until the chemical potentials of the solutions are equal inside and outside the structure (van Zoelen et al. 1976). While, the chemical potential in the outer media is given by the concentration of the components, those inside are determined by the concentration and the pressure exerted by the network of the bilayer on the solution. The viscoelastic properties of the bilayer may then contribute to the permeation processes. In the case of liposomes, the pressure can be exerted by the bilayer surrounding the internal solution. The cohesive forces maintaining the phospholipids in the structure contribute to this balance.

When permeant solutions are able to perturb the cohesive forces, more swelling would occur. At a 4 % volume increase in the swelling process of liposomes and vesicles results in a release of the liposomal content, such as ions which normally can not permeate the membrane, due to an increase in surface area (Evans and Skalak 1980). Thus, bilayer selectivity will vary according to the value of the reflection coefficient in the swelled structure a point also discussed by Nickels and Katsaras in Chap. 3.

The leakage produced at the critical volume increase is somehow congruent with the perturbation of the surface pressure produced when the area per lipid is above just 4 % larger than that corresponding to the hydration shell of the phospholipid head groups found in the cut –off.

Thus, variations of the membrane elasticity may be related to surface tension effects as inspected in monolayers and leakage cannot be necessary the result of membrane rupture. Instead, defects or fluctuations in the water lipid ratio along the membrane structure may explain the breakdown of the permeability barrier to ions (Deamer and Volkov 1995).

1.6 Interphase Water and Enzyme Activity

The studies by Damodaran in Chap. 10 extend further the relationship of monolayers with water activity, as a model experimental system for enzyme regulation.

A hypothesis is developed, which espouses that cells control activities of membrane-bound enzymes through manipulation of the thermodynamic activity of water in the lipid-water interfacial region. The hypothesis is based on the fact that the surface pressure of a lipid monolayer is a direct measure of the thermodynamic activity of water at the lipid-water interface. Accordingly, the surface pressure-dependent activation or inactivation of interfacial enzymes is directly related to changes in the thermodynamic activity of interfacial water.

Given that the physicochemical and structural properties of these enzymes are very different, it is inconceivable that all these enzymes would attain the catalytically active optimal conformation at a lateral mechanical force corresponding to a surface pressure of about 20 $mN{\bullet}m^{-1}$. This raises a very fundamental questions: Why at exactly this value? Is surface pressure really related to a *mechanical* lateral pressure exerted by the lipid layer or does it represent some other thermodynamic state or property of the interface that is essential for enzymes to adopt the catalytically most optimal structure?. Alternatively broader question are raised in terms of the evolutionary reasons for biology to choose the lipid-water boundary as the site for performing and/or controlling crucial biological reactions.

1.7 Life Without Water?

The question implicit in the origin of anhydrobiosis is if life can be preserved without water (Crowe, Chap. 11). It is known from a long time ago that one of the chief sites of damage in dehydration is the membrane, and some of them may be obviated by the production of certain sugars, particularly trehalose.

As discussed by Tristram-Nagle in Chap. 2, the main transition temperature of lipids (T_m) increased dramatically when the hydration level dropped below 20 wt.% water ($n_W = 12$). However, the lipid is maintained in the liquid-crystalline phase in the dry state if it is dried in the presence of trehalose.

Modifications of the gel state of hydrated phospholipids by trehalose can only be achieved if a drastic dehydration is performed in the presence of the sugar (Viera et al. 1993). The water activity in dimyristoylphosphatidylcholine (DMPC) decreases by 60 % when the lipid is dehydrated in the presence of trehalose concentrations higher than 0.02 M. Fourier transform infrared spectroscopy (FTIR) in these conditions denoted that trehalose binds to the carbonyl groups replacing 11 of 14 water molecules per lipid molecule. About four are displaced by changes in the water activity of the bulk solution, and seven by specific interactions with the phospholipids. In this last case, at least two of them are linked to the carbonyl group. This appears to be the cause of the decrease in the dipole potential of lipid monolayers spread on an air/water interface from 480 mV in pure water to 425 mV in 0.1 M trehalose (Luzardo et al. 2000).

Molecular dynamics simulations showed that trehalose binds to the phospholipid headgroups with its main axis parallel to the membrane normal. It establishes hydrogen bonds with the carbonyl and phosphate groups and replaces water molecules from the lipid headgroup. Notably, the number of hydrogen bonds (HBs) that the membrane made with its environment was conserved after trehalose binding. The HBs between lipid and trehalose have a longer lifetime than those established between lipid and water. The binding of the sugar does not produce changes either in the lipid area or in the lipid order parameter. The effect of trehalose on the dipole potential is in agreement with experimental results. The contribution of the different components to the membrane dipole potential was analyzed. It was observed that the binding of trehalose produces changes in the different components and the sugar itself contributes to the surface potential due to the polarization of its hydroxyl in the interface (Villarreal et al. 2004).

1.8 Contribution for a New Model for Lipid Membranes

Taken together, the approaches made by Crowe, Tsenkova and Damodaran extend the presence of water from the membrane interphase to the whole system. The details in the structure and properties of the membrane interphase is not constrained to a few nanometers near the lipids. Moreover, cells are now considered as crowded systems in which proximity between cell structure is linked by a few water layers. Thus, as an extension, the properties in the interphases can be visualized as the common interconnecting media. This new insight gives entrance to a holistic view of cells as entire complex systems in which membrane processes are essentials.

The traditional view of a membrane is an autonomous, rigid nonpolar slab sandwiched by bulk water phases implying isolation from the immediate surroundings. To fit membrane structure to the concept of complex systems, membrane should be considered as a non-autonomous phase. This means that the membrane (i.e. the lipid bilayer) is a phase whose properties depend on the phase which it is in contact with (i.e. water). This property has been denoted as responsive membranes by Sparr and Wennestrom (2001). This definition implies that the phenomena occurring at membrane level occurs with local or general changes in the membrane structure, that is, a response of the membrane to some component of the adjacent media, reflected in the thermodynamics and the kinetics of the protein–membrane interaction. The change in the membrane structure may be local with propagation to the whole structure in the plane and along the thickness with different degrees of cooperativity and synergism.

To decipher the biological phenomena in terms of membrane surface properties, the stability of the different arrays of water around the different membrane groups and its dynamical properties should be clarified. This includes namely: water as part of the membrane structure, the definition of the lipid interphases, the identification of the sites of hydration at the membrane surface; the synergism of their hydration and its modulation according to the lipid species (usually found in biological membranes in terms of head group and fatty acid chains).

Details of water location and its properties in these restricted domains are scarce. The great variety of lipid composition and the multiple combinations in mixtures acquires relevance and functional meaning. As far as the properties of water microenvironments may be changed by the protrusion of the different polar moieties into the water phase, lipid species may generate new different water species each of them identified by the type of interaction they may have with its neighbor water molecules and/or chemical groups of the lipids. Thus, at this point, lipidomics approaches to aquaomics. In other words, lipid species would give an in-print on water with specific thermodynamic features for membrane response.

It is hoped that the discussions and ideas put forward in this book can be a stimulating start to study membrane phenomena in a more integrated approach in which water, in its different states and arrangements, appears as a common and unavoidable intermediary in biological response.

References

Al-Awqati Q (1999) One hundred years of membrane permeability: does Overton still rule? Nat Cell Biol 1(8):E201–E202

Almaleck H, Gordillo GJ, Disalvo A (2013) Water defects induced by expansion and electrical fields in DMPC and DMPE monolayers: contribution of hydration and confined water. Colloids Surf B Biointerfaces 102:871–878

Antollini SS, Barrantes FJ (1998) Disclosure of discrete sites for phospholipid and sterols at the protein-lipid interface in native acetylcholine receptor-rich membrane. Biochemistry 37:16653–16662

Bagatolli LA, Ipsen JH, Simonsen AC, Mouritsen OG (2010) An outlook on organization of lipids in membranes: searching for a realistic connection with the organization of biological membranes. Prog Lipid Res 49(4):378–389

Ben-Shaul A (1995) Molecular theory of chain packing, elasticity and lipid-protein interaction in lipid bilayers. In: Lipowsky R, Sackmann E (eds) Handbook of biological physics. Elsevier Science, North-Holland

Berkowitz ML, Vácha R (2012) Aqueous solutions at the interface with phospholipid bilayers. Acc Chem Res 45:74–82

Berkowitz ML, Bostick DL, Pandit S (2006) Aqueous solutions next to phospholipid membrane surfaces: insights from simulations. Chem Rev 106:1527–1539

Bhide SY, Berkowitz ML (2005) Structure and dynamics of water at the interface with phospholipid bilayers. J Chem Phys 123:224702

Chaplin MF (1999) A proposal for structuring of water. Biophys Chem 83:211–221

Chapman D, Urbina J, Keough K (1974) Studies of lipid-water systems using differential scanning calorimetry. J Biol Chem 249(8):2512–2521

Chapmann D (1971) Liquid crystalline properties of phospholipids and biological membranes. Symp Faraday Soc 5:163–174

de Gier J (1989) Chapter 4: Osmotic properties of liposomes. In: Benga G (ed) Water transport in biological membranes, vol I. CRC Press, Boca Raton

de Gier J, Mandersloot JG, Hupkes JV, McElhaney RNM, Van Beek NP (1971). On the mechanism of non electrolyte permeation through lipid bilayers and through biomembranes. Biochim Biophys Acta 223:610–618

Deamer DW, Volkov AG (1995) Chapter 8: Proton permeability of lipid bilayers. In: Disalvo EA, Simon SA (eds) Permeability and stability of lipid bilayers. CRC Press, Boca Raton, pp 161–178

Debnath A, Mukherjee B, Ayappa KG et al (2010) Entropy and dynamics of water in hydration layers of a bilayer. J Chem Phys 133:174704

Disalvo EA (1986) Permeation of water and polar solutes in lipid bilayers. Adv Colloid Interf Sci 29:141–170

Disalvo EA, De Gier J (1983) Contribution of aqueous interphases to the permeability barrier of lipid bilayer for non-electrolytes. Chem Phys Lipids 32:39–47

Disalvo EA, Frías MA (2013) Water state and carbonyl distribution populations in confined regions of lipid bilayers observed by FTIR spectroscopy. Langmuir 29(23):6969–6974

Disalvo EA, Lairion F, Martini F, Tymczyszyn E, Frías M, Almaleck H, Gordillo GJ (2008) Structural and functional properties of hydration and confined water in membrane interfaces. Biochim Biophys Acta 1778:2655–2670

Disalvo EA, Bouchet AM, Frias MA (2013) Connected and isolated CH populations in acyl chains and its relation to pockets of confined water in lipid membranes as observed by FTIR spectrometry. Biochim Biophys Acta 1828:1683–1689

Evans EA, Skalak R (1980) Mechanics and thermodynamics of biomembranes. CRC Press, Boca Raton, pp 67–91

Flory PJ (1969) Statistical mechanics of chain molecules. Interscience, New York

Ge MT, Freed JH (2003) Hydration, structure, and molecular interactions in the headgroup region of dioleoylphosphatidylcholine bilayers: an electron spin resonance study. Biophys J 85:4023–4040

Goñi FM (2014) The basic structure and dynamics of cell membranes: an update of the Singer-Nicolson model. Biochim Biophys Acta 1838(6):1467–1476

Goñi FM, Arrondo JLR (1986) A study of phospholipid 410 phosphate groups in model membranes by Fourier transform infrared 411 spectroscopy. Faraday Discuss Chem Soc 81:117–126

Gordeliy VI (1996) Possibility of direct experimental check up of the theory of repulsion forces between amphiphilic surfaces via neutron and X-ray diffraction. Langmuir 12:3498–3502

Gordeliy VI, Cherezov VG, Teixeira J (1996) Evidence of entropic contribution to "hydration" forces between membranes Part II. Temperature dependence of the "hydration" force: a small angle neutron scattering study. J Mol Struct 383:117–124

Griffith OH, Dehlinger PJ, Van SP (1974) Shape of the hydrophobic barrier of phospholipid bilayers (evidence for water penetration in biological membranes). J Membr Biol 15:159–192

Haines T, Liebovitch LS (1995) Chapter 6: A molecular mechanism for the transport of water across phospholipid bilayers. In: Disalvo EA, Simon SA (eds) Permeability and stability of lipid bilayers. CRC Press, Boca Raton, pp 137–160

Heimburg T (2010) Lipid ion channels (review). Biophys Chem 150(1–3):2–22

Herrera FE, Bouchet A, Lairion F, Disalvo EA, Pantano S (2012) Molecular dynamics study of the interaction of arginine with phosphatidylcholine and phosphatidylethanolamine bilayers. J Phys Chem B 116:4476–4483

Ipsen JH, Mouritsen OG, Bloom M (1990) Relationships between lipid membrane area, hydrophobic thickness, and acyl- chain orientational order. Biophys J 57:405–412

Israelachvili JN (1977) Refinement of the fluid-mosaic model of membrane structure. Biochim Biophys Acta 469:221–225

Israelachvili J, Wennerström H (1996) Role of hydration and water structure in biological and colloidal interactions. Nature 379(6562):219–225

Jendrasiak GL, Hasty JH (1974) The hydration of phospholipids. Biochim Biophys Acta 337(1):79–91

Jendrasiak GL, Smith RL, Shaw W (1996) The water adsorption characteristics of charged phospholipids. Biochim Biophys Acta 1279:63–69

Kedem O, Katchalsky A (1958) A thermodynamic analysis of the permeability of biological membranes to non-electrolytes. Biochim Biophys Acta 27:229–246

Kiselev M, Lesieur P, Kisselev A et al (1999) DMSO-induced dehydration of DPPC membranes studied by X-ray diffraction, small-angle neutron scattering, and calorimetry. J Alloys Compd 286:195–202

Kodama M, Kato H, Aoki H (2001) Comparison of differently bound molecules in the gel and subgel phases of a phospholipid bilayer system. J Therm Anal Calorim 64:219–230

Kuntz ID, Kauzmann W (1974) Hydration of proteins and polypeptides. Adv Protein Chem 28:239–345

Luzardo MC, Amalfa F, Nuñez AM, Díaz S, Biondi De Lopez AC, Disalvo EA (2000) Effect of trehalose and sucrose on the hydration and dipole potential of lipid bilayers. Biophys J 78(5):2452–2458

MacCallum L, Bennett WF, Tieleman DP (2008) Distribution of amino acids in a lipid bilayer from computer simulations. Biophys J 94:3393–3404

Malaspina DC, Rodriguez Fris JA, Appignanesi GA, Sciortino F (2009) Identifying a causal link between structure and dynamics in supercooled water. Europhys Lett 88:16003

Mathai JC, Tristram-Nagle S, Nagle JF (2008) Structural determinants of water permeability through the lipid membrane. J Gen Physiol 131(1):69–76

McElhaney RN, de Gier J, van der Neut-Kok ECM (1973) The effect of alterations in fatty acid composition and cholesterol content on the nonelectrolyte permeability of Acholeplasma laidlawii B cells and derived liposomes. Biochim Biophys Acta 298:500–512

McIntosh TJ, Simon SA, Dilger JP et al (1989) Chapter 1: Location of water-hydrocarbon interface in lipid bilayers. In: Benga G (ed) Water transport in biological membranes, vol 1. CRC Press, Boca Raton

Murzyn K, Zhao W, Karttunen M et al (2006) Dynamics of water at membrane surfaces: effect of headgroup structure. Biointerphases 1:98–105

Nagle JF, Tristram-Nagle S (2000) Structure of lipid bilayers. Biochim Biophys Acta Rev Biomembr 1469(3):159–195

Overton E (1889) Über die allgemeinen osmotischen Eigenschaften der Zelle, ihre vermutlichen Ursachen und ihre Bedeutung für die Physiologie. Vierteljahrsschr Natur-forsch Ges Zürich 44:88–135

Parasassi T, Gratton E (1995) Membrane lipid domains and dynamics as detected by LAURDAN fluorescence. J Fluoresc 5:59–69

Parasassi T, Gratton E, Yu WM, Wilson P, Levi M (1997) Two-photon fluorescence microscopy of laurdan generalized polarization domains in model and natural membranes. Biophys J 72:2413–2429

Pinnick ER, Erramilli S, Wang F (2010) Computational investigation of lipid hydration water of L α 1-palmitoyl-2-oleoyl- sn -glycero-3-phosphocholine at three hydration levels. Mol Phys 108:2027–2036

Preston Moon C, Fleming KG (2011) Side-chain hydrophobicity scale derived from transmembrane protein folding into lipid bilayers. Proc Natl Acad Sci U S A 108(25):10174–10177

Simon SA, McIntosh TJ (1986) Depth of water penetration into lipid bilayers. Methods Enzymol 127:511–521

Singer SJ, Nicolson GL (1972) The fluid mosaic model of the structure of cell membranes. Science 175(23):720–723

Sovago M, Vartiainen E, Bonn M (2009) Observation of buried water molecules in phospholipid membranes by surface sum-frequency generation spectroscopy. J Chem Phys 131:161107–161111

Sparr E, Wennerström H (2001) Responding phospholipid membranes—interplay between hydration and permeability. Biophys J 81(2):1014–1028

Ti Tien H, Ottova AL (2001) The lipid bilayer concept and its experimental realization: from soap bubbles, kitchen sink, to bilayer lipid membranes. J Membr Sci 189:83–117

Träuble H (1971) The movement of molecules across lipid membranes: a molecular theory. J Membr Biol 4(1):193–208

Van Zoelen EJJ, Blok MC, De Gier J (1976) An improved method for the description of non-electrolyte permeation through liposomes, based on irreversible thermodynamics. Biochim Biophys Acta Biomembr 436(2):301–306

Viera LI, Alonso-Romanowski S, Borovyagin V, Feliz MR, Disalvo EA (1993) Properties of gel phase lipid-trehalose bilayers upon rehydration. Biochim Biophys Acta 1145(1):157–167

Villarreal MA, Díaz SB, Disalvo EA, Montich GG (2004) Molecular dynamics simulation study of the interaction of trehalose with lipid membranes. Langmuir 20:7844–7851

White SH (1976) The lipid bilayer as a "solvent" for small hydrophobic molecules. Nature 262:421–422

Wimley C, White SH (1996) Experimentally determined hydrophobicity scale for proteins at membrane interfaces. Nat Struct Biol 3:842–848

Yeagle PL (2004) The structure of biological membranes, 2nd edn. CRC Press, Boca Raton (FL).

Chapter 2
Use of X-Ray and Neutron Scattering Methods with Volume Measurements to Determine Lipid Bilayer Structure and Number of Water Molecules/Lipid

Stephanie Tristram-Nagle

Abstract In this chapter I begin with a historical perspective of membrane models, starting in the early twentieth century. As these membrane models evolved, so did experiments to characterize the structure and water content of purified lipid bilayers. The wide-spread use of the X-ray gravimetric, or Luzzati method, is critically discussed. The main motivation of the gravimetric technique is to determine the number of water molecules/lipid, n_W, and then derive other important structural quantities, such as area/lipid, A_L. Subsequent experiments from the Nagle/Tristram-Nagle laboratory using X-ray and neutron scattering, first determine A_L and then calculate n_W, using molecular lipid V_L and water V_W volumes. This chapter describes the details of our volume experiments to carefully measure V_L. Our results also determine n_W', the steric water associated with the lipid headgroup, and how our calculated value compares to many literature values of tightly-associated headgroup water.

Keywords Waters/lipid • Hydration • Lipid bilayer • X-ray scattering • Neutron scattering

Abbreviations

DSC	Differential scanning calorimetry
NMR	Nuclear magnetic resonance
EPR	Electron spin resonance
FTIR	Fourier transform infrared resonance
Tm	Main transition melting temperature
n_W	Number of waters/lipid

S. Tristram-Nagle (✉)
Biological Physics Group, Physics Department, Carnegie Mellon University, Pittsburgh, PA 15213, USA
e-mail: stn@cmu.edu

E.A. Disalvo (ed.), *Membrane Hydration*, Subcellular Biochemistry 71,
DOI 10.1007/978-3-319-19060-0_2

n_W'	Steric number of waters/lipid
A_L	Area/lipid
V_L	Molecular volume/lipid
V_W	Molecular volume/water
MLVs	Multilamellar vesicles
ULVs	Unilamellar vesicles
D, D-space	X-ray lamellar D-spacing
d, d-space	X-ray wide-angle chain spacing
η	Fluctuation parameter
K_C	Bending modulus
B	Bulk modulus
u_n	Vertical displacement
$I(q_z)$	X-ray intensity
$\|F(q_z)\|$	Form factor
MD simulation	Molecular dynamics simulation
$2D_C$	Hydrocarbon thickness
D_B	Bilayer thickness
D_H'	Headgroup thickness
interdig.	Interdigitated
DPPC	Dipalmitoylphosphatidylcholine
DSPC	Distearoylphosphatidylcholine
DHPC	Dihexadecanoyl-phosphatidylcholine
DLPE	Dilauroylphosphatidylethanolamine
DMPC	Dimyristoylphosphatidylcholine
DMPE	Dimyristoylphosphatidylethanolamine
DLPC	Dilauroylphosphatidylcholine
DOPC	Dioleoylphosphatidylcholine
DOPS	Dioleoylphosphatidylserine
EggPC	Egg phosphatidylcholine
POPC	Palmitoyloleoylphosphatidylcholine
SOPC	Stearoyloleoylphosphatidylcholine
diC22:1PC	Dierucoylphosphatidylcholine
18:0:22:5PC	Stearoyldocosapentaenoylphosphatidylcholine
18:0-22:6PC	Stearoyldocosahexaenoylphosphatidylcholine
diphytanoylPC	Diphytanoylphosphatidylcholine
DLPG	Dilauroylphosphatidylglycerol
DMPG	Dimyristoylphosphatidylglycerol
POPG	Palmitoyloleoylphosphatidylglycerol
SOPG	Stearoyloleoylphosphatidylglycerol
DOPG	Dioleoylphosphatidylglycerol
TMCL	Tetramyristoylcardiolipin
DMPS	Dimyristoylphosphatidylserine

2.1 Historical Perspective

The study of lipid membranes, the underlying structure of all cell membranes, began near the beginning of the twentieth century (Swart 1907). The phospholipid bilayer was first proposed by Gorter and Grendell in 1925, by comparing monolayer areas of extracted red blood cell lipids with calculated cell surface areas (Gorter and Grendel 1925). In 1935, Hugh Davson and James Danielli proposed a model of the cell membrane in which the phospholipid bilayer lies between two layers of globular protein (Danielli and Davson 1935). Twenty-five years later, J. David Robertson determined that the dark bands observed in electron microscopy on either side of the membrane were the lipid headgroups and associated proteins of two apposed lipid monolayers, comprising a rigid "unit cell" (Robertson 1960). This static picture of a membrane was rejected 10 years later due to experiments by Frey and Edidin who fused two fluorescently labeled cells and watched as the dye populations mixed (Frye and Edidin 1970). The results of this experiment were key in the development of the "fluid mosaic" model of the cell membrane by Singer and Nicolson, where membrane lipids and proteins are viewed as highly mobile (Singer and Nicolson 1972). More recently, the idea of transient or permanent membrane domains (rafts) containing cholesterol and detergent-resistant proteins diffusing within the fluid bilayer has become the leading paradigm (Simons and Toomre 2000). As these models evolved, many precise biophysical investigations of the thermotropic and lyotropic properties of the underlying lipid bilayer were carried out using NMR, differential scanning calorimetry, FTIR, X-ray diffraction, centrifugation and fluorescence. Since natural membranes exist in a fully hydrated environment, the water component is necessarily an important part of the lipid structure. Indeed, without water, life is impossible. In this chapter, I will discuss some early experiments to obtain the total number of water molecules/lipid in stacked membranes, n_W, and other experiments that attempted to obtain the number of tightly bound water molecules/lipid, n_W'. The main focus will be on the use of scattering techniques: early experiments with X-ray diffraction, and our more modern method of X-ray diffuse scattering, sometimes combined with neutron scattering, that first obtains the fluid phase area/lipid and then calculates n_W and n_W' from known equations.

2.2 Gravimetric, or Luzzati Method

Why should we be concerned about n_W, number of waters/lipid? The main reason is that if lipids become too dehydrated, their physical and structural properties change. I.e., the water is actually part of the structure of the hydrated lipid. Then the question is, how much water is needed to maintain the equilibrium lipid structure and properties? In the 1960s, a pioneer in this field was Dennis Chapman who used differential scanning calorimetry (DSC), nuclear magnetic resonance (NMR),

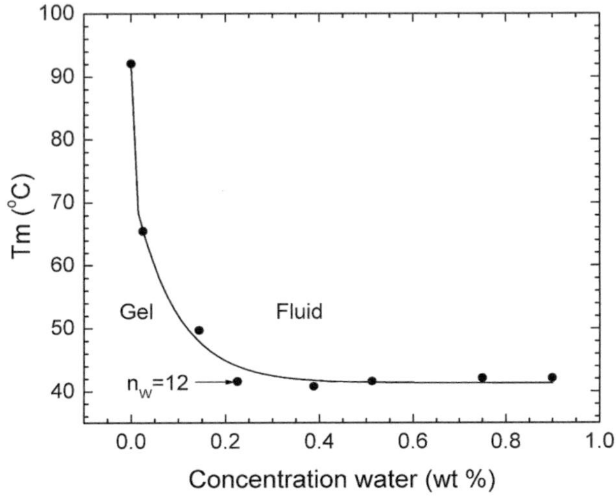

Fig. 2.1 Melting temperature (Tm) of DPPC vs. water concentration (wt %) obtained by DSC (Redrawn with permission from Fig. 3 (Ladbrook et al. 1968))

electron spin resonance (EPR), and Fourier transform infrared resonance (FTIR) to characterize the lyotropic behavior of lipid bilayers. In Fig. 2.1, Chapman's DSC data show the effect of the water content on the main melting transition temperatures (Tm's) of DPPC.

As shown in Fig. 2.1, when the hydration level dropped below 20 wt% water ($n_W = 12$), the measured Tm (main chain melting) increased dramatically. This indicates that the fluid phase of dipalmitoylphosphatidylcholine (DPPC) requires at least 12 water molecules for stability. Thus, in this relatively simple experiment, where careful weighing of water and lipid is required, large changes in lipid physical properties are observed at limiting water content, emphasizing the importance of adequate water in the lipid structure.

Another pioneer in the field of lipid bilayers is Vittorio Luzzati, who studied lipid bilayer structure using X-rays at about the same time that Chapman was using DSC and spectroscopic methods. When membranes are stacked to form an array, their X-ray diffraction pattern consists of Bragg reflections, which indicates that the array consists of membranes lying with their planes parallel, that are spaced relatively regularly. This occurs spontaneously when dried lipids are thoroughly mixed with water, forming onion-like structures, called multilamellar vesicles (MLVs). The D-spacing of the array is determined by the bilayer thickness plus the water between the bilayers. The amount of water between bilayers is determined by a balance of forces: in the gel phase there is an attractive van der Waals force and a repulsive hydration force, while in the fluid phase there is also a repulsive fluctuation force (Petrache et al. 1998a). A typical example of an X-ray pattern from an MLV consisting of the gel phase lipid, DPPC, is shown in Fig. 2.2.

The lamellar D-spacing is foundational to the gravimetric, or Luzzati method. This method entails: (1) drying the lipid completely before weighing, (2) weighing precise amounts of dried lipid with water, (3) equilibrating the lipid/water mixture,

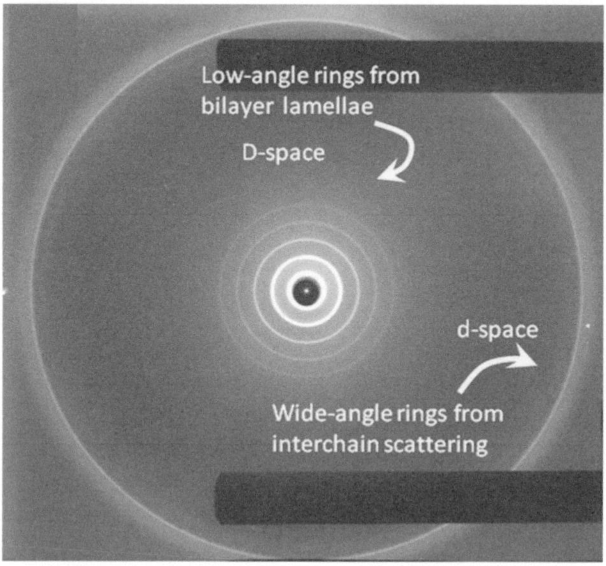

Fig. 2.2 X-ray diffraction pattern showing low-angle, central rings used to obtain the lamellar D-spacing, and wide-angle peripheral rings, using to obtain the d-spacing, distance between chains. Film data of DPPC at room temperature (*RT*) obtained by Prof. Roy Worthington at the Cornell High Energy Synchrotron Source (CHESS) in 1992. *Dark shadows* are from the film holder

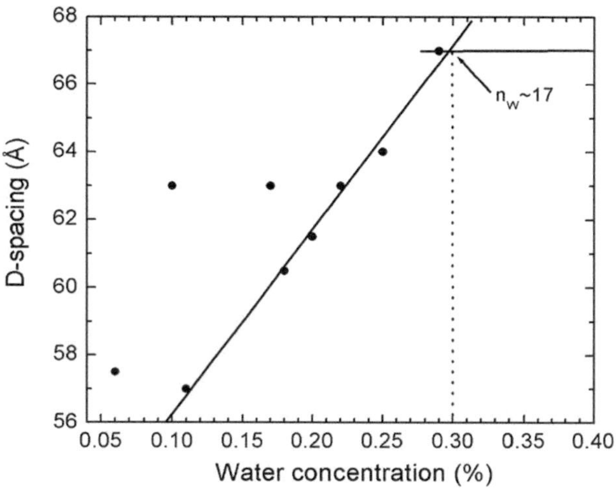

Fig. 2.3 D-spacing of DPPC at 20 °C vs. water concentration (wt %) obtained by X-ray diffraction (Redrawn with permission from Table 1 (Tardieu et al. 1973))

sometimes at high temperatures, to hydrate thoroughly, (4) X-raying the resulting multilamellar sample to obtain a D-spacing, or distance between layers, that includes both the lipid thickness and associated water, (5) plotting D-spacing vs. water concentration, (6) determining the fully hydrated water concentration where the D-spacing levels off by extrapolation, and (7) converting weight percent to n_W using the molecular weights of lipid and water. An example of results from the gravimetric method is shown in Fig. 2.3 for the same DPPC lipid as in Fig. 2.2.

Fig. 2.4 An illustration of the formation of water pools (*black*), in the center of MLVs and between MLVs (Reproduced with permission from Fig. 3 (Koenig et al. 1997))

There are several problems with the gravimetric method. The greatest problem has been pointed out by Gawrisch et al. (Koenig et al. 1997). Their previous studies on PC lipids in the Lα phase suggested that only the first 15 water molecules per lipid are homogeneously incorporated (Gawrisch et al. 1985). When more water is added, instead of locating neatly between the bilayers in a MLV, pools of water form, causing gradual changes in the morphology of the sample (Klose et al. 1988). This problem was discussed in (Nagle and Tristram-Nagle 2000) and is shown in the cartoon in Fig. 2.4. Formation of water pools can wreak havoc on the gravimetric method, since added water does not contribute to the measured D-spacing if the added water locates in a pool. Notice that the n_W determined in Fig. 2.3 is larger than that determined using DSC in Fig. 2.1 due to this artefact. Then, there are several other problems that add to the errors in this method. First, the lipid component must be completely anhydrous before weighing. It is difficult to remove the last 1–2 water molecules as has been measured (Jendrasiak and Hasty 1974), which normally is not a problem, but when accurate numbers are required, then this introduces an error of 10 %. Second, the lipid has to be precisely weighed, as does the water, so highly accurate balances are required. Human error can occur during this step, so many repetitions with different sample sizes are required to obtain accurate concentrations. Third, the sample must be hydrated by vortexing and temperature cycling. Evaporation of water can occur during this step if the sample is not properly sealed. Fourth, the sample must be transferred to a container that allows X-rays to pass through it. Usually this is a glass capillary, or the Luzzati

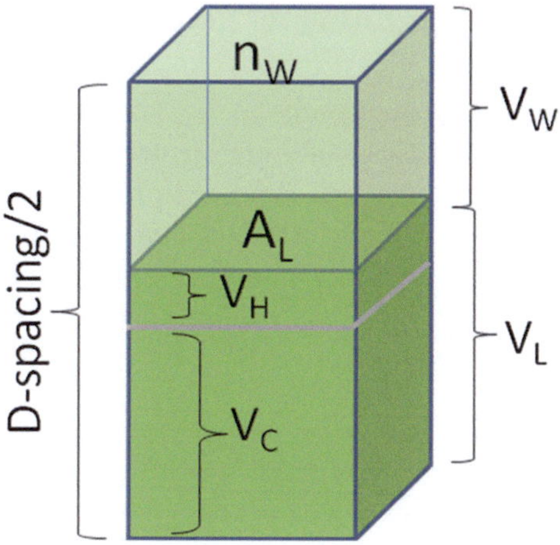

Fig. 2.5 Drawing showing V_L (*darker green*) with its associated water V_W (*transparent green*). V_H = volume/headgroup and V_C = volume/hydrocarbon. These volumes represent one monolayer in the bilayer which contributes to ½ lamellar D-spacing

cell where the sample is contained between mica or mylar windows. Evaporation of water can occur during the transfer and sealing of the X-ray container. Fifth, even with temperature cycling and vortexing, the MLV may not be at equilibrium with its surrounding water; the onion-skin structure can trap water molecules and prevent the maximum swelling from occurring, as if an osmotic pressure were being applied, especially in the fluid phase. Finally, in order to estimate the full water component using the gravimetric method, it is necessary to fit linearly the D-spacings on the approach to full hydration and in the plateau region, and find their intersection (see Fig. 2.3). There can be errors in the intersection value, due to noisy data, and an increasing slope after "full" hydration has been reached.

In the above methods, n_W is obtained first, and then the area/lipid (A_L) is calculated as the desired result. The equation that is used is:

$$A_L = 2\left(V_L + n_W V_W\right)/D \tag{2.1}$$

where V_L and V_W are the lipid and water molecular volumes (Å^3), respectively, at the temperature of the experiments, and D is the fully hydrated D-spacing (Å). This equation is shown pictorially in Fig. 2.5. In our laboratory we have always measured V_L using either the combined technique of differential scanning dilatometry (Nagle and Wilkinson 1978) and neutral flotation (Wiener et al. 1988), neutral flotation alone, or differential scanning densimetry with a modern Anton-Paar DMA 5000 M densimeter (Tristram-Nagle et al. 2010). Several other groups simply calculate V_L by adding measured (or calculated) volumes of the lipid's component groups. The latter practice is a source for error in obtaining n_W and A_L; V_L should be measured at the temperature of the experiments whenever possible. More details about our volume measurements will be given in Sect. 2.6.

2.3 Obtaining A_L: Gel Phase

Obtaining the area/lipid, A_L, is straightforward using X-ray diffraction techniques in the gel phase of a bilayer. In addition to low-angle scattering that produces the D-spacing discussed above, lipid bilayers also scatter in the wide-angle region due to their chain-chain correlation, with d-spacing 4–5 Å. Using the wide-angle d-spacings, A_L is calculated with the equation:

$$A_L = 2A_C/\cos\theta \tag{2.2}$$

where A_C is the area/chain (Å^2) and θ is the chain tilt angle. Details on how to obtain A_C from the wide-angle spacings can be found in Levine (1973) and Tristram-Nagle et al. (1993). Since the headgroup area is larger than the combined area of the two hydrocarbon chains beneath it, the chains tilt in order to maximize van der Waals interactions between them (Nagle 1976). In the gel phase, the lipid chains are in the *all-trans* state, which are essentially straight chains. The chain scattering is confined to Bragg rods in the wide-angle region; while the 2,0 rod lies on the equator, the 1,1 rod is lifted up due to the chain tilt. The chain tilt is obtained by first measuring the angle off the equator of the 1,1 wide-angle rod (see Fig. 2.6 for an example) in a fully hydrated, oriented gel phase sample. The chain tilt angle is calculated as in (Tristram-Nagle et al. 1993) using the angle η as drawn in Fig. 2.6. Notice that in an oriented gel phase pattern, low-angle scattering is along the q_z (vertical) axis, and wide-angle scattering has a q_r (horizontal) component, compared to the isotropic scattering in the MLV data (shown in Fig. 2.2).

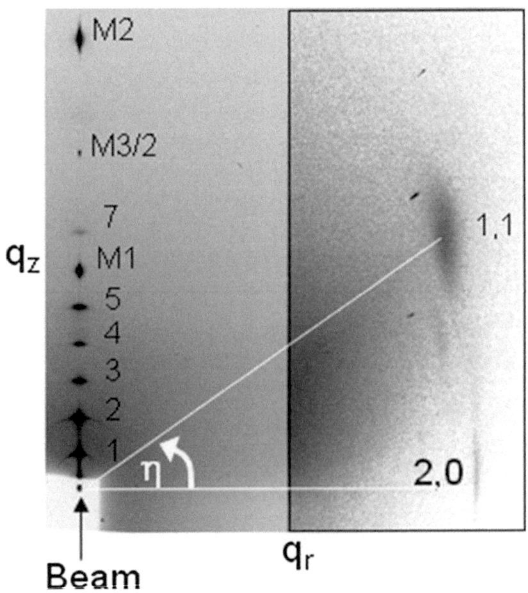

Fig. 2.6 Dimyristoylphosphatidylcholine (DMPC) low-angle lamellar orders are numbered in increasing q_z, while mica peaks are labeled with M. Wide-angle peaks (2,0 and 1,1) are shown to the right with a higher contrast. The angle η is directly related to the chain tilt angle θ (Reproduced with permission from Fig. 4 (Tristram-Nagle et al. 2002))

2.4 Obtaining A_L: Fluid Phase

In the fluid phase, the chains are disordered with 4 gauche rotamers per chain (Nagle 1980; Mendelsohn et al. 1991). A gauche rotamer is a rotation of 120° about a C-C single bond; when there are two or three gauche rotamers in a row, it is referred to as a kink in the chain. Although a theory from the liquid-crystal literature obtains A_L using a disordered wide-angle pattern (Levine 1973; Mills et al. 2008), the precision of this method is less than that of the gel phase A_L wide-angle determination described above. A more precise A_L is obtained by using low-angle X-ray scattering (LAXS) that results from fully hydrated, oriented lipid membranes as developed in the Nagle/Tristram-Nagle lab ((Tristram-Nagle and Nagle 2004), sometimes combined with neutron scattering from isotropic, unilamellar vesicles (ULV) (Kučerka et al. 2008). The basic principle of this method is to obtain the bilayer thickness, without its associated water. Then, by also measuring V_L, A_L is directly obtained using two equations, either separately or combined:

$$A_L = V_C/D_C \text{ (X-rays) or } A_L = 2V_L/D_B \text{ (neutrons)} \quad (2.3, 2.4)$$

where D_C is the hydrocarbon thickness and D_B is the bilayer thickness without water. V_C is determined by subtracting the headgroup volume, V_H, from V_L. Headgroup volume is determined using gel phase data as described in (Tristram-Nagle et al. 2002) and we assume that V_H doesn't change with temperature. D_C is determined by fitting fluid phase X-ray data to the Scattering Density Profile (SDP) fitting program. These thicknesses are depicted in Fig. 2.7c, which also shows the total electron density profile of a lipid bilayer (**b**) and its component group probabilities (**a**).

Neutrons use a different measure of bilayer thickness, D_B, which is defined as the Gibbs dividing surface of water. As shown in Fig. 2.7, D_B is in the headgroup region, while D_C is closer to the bilayer center. Using the SDP program, both X-ray and neutron scattering data can be fit simultaneously through the Fourier transform to a model of the bilayer with component groups. An example of these types of data are shown for the lipid diphytanoylPC in Fig. 2.8.

In that work $A_L = 80.5$ Å^2 and $n_W = 37$ (Tristram-Nagle et al. 2010). By combining Eqs. 2.3 and 2.4 during the fit to both X-ray and neutron data, a compromise A_L was determined, which is between a slightly larger A_L using X-ray data alone (83 Å^2) and a slightly smaller A_L using neutron data alone (78 Å^2). When studying another lipid, dioleoylphosphatidylcholine (DOPC), our lab similarly found a larger area (72.4 Å^2) using X-rays alone (Pan et al. 2008a; Kučerka et al. 2005b) compared to 66.5 Å^2 using neutrons alone (Gallova et al. 2008). When X-rays and neutron data were analyzed simultaneously using the SDP program, DOPC $A_L = 67.4$ Å^2 (Kučerka et al. 2008). Since areas are used to determine n_W, this variability represents a 16–20 % error. Although for these two examples, the neutron scattering data yielded smaller areas, for the case of DPPC in the fluid phase, the areas were the same for X-rays and neutrons (Kučerka et al. 2008).

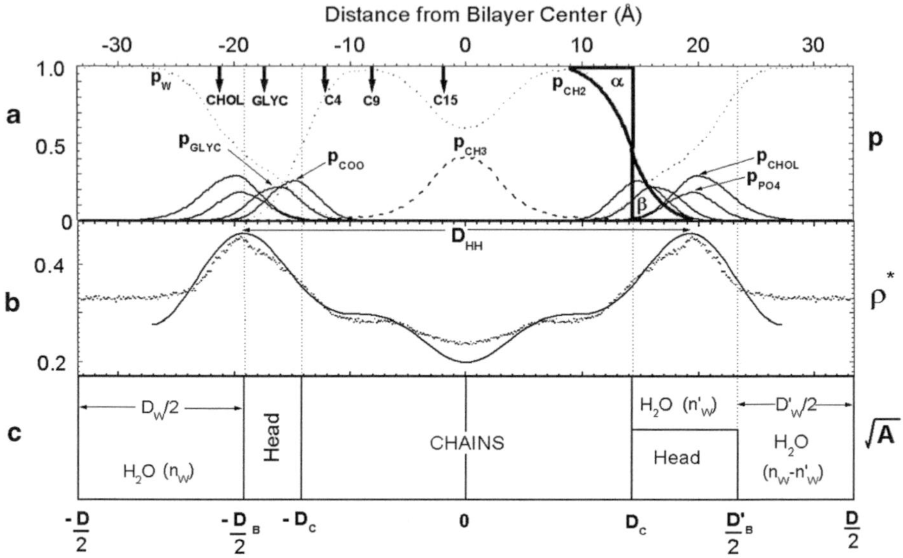

Fig. 2.7 Three representations of the structure of DPPC bilayers in the Lα fluid phase: (**a**) Probability distribution functions for component groups (from simulations). Abbreviations: *W* water, *Glyc* glycerol, *Chol* choline, *COO* carbonyl, *CH*$_2$ methylenes, *CH*$_3$ terminal methyl and *PO*$_4$ phosphate. The equality of the areas α and *β* determines the Gibbs dividing surface, or edge, of the hydrocarbon region. (**b**) Electron density profiles from X-ray data (*solid line*) and from simulations (*dotted line*). (**c**) Two volumetric pictures: *left*, a 3-compartment model and *right*, a more realistic model of the interfacial headgroup region (Reproduced with permission from Fig. 2 (Nagle and Tristram-Nagle 2000))

2.5 Details of the X-Ray Diffuse Scattering Method

While it is fairly straightforward to obtain neutron scattering data from ULV by methods described in (Lewis and Engelman 1983; Kučerka et al. 2004), it took our laboratory about 4 years to develop the data collection and analysis methods that use oriented, fully hydrated lipid membranes (Lyatskaya et al. 2001; Tristram-Nagle and Nagle 2004). Unlike gel phase X-ray diffraction where 7–10 lamellar orders are observed, X-ray scattering from fluid phase, fully hydrated MLV produces only two lamellar orders, with the h = 2 quite weak, due to the fluctuations and undulations caused by thermal energy. There is disorder both within a single bilayer as well as disorder caused by correlations between fluctuating bilayers (Guinier 1963). This causes loss of lamellar diffraction orders, as well as loss of intensity in the remaining two orders, which redistributes to diffuse scattering between peaks. This creates problems for structure determination, in that the resolution is quite poor. If a negatively charged lipid is present, there are no lamellar orders at all, since the lipid lamellae unbind from each other. What many labs have done is to dehydrate fluid phase bilayers, in order to produce more lamellar orders. While this method does

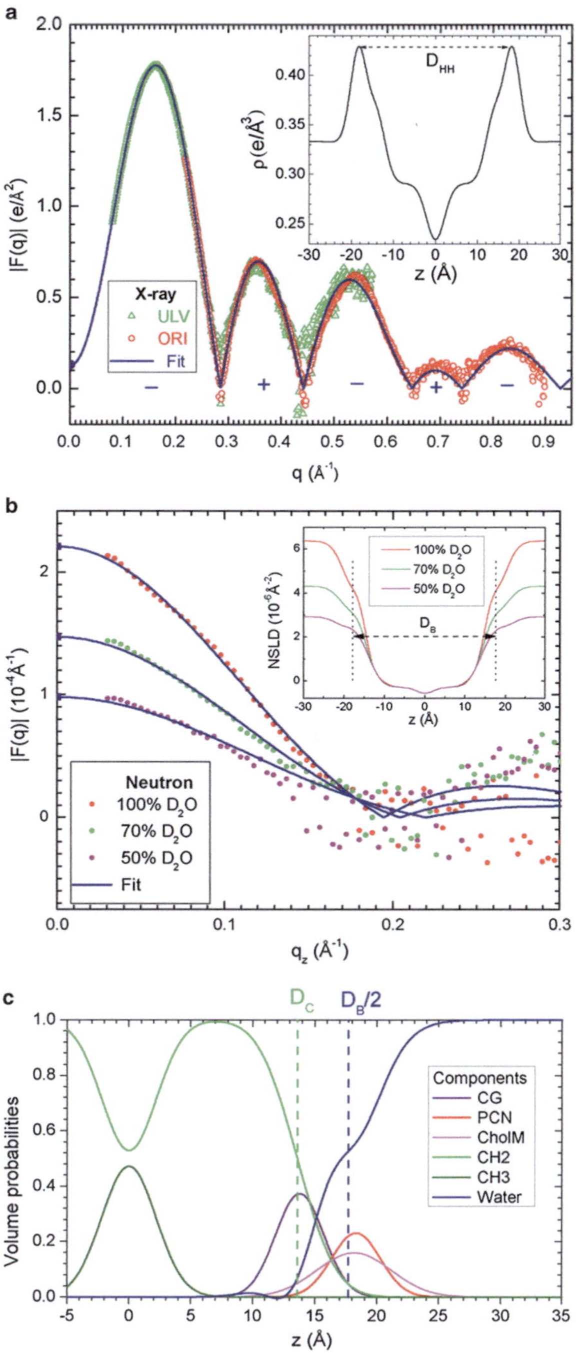

Fig. 2.8 (**a**) Absolute X-ray form factors for diphytanoylPC at 30 °C. ULV (*green*) and oriented (*red*) data are both fit to a model of a bilayer (*blue*) using the SDP program. *Inset* shows the corresponding electron density profile (Reproduced with permission from Fig. 6 (Tristram-Nagle et al. 2010)). (**b**) Absolute neutron form factors for ULV of diphytanoylPC at 30 °C in three concentrations of D_2O/H_2O. *Inset* shows the corresponding scattering length density profiles, with D_B indicated by *vertical, dotted lines* (Reproduced with permission from Fig. 7 (Tristram-Nagle et al. 2010)). (**c**) Volume probabilities for the components of the SDP model: CG (carbonyl + glycerol); PCN (phosphate $+2CH_2 + N$); CholM ($(CH_3)^3$ on N); CH_2 (chain methylenes); CH_3 (chain terminal methyls); and water. The probabilities are symmetric about the bilayer center at z = 0. The *dotted green line* is located at D_C, the Gibbs' dividing surface for the hydrocarbon region, and the *dashed blue line* is located at $D_B/2$, the Gibbs dividing surface for water (Reproduced with permission from Fig. 8 (Tristram-Nagle et al. 2010))

succeed at producing more orders, thermal fluctuations are usually lost, removing the lipids further from their biologically relevant state.

In fact, our lab started with this method, applying a correction to the obtained scattering, due to the fluctuations that are present when bilayers are only slightly dehydrated, but still close to full hydration (Zhang et al. 1994, 1995). We model the disorder based on liquid-crystal theory (Caillé 1972; de Gennes and Prost 1993), which, when compared to paracrystalline theory (Hosemann and Bagchi 1962), is a better fit to the diffuse X-ray scattering data that emanates from fluctuating bilayers (Zhang et al. 1996). In our original method, samples were MLVs in X-ray glass capillaries, producing an isotropic pattern, similar to that shown in Fig. 2.2 for gel phase DPPC, but with fewer lamellar orders due to fluctuations. Using a high-resolution synchrotron X-ray setup to resolve peak shape of lamellar Bragg orders, the thermal fluctuations were analyzed using the modified Caillé analysis (Zhang et al. 1996). With this we calculated η, which is a combined fluctuation parameter of the bending modulus, K_C, and the bulk modulus, B (Nagle and Tristram-Nagle 2000) (and references therein). While K_C is a measure of the energy required to bend a single membrane, B is a measure of interaction energy between bilayers in a stack (see Fig. 2.9). The theory is based upon a free energy function shown in Eq. 2.5.

The theory quantifies the displacement, u_n, of a fluctuating bilayer from its flat position, which is used in the correlation function that then obtains the structure factor of a fluctuating bilayer. While isotropic samples can only provide η, the combined fluctuation parameter, oriented samples can provide K_C and B separately (Lyatskaya et al. 2001). In the late 1990s, our lab succeeded in obtaining well-oriented thin membrane films produced by a method pioneered by Tristram-Nagle (Tristram-Nagle et al. 1993; Tristram-Nagle 2007). A diagram of this sample is shown in Fig. 2.10.

It was initially difficult to hydrate lipid samples to full hydration through the vapor, which was referred to as the vapor pressure paradox (Rand and Parsegian 1989). Even gel phase lipids, which require less water than fluid phase lipids, did

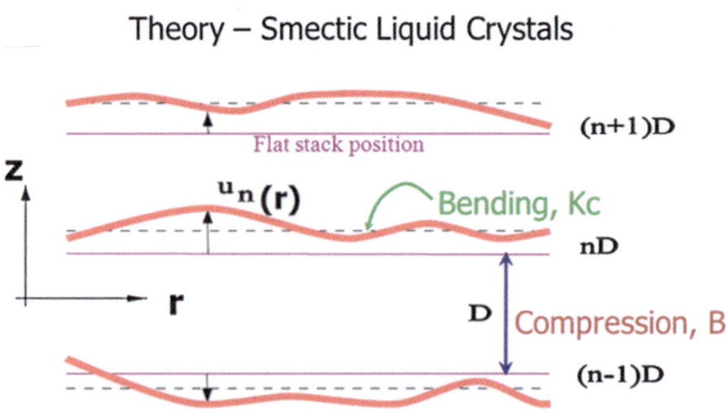

Fig. 2.9 Diagram of smectic liquid crystal theory for a fluctuating stack of bilayers and Eq. 2.5 showing how K_C and B are related to the displacement u_n in the free energy functional (de Gennes and Prost 1993). The displacement, u_n, is evaluated in the radial r direction

$$f_{fluc} = \frac{\pi}{NL^2} \int r dr \sum_{n=0}^{N-1} \left[K_c \left(\nabla_r^2 u_n(r) \right)^2 + B \left(u_{n+1}(r) - u_n(r) \right)^2 \right] \quad (2.5)$$

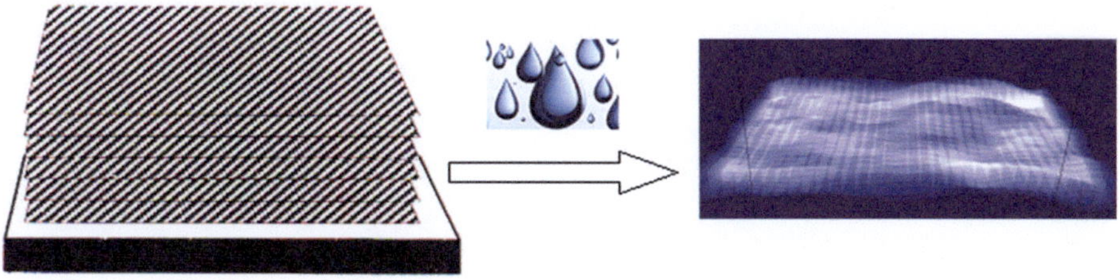

Fig. 2.10 An oriented stack *(left)*, 10 μ thick, consisting of 1800 bilayers, hydrates through the vapor in a well-insulated chamber producing thermal fluctuations near full hydration in the lipid fluid phase. Fluctuating bilayers were analyzed using a Monte Carlo program (Gouliaev and Nagle 1998) that was used to create this image *(right)*

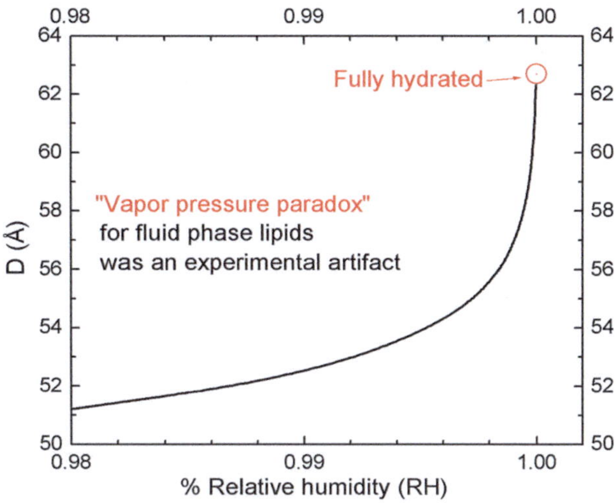

Fig. 2.11 Approximately 8 water molecules are taken up by DMPC in the final 0.2 % RH towards full hydration (Reproduced with permission from Fig. 3 (Chu et al. 2005))

not achieve the full n_W through the vapor (Jendrasiak and Hasty 1974; Torbet and Wilkins 1976). Our laboratory was successful in fully hydrating gel phase lipids through the vapor by placing the oriented sample on a Peltier cooling element within our X-ray hydration chamber, to gently condense all required water into the sample (Tristram-Nagle et al. 1993). Fluid phase lipid hydration through the vapor was much more difficult. See Fig. 2.11 for an example of the difficulty in attaining full hydration through the vapor.

This experimental hurdle was overcome with a well-insulated neutron hydration chamber (Katsaras 1998). Since that pioneering work, our laboratory has designed and built both a well-insulated X-ray (Kučerka et al. 2005a) and a well-insulated neutron chamber that each achieve full hydration of fluid phase lipids. The scattering from these samples is anisotropic similar to that shown in Fig. 2.6 for gel phase DMPC. K_C, the bending modulus, is an elastic parameter that characterizes the lipid membrane; it usually decreases when proteins are added (Tristram-Nagle and Nagle 2007; Shchelokovskyy et al. 2011; Pan et al. 2009a; Greenwood et al. 2008; Boscia

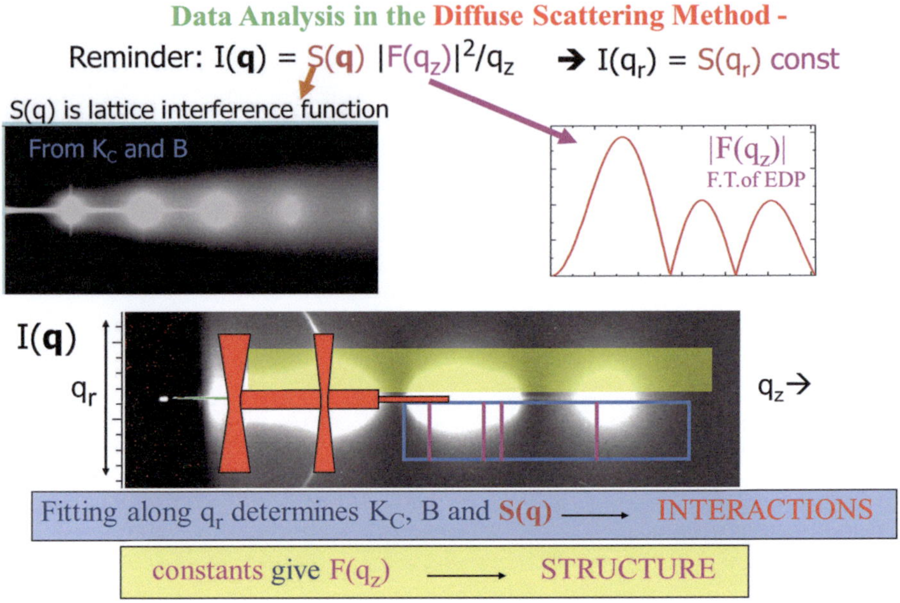

Fig. 2.12 Data analysis in the Diffuse Scattering Method first fits the X-ray diffuse data (*central image*) that results from fluctuating, fully hydrated bilayers in a stack to the liquid crystal theory using the free energy function shown in Fig. 2.9. This step obtains K_C, B and the structure factor shown in the *upper left*, by fitting the white intensity data in the diffuse lobes with 500 rays in q_r (some shown in magenta in the *grey box*). Then the q_z data under the yellow band are used to obtain the form factor by fixing the structure factor S(**q**). The intensity I(**q**) and the form factor $|(F(q_z)|$ are related by the equation shown at the top of this figure (Figure reproduced from Fig. 13 (Tristram-Nagle and Nagle 2004))

et al. 2013) and increases when cholesterol (Pan et al. 2008b, 2009b) or cardiolipin (Boscia et al. 2014) is added. A decrease in K_C indicates that it is easier to bend the membrane in the presence of a protein or peptide. K_C is determined *en route* to obtaining the structure factor, which then provides the form factor as is shown in Fig. 2.12.

Once the form factor is determined (brown curve in upper right of Fig. 2.12), the Scattering Density Profile (SDP) computer program is used to fit the form factor data to a model of a bilayer through the Fourier transform (Kučerka et al. 2008). The program models the component groups of a bilayer by using Gaussians for the two headgroup peaks, phosphocholine and glycerol-carbonyl, and for the methyl trough in the center of the bilayer, and by using error functions for the water and hydrocarbon electron densities. An example of an electron density profile determined by the SDP program is shown in Fig. 2.13 (solid lines). The profile also contains the bioflavinoid, BF, genistein. As shown, there is excellent agreement between the total and component electron densities when compared to the profile resulting from a CHARMM MD simulation carried out by Dr. Rich Pastor and Rick Venable at NHLBI (dotted lines).

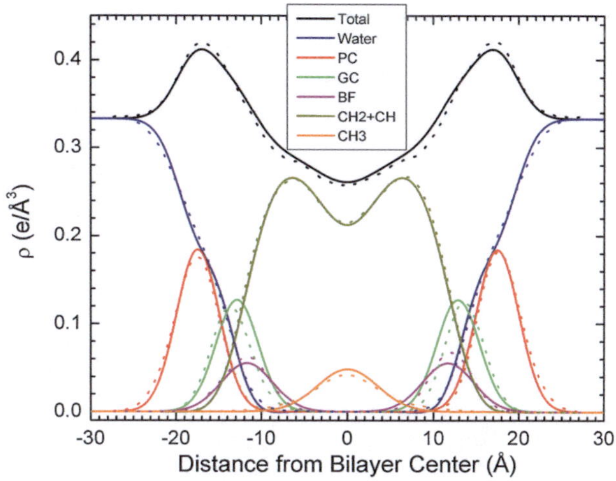

Fig. 2.13 Electron density profiles for DOPC with 20 mol % genistein resulting from SDP model fitting (*solid lines*) and MD simulations at $A_L = 83$ Å^2 (*dotted lines*). Component groups and bioflavinoid (BF) are identified in the legend (Reprinted (adapted) with permission from Fig. 4 (Raghunathan et al. 2012), American Chemical Society)

The SDP program also calculates bilayer thicknesses, $2D_C$ and D_B, as shown in the volume probabilities in Fig. 2.8c. Molecular volumes provide the underlying foundation upon which the SDP program is built; volume measurements will be discussed in Sect. 2.6. Using X-rays and, in some cases, X-rays plus neutrons, our laboratory has now determined many A_L's that have allowed the calculation of many n_W's using Eq. 2.1, shown in Table 2.1. n_W is 2–3 times larger in the fluid phase than in the gel phase, due to the larger A_L's and larger lamellar D-spacings (not shown). For fluid phase lipids, we rely more on the equilibrated D-spacings obtained by hydrating oriented samples through the vapor, than on the MLV capillary D-spacings. In many cases, the vapor-hydrated D-spacings are larger than the isotropic D-spacings, perhaps due to the artifact mentioned above caused by the constraining MLV geometry.

It is of interest to try to correlate A_L and n_W; this is done graphically in Fig. 2.14. As shown, n_W generally increases with A_L, both in the gel and fluid phases. The intuitive message is that larger lipid areas require more water. The increase is not linear, however, and is quite noisy since Eq. 2.1 involves other parameters besides A_L and n_W.

What about n_W', the steric number of water molecules associated with the headgroup molecule? For this, our group uses Eq. 2.6:

$$n_W' = \left(AD_H' - V_H\right)/V_W \tag{2.6}$$

where $A = A_L$, $D_H' = 9$ (PC's) or 8.5 (PE's) from neutron studies (Buldt et al. 1979), V_H = headgroup volume (Nagle and Tristram-Nagle 2000) and V_W is the volume of water. We can think of this water as headgroup structural water, shown in Fig. 2.7. It is considered to be a lower limit to measurements of tightly bound water. Using Eq. 2.6, n_W' is calculated for the lipids shown in Table 2.1. Values for n_W' are in the far right column in Table 2.1.

Again, n_W' is larger for fluid phase lipids than for gel phase lipids, but both are much smaller than n_W for their corresponding phase. n_W' shows a close correlation to A_L for fluid phase lipids when plotted in Fig. 2.15, with a nearly linear increase in n_W' between $A_L = 60–80\ Å^2$. This indicates that a small number of water molecules (6–13) are closely associated with the fluid phase headgroup, part of its structure. For gel phase, n_W' is 3 for several lipids, except cardiolipin (TMCL) where it is 6 for this 4-chain lipid. DHPC in the interdigitated phase is not included in Fig. 2.15, since its bilayer thickness is much smaller than the other lipids. Remember that n_W' is merely calculated from our structural results; inclusion of headgroup n_W' results in a third measure of bilayer thickness, D_B', shown in Fig. 2.7c. Many investigators have attempted to measure tightly-coupled headgroup water using various experimental techniques. These results appear in Table 2.2 together with our calculated values of n_W' for DMPC and DPPC.

Interestingly, bound water, when measured by many different experimental techniques, shows little dependence on the lipid phase. Although some of the smallest n_W' (<5) are for gel phase DMPC and DPPC lipids, there is one small $n_W' = 4$ for fluid phase Egg PC. In addition, both gel phase and fluid phase lipids

Table 2.1 Lipid areas and associated water molecules (Nagle/Tristram-Nagle lab)

Lipid	T (°C)	Area ± 0.5 ($Å^2$)	n_W	n_W'
Fluid Phase				
DPPC	50	62.9(±1.3)[a], 64.0[b], 64.3[c], 63.1[d]*	30[a]	8[a]
DHPC	48	65.1[e]	33[e]	9[e]
DLPE	35	51.2[b]	9[b]	6[b]
DMPC	30	59.7[f], 60.6[g]	27[f]	7[f]
DLPC	30	63.2[g]	31[f]	8[f]
DOPC	30	72.2[h], 72.5[b], 72.1[i], 72.4[j, k], 67.4[d]*	33[h], 28[d]	11[h], 9[d]
	(15)	(69.1[k])	(30[k])	(10[k])
	(45)	(75.5[k])	(35[k])	(12[k])
DOPS	30	65.3[l]	∞[l]	10[l]
EggPC	30	69.4[f, b]	35[b]	10[b]
POPC	30	68.3(±1.5)[j]	31[j]	9[j]
SOPC	30	67.0(±0.9)[m]	30[m]	9[m]
diC22:1PC	30	69.3[j]	30[j]	10[j]
18:0–22:5PC	24	68.7[n]	30[n]	10[n]
18:0–22:6PC	24	68.2[n]	29[n]	9[n]
DiphytanoylPC	30	80.5(±1.5)[r]	37[r]	13[r]
DLPG	30	65.6[s]	∞[s]	9[s]
DMPG	30	65.1[s]	∞[s]	9[s]
POPG	30	66.1[s]	∞[s]	9[s]
SOPG	30	66.7[s]	∞[s]	9[t]
DOPG	30	70.8[s]	∞[s]	10[s]
TMCL	50	108.6[t]	∞[t]	14[t]

(continued)

Table 2.1 (continued)

Lipid	T (°C)	Area ± 0.5 (Å²)	n_W	n_W'
Gel Phase				
DMPC	10	47.2[o]	12[o]	3[o]
DiC16PC,18,20,22,24	20	47.5[p, q]	12[p]	3[p]
DMPS	20	40.8[l]	∞[l]	3[l]
DLPE	20	41.0[b]	6[b]	3[b]
DHPC-Interdig	20	77.2[e]	24[e]	12[e]
DHPC-gel	20	46.9[e]	10[e]	3[e]
TMCL	35	81.5[t]	∞[t]	6[t]

[a]Nagle et al. (1996)
[b]Nagle and Tristram-Nagle (2000)
[c]Kučerka et al. (2006)
[d]Kučerka et al. (2008)
[e]Guler et al. (2009)
[f]Petrache et al. (1998b)
[g]Kučerka et al. (2005a)
[h]Tristram-Nagle et al. (1998)
[i]Liu and Nagle (2004)
[j]Kučerka et al. (2005b)
[k]Pan et al. (2008a)
[l]Petrache et al. (2004)
[m]Greenwood et al. (2008)
[n]Eldho et al. (2003)
[o]Tristram-Nagle et al. (2002)
[p]Tristram-Nagle et al. (1993)
[q]Sun et al. (1996)
[r]Tristram-Nagle et al. (2010)
[s]Pan et al. (2012)
[t]Boscia et al. (2014)
*Neutron data

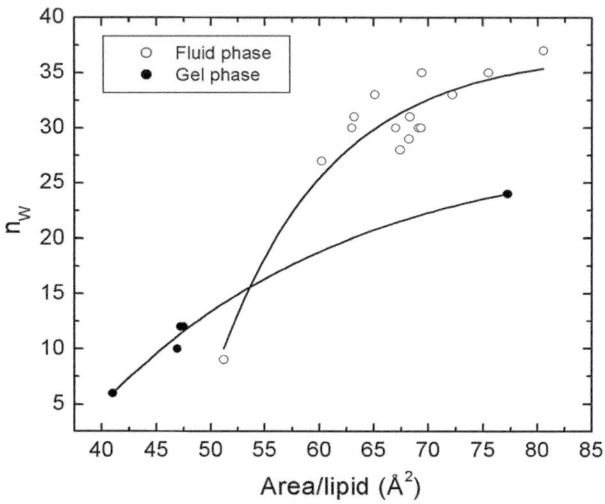

Fig. 2.14 Number of water molecules/lipid, n_W, vs. area/lipid, A_L

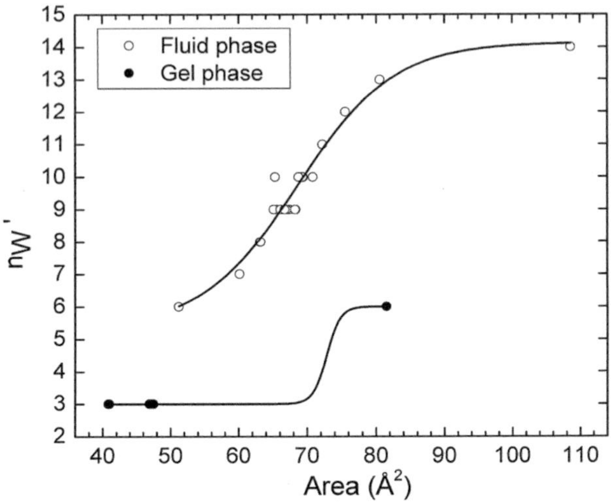

Fig. 2.15 Number of water molecules/lipid, n_W', vs. area/lipid, A_L

Table 2.2 Bound water, $n_{W'}$, determined experimentally

Authors (year)	Method	$n_{W'}$
Bach and Miller (1998)	Unfreezable water DSC (DMPC)	3.5
Salsbury et al. (1972) (& Chapman)	^{2}NMR linewidths in inner hydration shell (Egg PC)	4
Miller et al. (2002) (& Kodama)	Unfreezable water DSC (DMPC)	4.5
Bronshteyn and Steponkus (1993)	Unfreezable water DSC (DPPC)	5
Jendrasiak and Hasty (1974)	Inflection point in adsorption isotherm (DPPC)	5
Channareddy et al. (1997) (& Janes)	Radiolabel, sucrose & centrifugation (EggPC)	7
Binder et al. (1999) (& Heerklotz)	Isothermal titration calorimetry, gravimetry & FTIR (POPC)	~7
Pohle et al. (1998)	Adsorption isotherm using FTIR (DPPC)	7
Crowe et al. (1994)	Unfreezable water DSC (DPPC)	7.2
Nagle and Tristram-Nagle (2000)	X-ray fluid phase area, V_L, calculations (DMPC (30 °C) and DPPC (50 °C))	7, 8
Small (1967)	X-ray gravimetric, volumetric considerations (Egg PC)	8.3
Salsbury et al. (1972) (& Chapman)	^{2}NMR linewidths in inner hydration shell (gel phase DPPC)	9
Dufourcq et al. (1997)	Solid state NMR hydration regime (fluid phase DMPC)	9.7
Grdadolnik and Hadzi (1994)	FTIR bandshape and unfreezable water (DPPC)	10
Finer and Darke (1974)	^{2}NMR linewidths in inner hydration shell (Egg PC)	11
Hristova and White (1998)	Change in lipid structure using X-ray electron density profiles (DOPC)	12
Lairion et al. (2002) (& Disalvo)	Reverse micelles & turbidity (Egg PC)	12

have comparable values when $n_W' = 7$ or greater. Our lab's calculated steric values of 7 and 8 for fluid phase DMPC and DPPC, respectively, are near the middle of this range of experimental values. Some of the smallest values are for unfreezable water in DSC where the lipid is presumably in the gel phase, but one value is 7.2 (Crowe et al. 1994). Depending on the cooling rate, the lipid could be in the subgel phase, which requires less water than the gel phase. If the cooling rate is very slow, ice crystals can form in pools of water which create an osmotic pressure, lowering headgroup water. Unlike the gravimetric method, unfreezable water does not have to be sandwiched between bilayers, but simply associated with the phospholipid headgroup. In many of these studies, not only was an inner hydration shell found (reported in Table 2.2), but 1 or 2 outer hydration shells were found, before bulk water partitioned out. While these other hydration shells are of some interest, it has been shown using ^{2}H-NMR that hydration of DOPC is a smooth, continuous process, without separate hydration shells (Ulrich and Watts 1994). In that study, n_W' for DOPC = 22. The main point in showing values of headgroup water in Table 2.2 is to emphasize the large range of experimental values that have been reported in the literature. Perhaps it is not accurate to list these values in the same table, because they may be measuring different quantities.

2.6 Obtaining Molecular Volumes

As mentioned above, molecular volumes, V_L, are needed to calculate n_W and also to determine structure using the SDP modeling program. When I first joined the Nagle laboratory as a postdoc in 1982, I worked with a home-built instrument, the differential scanning dilatometer (DSD). It was constructed by John Nagle and his former postdoc Allan Wilkinson and was necessarily housed in the basement of Mellon Institute, to minimize vibrations. A grainy image of this instrument is shown in Fig. 2.16. The principle of the dilatometer is that expansion of a large volume of 0.2–1 g lipid in 10 ml water is compared to the expansion of 10 ml water as the temperature is increased. The lipid was in the form of MLVs, described in Sect. 2.2. The instrument was time-consuming to operate, since the glass stopcocks on top of the fragile glass "arms" had to be manually opened and closed often to release the pressure buildup after a few temperature increments, every 5 min. The 3D volume change in the lipid and water was transduced to a 2D vertical movement of the metal bellows shown in Fig. 2.16b. These bellows caused a precise movement of an optical flat; this movement was recorded using a laser that produced an interference pattern between the moving optical flat and a stationary optical flat below it (piezo-interferometry). The # fringes in the interference pattern were recorded manually by the operator, and a complete description of this instrument was published (Wilkinson and Nagle 1978). Although the dilatometer was difficult to operate, it was used to obtain many precise lipid volume measurements (Tristram-Nagle et al. 1987; Wiener et al. 1988; Wilkinson and McIntosh 1986; Wilkinson and

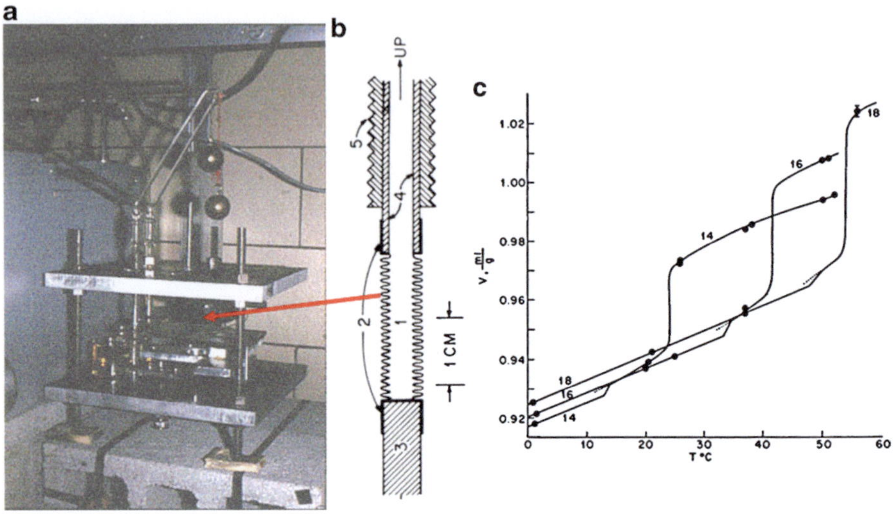

Fig. 2.16 (**a**) Differential scanning dilatometer with sample flasks extending into the grey water bath at *left of picture*. (**b**) Schematic of bellows containing a heavy hydrocarbon fluid in contact with expanding water. *Red arrow* shows their location in the dilatometer. (**c**) Specific volume data obtained using the dilatometer (*solid lines*) pinned to values obtained using neutral flotation (*solid circles*) (Data for DMPC (14), DPPC (16) and DSPC (18) reproduced with permission from Fig. 1 (Nagle and Wilkinson 1978))

Nagle 1979, 1981, 1984; Wilkinson et al. 1987; Yang and Nagle 1988). The DSD was also useful in obtaining α, the coefficient of thermal expansion. The change in volume vs. temperature reveals information regarding intermolecular interactions in the single phase regions (Wilkinson and Nagle 1982; Tristram-Nagle et al. 1986), as well as the total volume change during phase transitions (Nagle and Wilkinson 1978).

Besides its inconvenience, the differential dilatometer could only obtain relative, not absolute volumes. For this, the Nagle/Tristram-Nagle lab used neutral flotation (or neutral buoyancy) of lipids in D_2O/H_2O mixtures. The validity of neutral flotation for lipid bilayer dispersions was tested in (Wiener et al. 1988). The principle of this method is to prepare several D_2O/H_2O mixtures with specific densities (g/ml) (inverse specific volume) near to the expected lipid specific density. Volumes (1/densities) are first estimated by comparing to other lipids or by calculating the total volume using the component group volumes. An example of several mixtures prepared to measure the volume of DPPC at 21 °C is shown in Fig. 2.17a. The lipid will sink, swim or float as shown in Fig. 2.17b. Centrifugation can accelerate this process, but 2 days equilibrating in a temperature-controlled Incufridge is sufficient to produce the result shown in Fig. 2.17a. As shown, DPPC at 21 °C swims (is neutrally buoyant) at a density of 1.0705 g/ml. Density is converted to molecular volume using Eq. 2.7, where GMW is the gram molecular weight.

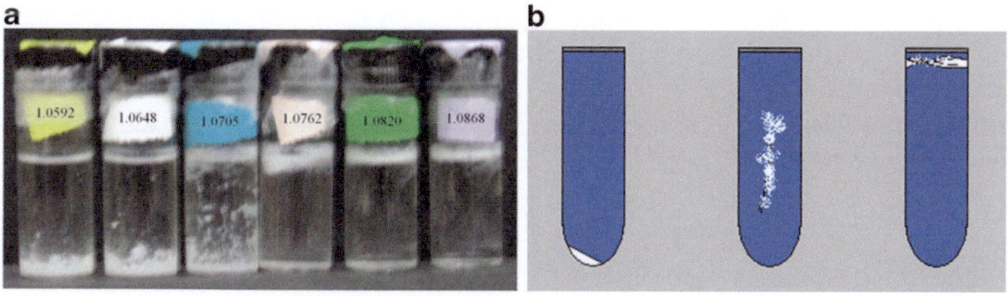

Fig. 2.17 (**a**) D_2O/H_2O solutions prepared with the densities shown contain DPPC (Reproduced with permission from Fig. 1 (Hallinen et al. 2012) from the PCCP Owner Societies). (**b**) Diagram showing three different lipid positions in a tube containing known densities

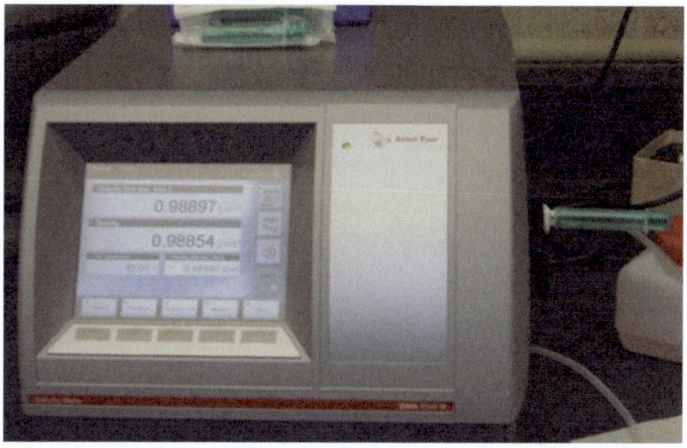

Fig. 2.18 Anton-Paar 5000M vibrating tube densimeter with sample syringe

$$\mathit{Density}\ (\mathrm{g/ml})\, x \left(\frac{1\ \mathrm{mole}}{GMW}\right) x \left(\frac{6.022\times 10^{23}\ \mathrm{molecules}}{1\ \mathrm{mole}}\right) x \left(\frac{1\ \mathrm{ml}}{10^{24}\mathring{A}^3}\right) = \left(\frac{1}{V_L\mathring{A}^3}\right) \tag{2.7}$$

The third method that our lab currently uses to obtain volumes is with a modern Anton-Paar DMA 5000 M densimeter. A picture of this instrument is shown in Fig. 2.18.

The operating principle of the densimeter is that the frequency of oscillation of a vibrating tube will be damped, i.e., the period τ will increase, when the solvent density ρ_S increases. Since it is a differential instrument, ρ_0, which is the density of a vibrating tube filled with air, is subtracted from ρ_S as shown in Eq. 2.8. K is an instrumental constant that depends on the barometric pressure in the laboratory.

$$\rho_S - \rho_0 = K\left(\tau_S^2 - \tau_0^2\right) \tag{2.8}$$

Normally fairly dilute MLV suspensions of lipids or lipid/protein mixtures are prepared (2–5 wt%). It is important that the concentration is known precisely since the density from this dilute solution is extrapolated to 100 % lipid or lipid/protein mixture. Generally 50 mg lipid in 1.2 ml MilliQ water is used for the densimeter experiment. The molecular volume is calculated using Eq. 2.9, where ρ_W is the density of water at a specific temperature, m_W = mass of water, m_L = mass of lipid or lipid/mixture and M_L = GMW.

$$V_L = \frac{M_L}{0.6022P_S}\left[1 + \frac{m_W}{m_L}\left(1 - \frac{\rho_S}{\rho_W}\right)\right] \tag{2.9}$$

For lipids that have been measured by both neutral flotation and densimetry, we have obtained quite good agreement provided that the densimeter sample is at least 5 wt% and that it is bubble-free. The samples are not degassed to avoid changes in concentration; however, they usually stand for one day prior to measurement to reduce bubble formation. For this same reason, measurements in the densimeter are confined to <50 °C. The densimeter is housed in a clear plastic box to reduce air drafts. We did encounter an artifact using the densimeter whereupon repeatedly scanning a lipid suspension up and down in temperature, the density appeared to increase. When we remixed the sample by removing and replacing it in the densimeter, the density returned to its original value (Hallinen et al. 2012). If care is taken with the sample preparation and loading procedure, then the Anton-Paar 5000 M densimeter obtains molecular volumes quite close to those obtained by neutral flotation. An example of this good agreement is shown in Fig. 2.19. The diphytanoylPC volume obtained using densimetry was 1425 Å^3 while that obtained using neutral buoyancy was 1435 Å^3.

As described above in Sect. 2.4, the headgroup volume V_H, is needed to determine A_L for X-ray data using Eq. 2.3. How do we know V_H? For PC lipids, we determined V_H by fitting well-determined DMPC gel phase data with a model of a bilayer (Tristram-Nagle et al. 2002). In that work, A_L was determined both using oriented WAXS as described in Sect. 2.3, and also by fitting 10 lamellar orders of

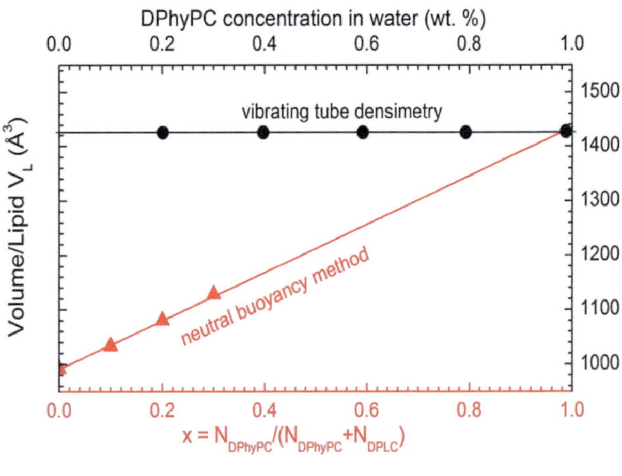

Fig. 2.19 Molecular volume of diphytanoylPC obtained using vibrating tube densimetry (*black circles*) compared to neutral buoyancy (*red triangles*) (Reproduced with permission from Fig. 2 (Tristram-Nagle et al. 2010))

LAXS. This study yielded 331 $Å^3$ for the PC headgroup which was close to an earlier value obtained by our lab of 319 $Å^3$ (Sun et al. 1994). We assume that V_H does not change with temperature because when the chain melting transitions (Tms) of lipids with n carbons/chain were plotted vs. 1(/n−3), a linear fit extrapolated to the melting temperature of polyethylene, indicating that the headgroup does not "melt" with increasing temperature (Nagle and Wilkinson 1978). For the PE lipids, we relied on the X-ray gel electron density profiles of McIntosh (Mcintosh and Simon 1986) and our own volumetric measurements (Wiener et al. 1988) to determine $V_H = 252$ $Å^3$. For PS lipids, we similarly used gel phase X-ray electron density profiles to determine that $V_H = 244$ $Å^3$ (Petrache et al. 2004). For PG lipids, we joined forces with Pan's team to determine that the headgroup volume is 291 $Å^3$, this time using X-ray and neutron scattering data to obtain precise structural parameters. Recently we determined that for ½ TMCL (4-chain cardiolipin), $V_H = 253$ $Å^3$, close to that of PE lipids.

To conclude, this chapter summarizes early membrane models and experiments to characterize the structure and thermodynamic properties of hydrated lipid bilayers. It discusses the problems with the gravimetric method to determine n_W. It describes in detail how the Nagle/Tristram-Nagle laboratory use LAXS and WAXS, and neutron scattering, to precisely determine A_L, which is then used to calculate n_W, the total number of water molecules/lipid, and n_W', the steric water near the lipid headgroup. Details of experiments to determine the volume, which is the foundation of the X-ray structural work, are also presented. Finally, a summary of many literature experiments to determine constrained headgroup water is presented in Table 2.2.

Acknowledgements The author would like to thank Ben Sauerwine for preparing the snapshot of fluctuating bilayers from the Monte Carlo simulation. Funding was from NIH GM 44976.

References

Bach D, Miller IR (1998) Hydration of phospholipid bilayers in the presence and absence of cholesterol. Biochim Biophys Acta 1368(2):216–224

Binder H, Kohlstrunk B, Heerklotz HH (1999) Hydration and lyotropic melting of amphiphilic molecules: a thermodynamic study using humidity titration calorimetry. J Colloid Interface Sci 220(2):235–249

Boscia AL, Akabori K, Benamram Z, Michel JA, Jablin MS, Steckbeck JD, Montelaro RC, Nagle JF, Tristram-Nagle S (2013) Membrane structure correlates to function of LLP2 on the cytoplasmic tail of HIV-1 gp41 protein. Biophys J 105(3):657–666

Boscia AL, Treece BW, Mohammadyani D, Klein-Seetharaman J, Braun AR, Wassenaar TA, Klosgen B, Tristram-Nagle S (2014) X-ray structure, thermodynamics, elastic properties and MD simulations of cardiolipin/dimyristoylphosphatidylcholine mixed membranes. Chem Phys Lipids 178:1–10

Bronshteyn VL, Steponkus PL (1993) Calorimetric studies of freeze-induced dehydration of phospholipids. Biophys J 65(5):1853–1865

Buldt G, Gally HU, Seelig J, Zaccai G (1979) Neutron-diffraction studies on phosphatidylcholine model membranes. 1. Head group conformation. J Mol Biol 134(4):673–691
Caillé A (1972) X-ray scattering by smectic-A crystals. CR Acad Sci Paris Sér B 274:891–893
Channareddy S, Jose SS, Janes N (1997) Direct determination of hydration in the lamellar to inverted hexagonal transition of phosphatidylethanolamine. J Am Chem Soc 119:2345–2347
Chu N, Kučerka N, Liu YF, Tristram-Nagle S, Nagle JF (2005) Anomalous swelling of lipid bilayer stacks is caused by softening of the bending modulus. Phys Rev E 71(4):041904
Crowe LM, Spargo BJ, Ioneda T, Beaman BL, Crowe JH (1994) Interaction of cord factor (alpha, alpha′-trehalose-6,6′-dimycolate) with phospholipids. Biochim Biophys Acta Biomembr 1194:53–60
Danielli JF, Davson H (1935) A contribution to the theory of permeability of thin films. J Cell Comp Physiol 5(4):495–508
de Gennes PG, Prost J (1993) The physics of liquid crystals, 2nd edn. Oxford University Press, New York
Dufourcq P, Louis H, Dandre F, Lavie J, Bonnet J, Lamaziere JM (1997) Phenotypic modification of arterial smooth muscle cells in response to medial dissection. Coron Artery Dis 8(3–4):163–170
Eldho NV, Feller SE, Tristram-Nagle S, Polozov IV, Gawrisch K (2003) Polyunsaturated docosahexaenoic vs docosapentaenoic acid – differences in lipid matrix properties from the loss of one double bond. J Am Chem Soc 125(21):6409–6421
Finer EG, Darke A (1974) Phospholipid hydration studied by deuteron magnetic resonance spectroscopy. Chem Phys Lipids 12(1):1–16
Frye LD, Edidin M (1970) The rapid intermixing of cell surface antigens after formation of mouse-human heterokaryons. J Cell Sci 7(2):319–335
Gallova J, Uhrikova D, Kučerka N, Teixeira J, Balgavy P (2008) Hydrophobic thickness, lipid surface area and polar region hydration in monounsaturated diacylphosphatidylcholine bilayers: SANS study of effects of cholesterol and beta-sitosterol in unilamellar vesicles. Biochim Biophys Acta 1778(11):2627–2632
Gawrisch K, Richter W, Mops A, Balgavy P, Arnold K, Klose G (1985) The influence of water concentration on the structure of egg-yolk phospholipid water dispersions. Stud Biophys 108(1):5–16
Gorter E, Grendel F (1925) On bimolecular layers of lipoids on the chromocytes of the blood. J Exp Med 41(4):439–443
Gouliaev N, Nagle JF (1998) Simulations of interacting membranes in the soft confinement regime. Phys Rev Lett 81(12):2610–2613
Grdadolnik J, Hadzi D (1994) Conformational effects of metal salt binding to the polar head of phosphatidylcholines investigated by FTIR spectroscopy. Chem Phys Lipids 65:121–132
Greenwood AI, Pan JJ, Mills TT, Nagle JF, Epand RM, Tristram-Nagle S (2008) CRAC motif peptide of the HIV-1 gp41 protein thins SOPC membranes and interacts with cholesterol. Biochim Biophys Acta Biomembr 1778(4):1120–1130
Guinier A (1963) X-ray diffraction in crystals, imperfect crystals, and amorphous bodies. W. H. Freeman, San Francisco
Guler SD, Ghosh DD, Pan JJ, Mathai JC, Zeidel ML, Nagle JF, Tristram-Nagle S (2009) Effects of ether vs. ester linkage on lipid bilayer structure and water permeability. Chem Phys Lipids 160(1):33–44
Hallinen KM, Tristram-Nagle S, Nagle JF (2012) Volumetric stability of lipid bilayers. Phys Chem Chem Phys 14(44):15452–15457
Hosemann R, Bagchi SN (1962) Direct analysis of diffraction by matter. North-Holland, Amsterdam
Hristova K, White SH (1998) Determination of the hydrocarbon core structure of fluid dioleoylphosphocholine (DOPC) bilayers by x-ray diffraction using specific bromination of the double-bonds: effect of hydration. Biophys J 74(5):2419–2433
Jendrasiak GL, Hasty JH (1974) The hydration of phospholipids. Biochim Biophys Acta 337(1):79–91

Katsaras J (1998) Adsorbed to a rigid substrate, dimyristoylphosphatidylcholine multibilayers attain full hydration in all mesophases. Biophys J 75(5):2157–2162
Klose G, Konig B, Meyer HW, Schulze G, Degovics G (1988) Small-angle X-ray-scattering and electron-microscopy of crude dispersions of swelling lipids and the influence of the morphology on the repeat distance. Chem Phys Lipids 47(3):225–234
Koenig BW, Strey HH, Gawrisch K (1997) Membrane lateral compressibility determined by NMR and X-ray diffraction: effect of acyl chain polyunsaturation. Biophys J 73(4):1954–1966
Kučerka N, Kiselev MA, Balgavy P (2004) Determination of bilayer thickness and lipid surface area in unilamellar dimyristoylphosphatidylcholine vesicles from small-angle neutron scattering curves: a comparison of evaluation methods. Eur Biophys J Biophys 33(4):328–334
Kučerka N, Liu YF, Chu NJ, Petrache HI, Tristram-Nagle ST, Nagle JF (2005a) Structure of fully hydrated fluid phase DMPC and DLPC lipid bilayers using X-ray scattering from oriented multilamellar arrays and from unilamellar vesicles. Biophys J 88(4):2626–2637
Kučerka N, Tristram-Nagle S, Nagle JF (2005b) Structure of fully hydrated fluid phase lipid bilayers with monounsaturated chains. J Membr Biol 208(3):193–202
Kučerka N, Tristram-Nagle S, Nagle JF (2006) Closer look at structure of fully hydrated fluid phase DPPC bilayers. Biophys J 90(11):L83–L85
Kučerka N, Nagle JF, Sachs JN, Feller SE, Pencer J, Jackson A, Katsaras J (2008) Lipid bilayer structure determined by the simultaneous analysis of neutron and x-ray scattering data. Biophys J 95(5):2356–2367
Ladbrook BD, Williams RM, Chapman D (1968) Studies on lecithin-cholesterol-water interactions by differential scanning calorimetry and X-ray diffraction. Biochim Biophys Acta 150(3):333–340
Lairion F, Filler R, Disalvo EA (2002) Reversed micelles as model systems to study interfacial properties of lipid bilayers. Colloid Surf B 25(4):369–371
Levine YK (1973) X-ray diffraction studies of membranes. Prog Surf Sci 3:279–352
Lewis BA, Engelman DM (1983) Lipid bilayer thickness varies linearly with acyl chain-length in fluid phosphatidylcholine vesicles. J Mol Biol 166(2):211–217
Liu YF, Nagle JF (2004) Diffuse scattering provides material parameters and electron density profiles of biomembranes. Phys Rev E 69(4):040901
Lyatskaya Y, Liu YF, Tristram-Nagle S, Katsaras J, Nagle JF (2001) Method for obtaining structure and interactions from oriented lipid bilayers. Phys Rev E 63(1):011907
Mcintosh TJ, Simon SA (1986) Area per molecule and distribution of water in fully hydrated dilauroylphosphatidylethanolamine bilayers. Biochemistry 25(17):4948–4952
Mendelsohn R, Davies MA, Schuster HF, Xu ZC, Bittman R (1991) Cd2 rocking modes as quantitative infrared probes of one-bond, 2-bond, and 3-bond conformational disorder in dipalmitoylphosphatidylcholine and dipalmitoylphosphatidylcholine cholesterol mixtures. Biochemistry 30(35):8558–8563
Miller IR, Bach D, Wachtel EJ, Eisenstein M (2002) Interrelation between hydration and interheadgroup interaction in phospholipids. Bioelectrochemistry 58(2):193–196
Mills TT, Toombes GES, Tristram-Nagle S, Smilgies DM, Feigenson GW, Nagle JF (2008) Order parameters and areas in fluid-phase oriented lipid membranes using wide angle x-ray scattering. Biophys J 95(2):669–681
Nagle JF (1976) Theory of lipid monolayer and bilayer phase-transitions – effect of headgroup interactions. J Membr Biol 27(3):233–250
Nagle JF (1980) Theory of the main lipid bilayer phase transition. Annu Rev Phys Chem 31:157–195
Nagle JF, Tristram-Nagle S (2000) Structure of lipid bilayers. Biochim Biophys Acta Rev Biomembr 1469(3):159–195
Nagle JF, Wilkinson DA (1978) Lecithin bilayers – density measurements and molecular interactions. Biophys J 23(2):159–175
Nagle JF, Zhang RT, Tristram-Nagle S, Sun WJ, Petrache H, Suter RM (1996) X-ray structure determination of fully hydrated L(alpha) phase DPPC bilayers. Biophys J 70(2):1419–1431

Pan J, Tristram-Nagle S, Kučerka N, Nagle JF (2008a) Temperature dependence of structure, bending rigidity, and bilayer interactions of dioleoylphosphatidylcholine bilayers. Biophys J 94(1):117–124

Pan JJ, Mills TT, Tristram-Nagle S, Nagle JF (2008b) Cholesterol perturbs lipid bilayers nonuniversally. Phys Rev Lett 100(19)

Pan JJ, Tieleman DP, Nagle JF, Kučerka N, Tristram-Nagle S (2009a) Alamethicin in lipid bilayers: combined use of X-ray scattering and MD simulations. Biochim Biophys Acta Biomembr 1788(6):1387–1397

Pan JJ, Tristram-Nagle S, Nagle JF (2009b) Effect of cholesterol on structural and mechanical properties of membranes depends on lipid chain saturation. Phys Rev E 80(2):021931

Pan JJ, Heberle FA, Tristram-Nagle S, Szymanski M, Koepfinger M, Katsaras J, Kučerka N (2012) Molecular structures of fluid phase phosphatidylglycerol bilayers as determined by small angle neutron and X-ray scattering. Biochim Biophys Acta Biomembr 1818(9):2135–2148

Petrache HI, Gouliaev N, Tristram-Nagle S, Zhang RT, Suter RM, Nagle JF (1998a) Interbilayer interactions from high-resolution x-ray scattering. Phys Rev E 57(6):7014–7024

Petrache HI, Tristram-Nagle S, Nagle JF (1998b) Fluid phase structure of EPC and DMPC bilayers. Chem Phys Lipids 95(1):83–94

Petrache HI, Tristram-Nagle S, Gawrisch K, Harries D, Parsegian VA, Nagle JF (2004) Structure and fluctuations of charged phosphatidylserine bilayers in the absence of salt. Biophys J 86(3):1574–1586

Pohle W, Selle C, Fritzsche H, Binder H (1998) Fourier transform infrared spectroscopy as a probe for the study of the hydration of lipid self-assemblies. I. Methodology and general phenomena. Biospectroscopy 4(4):267–280

Raghunathan M, Zubovski Y, Venable RM, Pastor RW, Nagle JF, Tristram-Nagle S (2012) Structure and elasticity of lipid membranes with genistein and daidzein bioflavinoids using X-ray scattering and MD simulations. J Phys Chem B 116(13):3918–3927

Rand RP, Parsegian VA (1989) Hydration forces between phospholipid bilayers. Biochim Biophys Acta 988(3):351–376

Robertson JD (1960) The molecular structure and contact relationships of cell membranes. Prog Biophys Mol Biol 10:343–418

Salsbury NJ, Chapman D, Darke A (1972) Deuteron magnetic-resonance studies of water associated with phospholipids. Chem Phys Lipids 8(2):142–151

Shchelokovskyy P, Tristram-Nagle S, Dimova R (2011) Effect of the HIV-1 fusion peptide on the mechanical properties and leaflet coupling of lipid bilayers. New J Phys 13:025004

Simons K, Toomre D (2000) Lipid rafts and signal transduction. Nat Rev Mol Cell Biol 1(1):31–39

Singer SJ, Nicolson GL (1972) The fluid mosaic model of the structure of cell membranes. Science 175(4023):720–731

Small DM (1967) Phase equilibria and structure of dry and hydrated egg lecithin. J Lipid Res 8(6):551–557

Sun W, Suter RM, Knewtson MA, Worthington CR, Tristram-Nagle S, Zhang R, Nagle JF (1994) Order and disorder in fully hydrated unoriented bilayers of gel-phase dipalmitoylphosphatidylcholine. Phys Rev E Stat Phys Plasmas Fluids Relat Interdiscip Topics 49(5):4665–4676

Sun WJ, Tristram-Nagle S, Suter RM, Nagle JF (1996) Structure of gel phase saturated lecithin bilayers: temperature and chain length dependence. Biophys J 71(2):885–891

Swart SP (1907) About the permeability of artificial lipid membranes for proferments. Biochem Z 6:358–365

Tardieu A, Luzzati V, Reman FC (1973) Structure and polymorphism of the hydrocarbon chains of lipids: a study of lecithin-water phases. J Mol Biol 75(4):711–733

Torbet J, Wilkins MHF (1976) X-Ray diffraction studies of lecithin bilayers. J Theor Biol 62(2):447–458

Tristram-Nagle SA (2007) Preparation of oriented, fully hydrated lipid samples for structure determination using X-ray scattering. Methods Mol Biol 400:63–75

Tristram-Nagle S, Nagle JF (2004) Lipid bilayers: thermodynamics, structure, fluctuations, and interactions. Chem Phys Lipids 127(1):3–14

Tristram-Nagle S, Nagle JF (2007) HIV-1 fusion peptide decreases bending energy and promotes curved fusion intermediates. Biophys J 93(6):2048–2055
Tristram-Nagle S, Yang CP, Nagle JF (1986) Thermodynamic studies of purple membrane. Biochim Biophys Acta 854(1):58–66
Tristram-Nagle S, Wiener MC, Yang CP, Nagle JF (1987) Kinetics of the subtransition in dipalmitoylphosphatidylcholine. Biochemistry Us 26(14):4288–4294
Tristram-Nagle S, Zhang R, Suter RM, Worthington CR, Sun WJ, Nagle JF (1993) Measurement of chain tilt angle in fully hydrated bilayers of gel phase lecithins. Biophys J 64(4):1097–1109
Tristram-Nagle S, Petrache HI, Nagle JF (1998) Structure and interactions of fully hydrated dioleoylphosphatidylcholine bilayers. Biophys J 75(2):917–925
Tristram-Nagle S, Liu YF, Legleiter J, Nagle JF (2002) Structure of gel phase DMPC determined by X-ray diffraction. Biophys J 83(6):3324–3335
Tristram-Nagle S, Kim DJ, Akhunzada N, Kučerka N, Mathai JC, Katsaras J, Zeidel M, Nagle JF (2010) Structure and water permeability of fully hydrated diphytanoylPC. Chem Phys Lipids 163(6):630–637
Ulrich AS, Watts A (1994) Molecular response of the lipid headgroup to bilayer hydration monitored by H2-Nmr. Biophys J 66(5):1441–1449
Wiener MC, Tristram-Nagle S, Wilkinson DA, Campbell LE, Nagle JF (1988) Specific volumes of lipids in fully hydrated bilayer dispersions. Biochim Biophys Acta 938(2):135–142
Wilkinson DA, McIntosh TJ (1986) A subtransition in a phospholipid with a net charge, dipalmitoylphosphatidylglycerol. Biochemistry 25(2):295–298
Wilkinson DA, Nagle JF (1978) Differential dilatometer. Anal Biochem 84(1):263–271
Wilkinson DA, Nagle JF (1979) Dilatometric study of binary mixtures of phosphatidylcholines. Biochemistry 18(19):4244–4249
Wilkinson DA, Nagle JF (1981) Dilatometry and calorimetry of saturated phosphatidylethanolamine dispersions. Biochemistry 20(1):187–192
Wilkinson DA, Nagle JF (1982) Specific heats of lipid dispersions in single-phase regions. Biochim Biophys Acta 688(1):107–115
Wilkinson DA, Nagle JF (1984) Metastability in the phase behavior of dimyristoylphosphatidylethanolamine bilayers. Biochemistry 23(7):1538–1541
Wilkinson DA, Tirrell DA, Turek AB, McIntosh TJ (1987) Tris buffer causes acyl chain interdigitation in phosphatidylglycerol. Biochim Biophys Acta 905(2):447–453
Yang CP, Nagle JF (1988) Phase transformations in lipids follow classical kinetics with small fractional dimensionalities. Phys Rev A 37(10):3993–4000
Zhang R, Suter RM, Nagle JF (1994) Theory of the structure factor of lipid bilayers. Phys Rev E Stat Phys Plasmas Fluids Relat Interdiscip Topics 50(6):5047–5060
Zhang R, Sun WJ, Tristram-Nagle S, Headrick RL, Suter RM, Nagle JF (1995) Critical fluctuations in membranes. Phys Rev Lett 74(14):2832–2835
Zhang R, Tristram-Nagle S, Sun WJ, Headrick RL, Irving TC, Suter RM, Nagle JF (1996) Small-angle X-ray scattering from lipid bilayers is well described by modified Caillé theory but not by paracrystalline theory. Biophys J 70(1):349–357

Chapter 3
Water and Lipid Bilayers

Jonathan D. Nickels and John Katsaras

Abstract *Water is crucial to the structure* and function of biological membranes. In fact, the membrane's basic structural unit, i.e. the lipid bilayer, is self-assembled and stabilized by the so-called hydrophobic effect, whereby lipid molecules unable to hydrogen bond with water aggregate in order to prevent their hydrophobic portions from being exposed to water. However, this is just the beginning of the lipid-bilayer-water relationship. This mutual interaction defines vesicle stability in solution, controls small molecule permeation, and defines the spacing between lamella in multi-lamellar systems, to name a few examples. This chapter will describe the structural and dynamical properties central to these, and other water- lipid bilayer interactions.

Keywords Permeation • Water distribution • Dynamics

3.1 Water at the Interface of Model Membranes

Lipid-water interactions are ubiquitous in biological systems; our goal is to discuss lipid bilayer-water interactions described in the literature, and more precisely, the ways in which water and lipid bilayers mutually define the structure and dynamics of the lipid-water interface. In fact, the importance of water is such, that bilayer-bilayer interactions have been modeled as the interaction of their associated water shells (Leikin et al. 1993). Keeping this in mind, we begin our discussion with the properties of water at the interface of model membranes.

The structure of interfacial water can be described in a number of ways – depending on how one chooses to approach the system. The classical double layer (Debye and Hückel 1923) description of a lipid bilayer in an aqueous solution, the

J.D. Nickels (✉)
Joint Institute for Neutron Sciences, Oak Ridge National Laboratory, Oak Ridge, TN, USA
e-mail: nickelsjd@ornl.gov

J. Katsaras (✉)
Biology & Soft Matter and Biosciences Division, Oak Ridge National Laboratory, Oak Ridge, TN, USA
e-mail: katsarasj@ornl.gov

E.A. Disalvo (ed.), *Membrane Hydration*, Subcellular Biochemistry 71,
DOI 10.1007/978-3-319-19060-0_3

solvent is treated as a continuum and the ions in solution as point charges. Some of these ions are attracted to the bilayer surface, where they act as immobile point charges, screening the membrane. The free ions and solvent molecules near the membrane's surface make up the so-called diffuse layer, and are influenced by the bilayer's net electric field. Within this diffuse layer is the slip plane, a boundary separating the region dynamically associated with the lipid bilayer surface from the mobile region which intermingles with the bulk solvent. The electrical potential at the slip plane – the bulk boundary interface – is known as the zeta potential (ζ). Although ζ is not a true measure of surface charge, it is useful in estimating particle stability. In the case of lipid vesicles, ζ is usually in the range of +/−100 meV, with highly charged particles (-25 mV $> \zeta$ or $\zeta > 25$ mV) being more stable in solution.

The spatial extent to which a lipid bilayer electrostatically influences an aqueous solution is defined as the Debye length, κ^{-1}, and is expressed as:

$$\kappa^{-1} = \sqrt{\frac{\varepsilon_r \varepsilon_0 k_B T}{2 N_A E^2 I}}, \tag{3.1}$$

where ε_r is the dielectric constant, ε_0 is the permittivity constant, I is the ionic strength, N_A is Avogadro's number, $k_B T$ is the temperature and Boltzman constant, and E is the elementary unit of charge. The Debye length is typically on the order of one to several nanometers for vesicle and bilayer systems. This is an important consideration when it comes to understanding the effects of ions in screening the surface charge of lipid bilayers. For example, large amounts of salt increase the ionic strength, thus decreasing the Debye length. This decrease in κ^{-1} allows adjacent bilayers to approach closer to each other prior to experiencing electrostatic repulsion. This notion has been borne-out by numerous experiments studying multiple lamellae and vesicle suspensions.

The resulting forces (e.g., van der Waals attractive) between charged surfaces interacting through a liquid medium are quantitatively described by the DLVO theory (Verwey et al. 1948). In the case of two vesicles of radius, r, and separated by a distance, D, one can calculate the interaction energy between the two vesicles through the approximation of the attractive van der Waals force as follows:

$$\Phi_{\mathrm{vdW}}(\mathrm{D}) = \frac{-\mathrm{Ar}}{12\mathrm{D}}, \tag{3.2}$$

where A is the Hamaker constant. Similarly, one can define the repulsive electrostatic force as:

$$\Phi_{\mathrm{R}}(\mathrm{D}) = 2\pi \varepsilon_r \varepsilon_0 \mathrm{r} \mathrm{E}^2 \mathrm{e}^{-\kappa \mathrm{D}}. \tag{3.3}$$

By combining these two equations, we get the interaction energy between the two spheres:

$$\Phi(\mathrm{D}) = \Phi_{\mathrm{vdW}}(\mathrm{D}) + \Phi_{\mathrm{R}}(\mathrm{D}). \tag{3.4}$$

This treatment is useful in providing an intuitive feel for the role of ions and water, and how they interact at the interface of a lipid bilayer, influencing its stability and enabling the formation of other aggregate morphologies (e.g., stalks, vesicles, etc.).

Beyond this, ions screening the bilayer surface are also known to bind to the head group carbonyl oxygen, as was shown in POPC (16:0–18:1 phosphatidylcholine) bilayers, creating larger interacting lipid complexes (Böckmann et al. 2003). This aggregation of lipid molecules increases the ordering of the acyl chains, and results in thicker bilayers with reduced lipid in-plane mobility. Ions in solution have also been proposed to affect water structure and diffusional rates within the bilayer (Kausik and Han 2011), consistent with what has been observed in the bulk (Ishai et al. 2013). Perhaps the most notable example of this is the role of the water-ion complex in the selectivity of ion channels (Doyle et al. 1998).

It is clear, however, that descriptions treating the interface in a manner similar to the bulk fail to describe many of the molecular details regarding ions, lipids, and more importantly, the solvent. Alternative perspectives of understanding the interface, such as the introduction of water sub-phases to which one can ascribe anomalous properties (Disalvo et al. 2008), can be useful. In doing so, one can develop models which better predict the structural changes of water and the resultant modified hydrogen bond network (Lelkes and Miller 1980; Gawrisch et al. 1992; König et al. 1994; Zhou and Schulten 1995; Swenson et al. 2008).

In light of these developments, others have sought to understand the repulsive force between bilayers microscopically, i.e., based on the local restructuring of water by the bilayer. Experimental evidence of structured water was first observed using NMR in the studies of Finer and Darke (1974), and was later theoretically interpreted by the order parameter model of Marčelja and Radić (1976). This model uses a free energy approach, relating the imposed structural order of water at the interface to disordered bulk water to the repulsive hydration pressure, P_h, between two surfaces. This is written as;

$$P_h = -\partial G\left(d_f\right) / \partial\left(d_f\right) \tag{3.5}$$

where d_f is the fluid space between adjacent bilayers and $G(d_f)$ is the excess Gibbs free energy per unit volume. The minimum of this expression for repulsive pressure can be used to define the equilibrium spacing of adjacent bilayers when evaluated fully (Rand and Parsegian 1989).

The order parameter is defined by the geometrical and dynamical restrictions imposed by the bilayer, thereby bringing in to play the presence of hydrogen bond partners, hydrogen bond lifetimes, and dynamical retardation factors. Interestingly, this result is equivalent to the Debye model when evaluated at distances much greater than the order parameter, λ, which is expressed below using the nomenclature of McIntosh and Simon:

$$P_h \approx 4P_o \mathrm{e}^{-d_f/\lambda} \tag{3.6}$$

Here λ is the length scale of the order parameter (McIntosh and Simon 1994) and P_0 is a function of the solvent and surface properties. P_0 can be expanded as follows:

$$P_o = 2\chi \frac{{\Psi_h}^2}{\lambda} \tag{3.7}$$

where Ψ_h is the hydration potential and χ is the orientational susceptibility of the solvent. Hydrated lipid bilayers exhibit a variety of interactions (e.g., water-water, lipid-water, and lipid-lipid), and this model can be extended to include repulsive pressures such as those arising from water-water interactions, as well as steric lipid-lipid interactions, including thermally induced undulations (Helfrich and Servuss 1984; Kornyshev and Leikin 1989; Cevc 1991; Parsegian and Rand 1991; Israelachvili 2011).

Early investigators lacked information about the actual energies and geometries of hydration water, and initially the structure of ice was seen as a system from which one was able to glean information regarding the orientational biases of water (Bernal and Fowler 1933; Jorgensen et al. 1983). However, ice is far from being an accurate description of hydration water, failing to include more complicated collective features that had been reported in bulk water (Frank and Wen 1957; Fecko et al. 2003; Tielrooij et al. 2010) (*Note of the Editor: see Chap. 7 by Alarcon et al.).

New models are still being proposed to explain the reorientation of water (Torre et al. 2004; Laage and Hynes 2006; Taschin et al. 2013), which will undoubtedly provide some new insights about the nature of hydration water at the bilayer surface.

We must also recognize that in biological membranes, the system is not uniformly distributed. Complex lipid and sterol compositions, the inclusions of proteins, glycosylation, as well as the formation of lipid domains all result in inhomogeneous local properties of the bilayer and the solvent. This complexity is critical for the function of biological membranes, but it also makes the analysis of real systems non-trivial.

3.2 Water Distribution Within the Bilayer

Water not only associates with lipid head groups (Griffith et al. 1974), but also with many of the membrane's other functional groups – in addition to occupying free volume. Early measurements of lecithin bilayers by Zaccai et al., mapped out the distribution of water across the bilayer (Zaccai et al. 1975). They were able to quantify water molecules in equilibrium near the glycerol and phosphorylcholine (PC) head group regions, even at low hydration and below the gelation temperature.

Up to this point we have focused on water outside the formal bilayer volume. However, there is a finite amount of water that can reside deep within the lipid-water interface. In this region, water can directly affect the membrane's structural and dynamical ordering. Ultimately, we will consider the role that the hydrocarbon region plays in defining the amount of water associated with the head group, and consider the limited, but necessary presence of water inside the bilayer's hydrocarbon core.

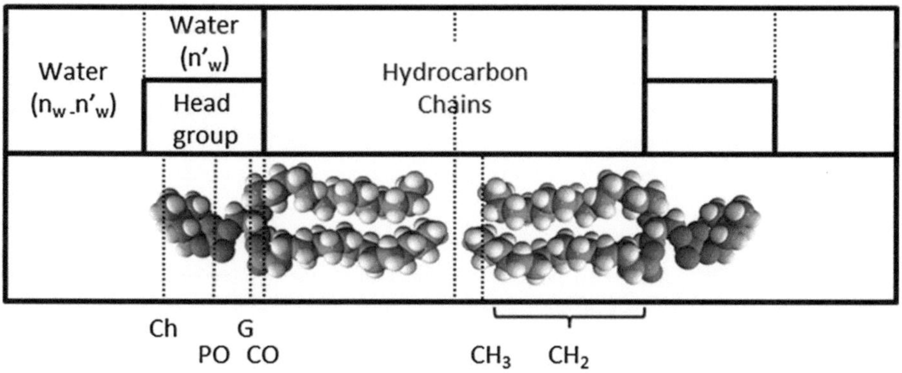

Fig. 3.1 The *upper panel* shows the parsing scheme of a DPPC bilayer in the L_α fluid phase and its waters of hydration, as depicted by Nagle and Wiener (1988). In the *lower panel*, the relative positions of the different head group moieties are illustrated, with the abbreviations Ch, PO, G, CO, CH_2, and CH_3 representing the choline, phosphate, glycerol, carbonyl, methylene, and methyl groups, respectively

The upper panel of Fig. 3.1 shows an idealized lipid bilayer (Nagle and Wiener 1988), emphasizing the substantial presence of water in the region of the polar head group, as well as recognizing that approximately 40 % of the total bilayer thickness is due to this hydrated region (*see Chap. 2 by Tristram-Nagle). (For illustrative purposes we will use the structure of a DPPC bilayer, which has been extensively studied over a number of decades, using both neutron and x-ray scattering techniques; and more recently, molecular dynamics simulations). For all intents and purposes, the DPPC bilayer is considered the prototypical model membrane system (Büldt et al. 1978; Katsaras 1995; Nagle et al. 1996; Petrache et al. 1997; Nagle and Tristram-Nagle 2000; Kučerka et al. 2008a).

Starting from the bilayer center, the terminal methyl groups are followed by 28 methylene groups – 14 in each of DPPC's hydrocarbon chains. Taken together, the methyl and methylene groups represent the bilayer's hydrophobic core. Interactions between hydrocarbon chains are predominantly van der Waals, and there is no possibility for hydrogen bonding with neighboring chains.

The two acyl chains and the PC head group are attached, via ester bonds, to the glycerol backbone's sn-1 and sn-2, and sn-3 positions, respectively (Fig. 3.1). Variations to this structure involve substitutions to the choline head group or the acyl chains. Typical head groups include serine, ethanolamine, glycerol, and inositol, to name a few. Acyl chains can also vary in length and degree of unsaturation (number of double bonds). In fact, biologically relevant lipids contain a saturated hydrocarbon chain at the sn-1 position and a longer unsaturated chain at the sn-2 position (e.g., POPC). These variations in acyl chain length and unsaturation have a major impact on the thickness and packing density of the bilayer's hydrophobic core, with longer saturated chains trending to thicker bilayers, and increased unsaturation leading to more disordered hydrophobic cores.

The lipid head group region intimately associates water, as we have discussed in the previous section. The carbonyl oxygen atoms participate, on average, one hydrogen bond with water. Non-ether phosphate oxygen atoms average about 4 hydrogen bonds per lipid, of which 1.7 are involved in bridging two lipid molecules together (Pasenkiewicz-Gierula et al. 1997). These water molecules are considered to be "tightly" bound, and are structurally and dynamically distinct from bulk water (Ghosh et al. 2007; Nagata and Mukamel 2010).

Simulations have studied the role that head group chemistry plays in determining the structure of water in the vicinity of the bilayer (Murzyn et al. 2006). For example, phosphatidylethanolamine (PE) and phosphotidylglycerol (PG) head groups hydrogen bond with only 1 or 2 water molecules each, while PC head groups affect ~11 water molecules by inducing a "clathrate" structure to accommodate the three methyl groups of choline. This water structuring scenario is consistent with experimental data from amphiphilic molecules in solution (Perticaroli et al. 2011).

In DPPC bilayers, up to 16 water molecules per lipid have been found to be affected by the presence of the bilayer (i.e., $\sim 5H_2O$ molecules tightly bound to CO and PO, $\sim 11H_2O$ molecules associated with the choline methyls). Åman and Lindahl suggest that a second shell of ~6 additional water molecules exhibit some lesser ordering imposed from contact with the more ordered inner shell of water (Åman et al. 2003). This is consistent with the picture of a double layer. In the context of the order parameter model, it is the work required to orient these additional water molecules that gives rise to the so-called 'hydration force' which manifests itself when two, apposing, hydrated bilayers approach each other (Rand and Parsegian 1989; Leikin et al. 1993).

We will now go on to consider how the bilayer's hydrophobic core contributes to the water structure at the bilayer surface. Primarily this affects water in the choline region and the 'second shell', which we referred to in the previous paragraph. The latter population of waters appears to be highly sensitive to the area per lipid (Table 3.1), a parameter known to vary, for a given phase (e.g., gel or liquid crystalline) and as a function of acyl chain length. Thickening of the bilayer with increasing hydration or a thermodynamic phase change (e.g., gel to liquid crystalline) are two ways that this effect can be observed.

In purple membranes for example, upon full hydration the bilayers thicken by about 0.5 nm, even below the gel transition (Fitter et al. 1999). In this rigid structure, the water molecules associate with the lipid's glycerol backbone, and each water molecule is able to hydrogen bond with multiple neighboring phospholipids. Upon heating through the gel transition, bilayer thickness expands from 5.3 to 6.8 nm. This change is associated with incoming hydration water in to the newly available volume around the head group. This picture of bilayer thickening with increasing levels of hydration has been found to be consistent with results from NMR (Ulrich and Watts 1994) and infrared spectroscopy (Binder 2007). Changes in hydration have been shown to play a role in determining the phase of gel DMPC (Sirota et al. 1988) and DPPC (Raghunathan and Katsaras 1995) bilayers, where increasing humidity was shown to lower the transition temperature for the L_α phase as well as the transition temperatures of the other sub phases (Katsaras et al. 2000).

Table 3.1 Number of water molecules associated with the lipid head group region, n_w' and the area per lipid molecule

	n_w'	Area [Å^2]	Reference
DOPC	10.7	72.4	Kučerka et al. (2006)
DOPC	11.0	72.2	Tristram-Nagle et al. (1998)
DOPC	11.1	72.5	Nagle and Tristram-Nagle (2000)
POPC	9.4	68.3	Kučerka et al. (2006)
di22:1PC	9.7	69.3	Kučerka et al. (2006)
DMPC	7.2	60.3	Kučerka et al. (2006)
DMPC	7.2	59.6	Nagle and Tristram-Nagle (2000)
EPC	10.2	69.4	Nagle and Tristram-Nagle (2000)
DPPC (Gel)	3.7	47.9	Nagle and Tristram-Nagle (2000)
DPPC	8.6	64.0	Nagle and Tristram-Nagle (2000)
DPPC	11.9	62.9	Kučerka et al. (2004)
DLPE (Gel)	2.0	41.0	Nagle and Tristram-Nagle (2000)
DLPE	4.7	51.2	Nagle and Tristram-Nagle (2000)
DLPE (Gel)	7	41.0	McIntosh and Simon (1986)
DLPE	9	49.1	McIntosh and Simon (1986)

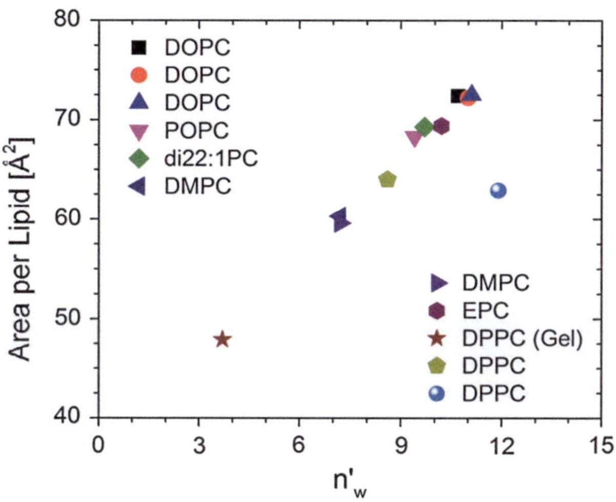

Fig. 3.2 Here we illustrate the number of head group associated water molecules, n_w', versus area per lipid for the different PC lipids in Table 3.1. We see that the available volume between head groups, as implied from area per lipid, is directly correlated to the water content in the head group region

A basic parameter that is used to describe interfacial water is n_w', the number of water molecules within the volume defined by the head group – i.e., the head group thickness, D_{HH}. This parameter is obtained through X-ray and neutron scattering studies of bilayers (Nagle and Tristram-Nagle 2000). It should be noted that new joint refinement models based on molecular volumes do not explicitly compute n_w' (Kučerka et al. 2008a). Also, as summarized in Table 3.1, a strong correlation can be implied between the number of water molecules and the area per lipid (see Fig. 3.2).

It is interesting that the number of water molecules associated with each PC head group is correlated to the packing of acyl chains in the hydrophobic core. This means that the free volume in the head group region is a determining factor in the number of hydration waters. It follows then that this relationship would also be observed when pressure is used to modulate area per lipid. In the study by Tristram-Nagle et al., the number of water molecules per DOPC lipid indeed decreased as a function of increasing hydrostatic (or osmotic stress) pressure – unsurprisingly, the area per lipid molecule also decreased with increasing pressure (Tristram-Nagle et al. 1998). This has implications in how one considers permeation of water through the bilayer (Mathai et al. 2008), as will be discussed in a later section of this chapter.

It should also be pointed out that the experimentally reported head group associated water molecules are fewer than those reported from simulations. At this point we should mention a few caveats regarding the water estimates from liquid crystallographic studies. Firstly, the number of head group associated water molecules, n_w', are determined from the excess of the fit functions used for lipid volumes, i.e. there is no explicit term in these models for water. This brings us to the second caveat, namely that the volumes of all water molecules (associated and bulk) are the same, thus enabling one to determine n_w'. However, it is clear that the order parameters (and hence the volume) of hydration water is different from those in the bulk. These obvious deficiencies clearly show the need to further refine the structural models used to interpret scattering data.

These deficiencies are important to keep in mind when one considers water within the hydrophobic region. So, while NMR (Griffith et al. 1974) and a few scattering studies (Kučerka et al. 2008b) have shown water penetration into the bilayer's hydrophobic core, the absence of a measureable probability for water in most scattering models is simply a reflection of how the structural model deals with the exceedingly low statistical probability of finding water in this region.

Generically, trace water content decreases from the acyl chain carbonyls to the bilayer center, with a slight increase at the center of the bilayer, where there is an increase in free volume. This is illustrated by the free energy calculations of Marrink and Berendsen for a number of small molecules in DPPC (Marrink and Berendsen 1994). The penetration of water into the bilayer is also dependent on acyl chain length, degree of unsaturation, and the presence of cholesterol (Subczynski et al. 1994). For example, Subczynski et al. demonstrated increased hydrophobicity with increasing fatty acid chain length and the introduction of a double bond at C9–C10 position. The introduction of cholesterol, however, increased water penetration to a depth of C7–C9, the approximate depth of the cholesterol steroid ring.

3.3 Permeation

Water and other small neutrally charged solute molecules are known to passively cross lipid bilayers. As discussed in Sect. 3.2, water is present, to some degree, at all depths within the bilayer. It is therefore unsurprising that water is transiting the bilayer. In fact, this is a vital feature of biological membranes, allowing for osmotic equilibrium, while at the same time, maintaining the separation of ions, electrolytes, and large biomolecules.

Classically, Overton's rule informs us that the permeability of a small molecule, such as water, through a hydrocarbon layer is defined by the partition coefficient (Overton 1899). However, Overton's rule is known to overestimate the thickness of a bilayer by nearly a factor of four (Finkelstein 1987; Nagle and Tristram-Nagle 2000), inviting new assumptions (Mathai et al. 2008) and models (Deamer and Bramhall 1986) in order to ameliorate this overestimate.

Kedem and Katchalsky (1958) quantitatively describe the passive transport of water, and any uncharged permeable solute, in terms of classical diffusion plus a convection model. (We must recognize that the presence of a convective term highlights the importance of water permeation in defining all types of transport across the bilayer.) The three parameters that describe the transport of water across a bilayer are: (i) hydraulic conductivity, L_P; (ii) solute permeability, ω; and (iii) the reflection coefficient, σ. The hydraulic conductivity, L_P, is defined as the volume flux of water, J_V, per pressure differential across the membrane, ΔP, and is evaluated for an infinitely thin membrane per unit area. This is expressed as follows:

$$L_P \equiv \left(\frac{J_V}{\Delta P}\right)_{\Delta x=0} \tag{3.8}$$

The bilayer solute permeability, ω, which is defined in terms of the solvent flux, J_s, and the osmotic pressure, $\Delta\pi$, at zero flux of water (per unit area) can be written as:

$$\omega \equiv \left(\frac{J_s}{\Delta \pi}\right)_{J_v=0} \tag{3.9}$$

where the osmotic pressure is,

$$\Delta\pi \equiv RT\Delta c_s \tag{3.10}$$

where R is the gas constant, T the temperature, and Δc_S is the concentration difference of the solute across the membrane. Finally, the reflection coefficient, σ, is used to characterize interactions of the solute with the bilayer, or solvent, that reduce transport. It is confined to values $0 \leq \sigma \leq 1$.

By combining these relations with the appropriate driving forces, i.e., ΔP, $\Delta\pi$, and Δc_S, we can formulate the Kedem-Katchalsky equations which describe the net solvent volume flux, J_v, and the net solute flux, J_s across a lipid bilayer;

$$J_v = L_P\Delta P - \sigma L_P \Delta\pi \tag{3.11}$$

$$J_s = \overline{c}_s\left(J_v - \sigma L_P \Delta P\right) + \omega\Delta\pi \tag{3.12}$$

Expressed in this way, it becomes clear that water traversing a bilayer plays a critical role in defining the net flux of most molecules going through a membrane.

Essentially there are two perspectives from which to consider the passive transport of water across a bilayer. The solubility diffusion model considers water to be homogenously solubilized in the lipid bilayer, and implies that solute partitioning and diffusivity are responsible for the rates of permeation of solute molecules across the bilayer (Hanai and Haydon 1966; Träuble 1971; Finkelstein 1976; Nagle et al. 2008). Alternatively, water can cross a membrane through nanometer scale pores spanning the thickness of the membrane (Finkelstein 1987; Tepper and Voth 2005), or transient pore diffusion, where clusters of bulk-like water within the bilayer diffuse across (Deamer and Bramhall 1986; Jansen and Blume 1995; Leontiadou et al. 2004; Kausik and Han 2011). These pore-based models essentially connect permeability to bilayer stability. In biological systems, protein structures such as aquaporin, also allow for the passive transport of water (Murata et al. 2000; de Groot and Grubmüller 2001; Sui et al. 2001).

The solubility-diffusion model is based on a few key assumptions. For example, it considers the membrane as a homogenous slab, or series of homogeneous slabs in which water is partitioned and diffuses through. Additionally, it assumes that the slab is homogeneous in the lateral dimension, a condition that is not true for lipid compositions which spontaneously form raft domains. In such systems, not only does lipid composition change laterally, and hence the diffusivity and partition coefficients, but bilayer thickness is also known to vary as well (Heberle et al. 2013).

In the context of the Kedem-Katchalsky definitions, the permeability can be equated to hydraulic conductivity. The permittivity of each slab can then be expressed as:

$$L_P = P = \frac{KD_C}{d_c} \tag{3.13}$$

where K is the partition coefficient, D_C is the diffusion coefficient in the membrane, and d_C is the bilayer thickness. From this, it is clear to see how bilayer thickness is overestimated when using Overton's rule.

As one might suspect, real systems cannot be fully described by the single slab model, as it fails to capture the membrane's structural and chemical heterogeneity. The head group region and hydrocarbon core, for example, have vastly different properties as they relate to water. Both simulation (Marrink and Berendsen 1994; Bemporad et al. 2004) and experimental results (Kausik and Han 2011) illustrate this point by showing that D_C varies along the bilayer; as does the amount of water (Xiang and Anderson 1998; Bemporad et al. 2004).

Using simulations, it is possible to evaluate the integral definition of the permeability, which is given by:

$$P = \int_0^d \frac{K(Z)\ D_c(Z)}{dz} \tag{3.14}$$

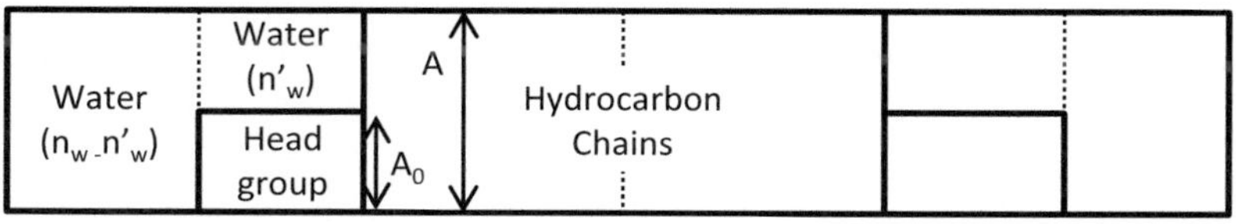

Fig. 3.3 A representation of the reduction in lipid area as a result of the head group encountering a diffusing water molecule as it permeates the bilayer (Nagle and Wiener 1988)

where the partition and diffusion coefficients can be evaluated as functions of the their position along a bilayer of thickness d (Fig. 3.3).

It is convenient to parse the bilayer into multiple slabs with distinct properties. The rational for this approach is laid out in detail by Nagle and coworkers (Nagle et al. 2008), where they advocate that the bilayer be divided into three slabs, namely: (1) two slabs describing the head group regions (one on each side); and (2) a slab representing the acyl chain region. In this model, the permeability of the two head group regions, P_H, and hydrocarbon region, P_C, are related as follows:

$$\frac{1}{P_f} = \frac{2}{P_H} + \frac{1}{P_C} \tag{3.15}$$

From this, one can account for water partitioning, diffusion constants, and relative bilayer thickness which result from inhomogeneities of the membrane. Nagle et al. (2008) also advocate for a fractional area term, $(A - A_0)/A$, when calculating the permeability of the head group region, P_H, where A_0 is then the head group barrier area (Mathai et al. 2008). As a result, permeability tends to zero as A_0 tends to A, as is evident from:

$$P_H = \left(\frac{K_H D_H}{d_H}\right)\left(\frac{A - A_0}{A}\right) \tag{3.16}$$

where D_H and K_H are, respectively, the diffusion and partition coefficients in the head group region, and d_H is the thickness of the head group region. This is somewhat reminiscent of the reflection coefficient used by Kedem and Katchalsky, except that in the Nagle et al. case the correction accounts for water/head group interactions, rather than water/solute interactions. In the hydrocarbon chain region, area is not thought to be important. As such, P_C is defined as:

$$P_C = \frac{K_C D_C}{2\, d_C} \tag{3.17}$$

where D_C and K_C are, respectively, the diffusion and partition coefficients in the bilayer's hydrophobic core, and d_C is the hydrophobic core thickness. These two relations make it clear that the rate controlling step in this model is the effective partition coefficient in the head group region, which is being modulated by the area of the lipid head group relative to the area per lipid (Mathai et al. 2008). This is an important insight which arises from a deep structural understanding of lipid hydration mentioned in a previous section. From those studies, it was clear that the area per lipid scales with the number of water molecules associated with the head group region. It is only natural then, to apply this knowledge to modeling water permittivity.

Pore models describe water transport by very different physical phenomena. To understand pore mediated water transport, one is not concerned with how much water exists in the bilayer at equilibrium, and how it diffuses, but rather in the following: (1) the number of water containing pores; (2) how long do they last; and (3) how does water diffuse within the pores during their lifetime. These notions can be expressed as a partition coefficient for the bilayer as follows:

$$P = \frac{n}{A} P_p \tag{3.18}$$

where P_p is the permeability per pore, and *n/A* is the number of pores per unit area of bilayer. We can further define P_p as:

$$P = \frac{v_w N D_w}{L^2} \tag{3.19}$$

where v_w is the molar volume of water, D_w and N are the diffusion constant of water and the average number of water molecules in a pore, respectively, and L is the pore length.

Thermodynamically, pore formation and stability can be thought of in terms of free energy. To form a pore of radius, r, in a bilayer, one considers the line tension, γ, which stabilizes the bilayer, and the surface tension, Γ, which stabilizes a pore. These quantities can be expressed as:

$$E(r) = 2\pi r\gamma - \pi r^2 \Gamma \tag{3.20}$$

The differences between these competing forces give us some understanding of pose size and the frequency of pore formation. Line tension and surface tension are parameters relating to bilayer stability, implying that pores appear more frequently in unstable bilayers. Indeed, large hydrophilic pores require a significant structural rearrangement of the bilayer, and can approach a nanometer in diameter, frequently spanning the entire bilayer (Leontiadou et al. 2004).

Hydrophobic pores, on the other hand, are thought to be smaller and more transient, with sizes smaller than a nanometer in diameter and lifetimes on the order of tens of picoseconds (Kausik and Han 2011). Some experimental evidence for this notion has recently been put forth, which was based on the mismatch in the temperature dependence of the activation energy of water diffusion and the observed internal water diffusion rate. One would expect these two quantities to scale with each other above and below the phase transition. However, the local diffusion constant of internal water was found to vary only slightly in the different phases, while the activation energy experienced a discrete change at the phase transition (Kausik and Han 2011). The reasoning for this was that water diffuses freely within a pore, but that the pore diffuses much slower in the plane of the bilayer below the gel temperature. This is good evidence for the lack of dynamical coupling between the water and the bilayer, as implied by the solubility-diffusion model. However, the use of a spin probe in these studies may have also perturbed the system, as has been previously observed for fluorescent probes (Ackerman et al. 2013). Such a perturbation can affect bilayer properties, allowing for local, dynamically decoupled water regions near the spin probe.

Most physical evidence points to solubility-diffusion as the predominant mechanism for water permeation (Al-Awqati 1999). This was demonstrated in the dependence of the water permeation rate as a function of bilayer thickness. One would expect a dependence of d_C^{-1} in the case of the solubility diffusion model, but a much stronger, exponential dependence as a function of bilayer thickness in a pore model. Indeed this observation is borne-out by experiments from Paula et al. (1996), where they showed that water permeation does not follow the predicted exponential dependence. Instead, their data were consistent with transport through pores of charged and ionic solutes. This seems reasonable as such molecules must traverse the bilayer as a water cluster, and clearly indicates that pores are not the predominant mode of action for water permeation. Another possibility is that both mechanisms are taking place, but that the majority of water permeation takes place through a solubility-diffusion mechanism.

3.4 Dynamics

As has been discussed in previous sections of this chapter, water molecule dynamics are highly sensitive to the relative location of the molecule to the lipid bilayer. For example NMR studies using a spin labeled PC system at 300 K demonstrated a fourfold reduction in the diffusion constant of choline associated water (0.69×10^{-9} m^2 s^{-1}), compared to the bulk (2.3×10^{-9} m^2 s^{-1}) (Kausik and Han 2011). This retardation of the water's diffusion constant was increased by an additional 20 % in the vicinity of the glycerol/carbonyl, and water motions slowed yet further in the acyl chain region. These results are in agreement with earlier experimental (Griffith et al. 1974; Marsh 2002) and simulation studies (Marrink and Berendsen 1994; Pasenkiewicz-Gierula et al. 1997; Åman et al. 2003). Diffusion, however, is not the only relevant dynamical feature of water. Many of its vibrational, librational, stretching, and bending modes are also modified when in the presence of a lipid bilayer. These reorientational motions of water can be highly informative when it comes to order parameters, as retardation of this process directly relates to the change in free energy of the bound molecule versus the bulk (Åman et al. 2003).

Current models of water dynamics seek to bring together the translational and reorientational motions of water through collective or concerted mechanisms related to hydrogen bond cleavage and large amplitude angular jumps (Torre et al. 2004; Laage and Hynes 2006; Taschin et al. 2013). While these models have been evaluated primarily for bulk water, or in the presence of some small solute molecules, they may be informative regarding water interactions with a lipid bilayer. So what insights do these models provide? The rotational and translational diffusion coefficients may be governed by the rate of hydrogen bond cleavage and reformation. In support of this, is the example of protein hydration water, where the rotational retardation factors ($D_{R,bulk}/D_{R,hyd}$) are comparable to translational retardation ($D_{T,bulk}/D_{T,hyd}$), but with the added advantage that rotational relaxations can be probed locally using NMR techniques (Marchi et al. 2002). This is quite elegantly described for protein hydration dynamics (Halle 2004), and is applicable to lipids. This similarity in retardation data between rotational and translational motions means that both processes use the lifetime of the hydrogen bond as a rate limiting step. When compared to the bulk, the hydrogen bond lifetime of hydration water is more than 5 times longer (Balasubramanian et al. 2002), and even structural perturbations imposed by non-hydrogen bonding head groups, such as choline, reduce reorientation times of local water by a factor of 4 (Murzyn et al. 2006). Having said this, an extended residence time for water near the bilayer is not indicative of strong bonding interactions, but could also be the result of a local topography/geometry which prevents water participation in the cooperative rearrangement of the bulk (Laage et al. 2009). The importance of a topographical/geometric restriction interfering with the cooperative mechanism becomes clearer when one considers the restricted geometry near the carboxyl region of the bilayer.

Before moving on, we would like to briefly return to the local translational diffusion rate within the hydrophobic core of the bilayer, where water motions are thought to be distinctly different. This is a limiting case where there is a lack of hydrogen bonding and a very low barrier to reorientation, resulting in rapid rotational diffusion, but a restricted translational diffusion – based on the available free space (Kausik and Han 2011). A diffusion mechanism in this environment requires free volume (Almeida et al. 1992), a volume that happens to be occupied by the acyl chains. As a result, water motions in the bilayer become connected to the motions of the acyl chains despite the lack of a shared hydrogen bond network. This part of the lipid bilayer undergoes a range of processes from structural relaxations, to vibrations, rotations and lateral translations (Rheinstädter 2012). Collective dispersions also appear to play a prominent role in the mechanisms governing the dynamical processes of the bilayer (Chen et al. 2001; Rheinstädter et al. 2004).

Experimentally, a wide range of techniques have been applied to the water-interface problem. It must be emphasized that different techniques probe different physical properties; i.e. spin, polarization, dipole, neutron scattering cross section, etc. As such, the identity of the perturbation, the influence of length scale, and the sensitivity to the bilayer will be highly technique dependent. For a more detailed discussion regarding this, the reader is referred to the excellent review by B. Halle (2004).

NMR has long been used to study the structure of water at the interface (Seelig 1977), and a number of NMR dynamical techniques have been applied to this problem, such as pulsed field gradient (Devaux and McConnell 1972; Volke et al. 1994; Wassall 1996), Overhauser dynamic nuclear polarization (Armstrong and Han 2009; Kausik and Han 2011), and residual magnetic relaxation profile (Victor et al. 2013). This powerful tool has yielded important information on highly localized dynamical processes on the picosecond to nanosecond timescale. Depolarized light scattering techniques, Raman, Kerr-Effect, IR and 2D-IR are all also suited to studying the dynamics of water (Auer et al. 2007; Binder 2007; Mazur et al. 2010). However, separating the water signal at the interface from that of the bulk is often difficult and requires careful fitting of the data in order to properly assign spectral features to corresponding hydration numbers and retardation factors (Perticaroli et al. 2013). New techniques based on two dimensional infrared methods (Kolano et al. 2006) provide detailed information about the vibrational modes of water and how they are affected in the presence of a lipid bilayer (Nagata and Mukamel 2010). Dielectric spectroscopy has also been used to investigate lipid bilayers, with time-domain terahertz techniques showing promise in the study of reorientational dynamics in aligned bilayers (Tielrooij et al. 2009).

Quasielastic neutron scattering (QENS) is a probe-free, ensemble technique (Bee 1988) that has been used to directly study the dynamics of bilayer associated water. (* N of E: for details see Chap. 4 by Pfeiffer) The advantages of neutron scattering are twofold: (1) the incoherent scattering cross-section of hydrogen and deuterium differ by a factor of ~40 (Sears 1992), making isotopic substitution of the bilayer an attractive strategy for separating the dynamics of water from those of the bilayer. (2) Neutron scattering offers the possibility to directly observe dynamics as a function of scattering wave vector, Q. This important feature allows one to assign the dynamic process being observed, i.e., a true diffusive process scales as the length scale squared,

$$D_w = \frac{\Gamma\,(Q)}{Q^2} \tag{3.21}$$

where D_w is the diffusion coefficient of water and $\Gamma(Q)$ is the decay constant in units of s^{-1}, which can be determined from a fit of the observed dynamic structure factor, S(Q,E). This fitting of neutron data is typically performed by convoluting Lorentzians, representing the contribution of particles in motion, with delta functions, representing stationary particles on the timescale of the instrumental resolution (Fitter et al. 1999). It is also possible, and arguably more appropriate, to utilize the dynamic susceptibility formalism of the neutron scattering spectrum, and then fit the data with a Cole-Cole distribution to account for the stretching of water relaxation spectra at the interface of a weakly hydrated biological sample

(Settles and Doster 1996; Nickels et al. 2012). The conversion to dynamic susceptibility is accomplished by removing the Bose occupation number, $n_B(E)$, from the observed dynamic structure factor,

$$\chi''(Q, E) \propto \frac{S(Q, E)}{n_B(E)} = \frac{S(Q, E)}{\left(e^{E/k_B T} - 1\right)^{-1}} \tag{3.22}$$

The Cole-Cole distribution is typically fit in the frequency domain, which is accessible through Plank's constant, $\hbar$;

$$\frac{\chi''(Q, E)}{\hbar} = \chi''(Q, \omega) \tag{3.23}$$

and can then be used to fit at all measured values of Q according to:

$$\chi''(Q, \omega) = \frac{\chi_0(\omega\tau)^{1-\alpha} \sin\left(\frac{\alpha\pi}{2}\right)}{1 + 2(\omega\tau)^{1-\alpha} \cos\left(\frac{\alpha\pi}{2}\right) + (\omega\tau)^{2-2\alpha}} \tag{3.24}$$

where α is the stretching parameter, χ_0 is the scaling factor, and τ is the central relaxation time.

Using QENS, and other physical characterization techniques, a picture of dynamical decoupling for the most tightly associated water molecules emerges that is consistent with the removal of the cooperative relaxation mechanism, in low hydration samples, and the carboxyl associated water molecules. Swenson and coworkers (Swenson et al. 2008) have built on the works by König and Pfeiffer (Pfeiffer et al. 1989; König et al. 1994) to experimentally determine the slowdown in translational diffusion of hydration water. They looked at the dynamics of interfacial water in low-hydration, aligned DMPC and DPPC bilayers, and noted that the onset of water translation begins at ~295 K, which coincides with the gel-to-liquid transition of the DMPC lipid bilayer – 20 K higher than bulk water (Swenson et al. 2008). It must be pointed out, however, that these QENS observations of water dynamics were made at hydration levels far below what is considered "full" hydration – i.e., the amount of water where the lamellar repeat spacing ceases to change with the further addition of water (Katsaras 1997; Fitter et al. 1999). This implies a predominant role of acyl chain dynamics in facilitating water translation in the low hydration condition, i.e. in the tightly associated water region.

The dramatic slowdown of water in the hydration layer is not an isolated result. A study focusing on the hydration water of purple membranes at higher concentration of water also illustrated the dynamical decoupling of hydration water from the membrane (Wood et al. 2007). Wood et al. demonstrated that transitions in hydration water had little or no influence on the dynamics of the membrane itself. Separate transitions, assigned to the onset of collective dynamics of the acyl region and the freezing of bulk-like heavy water, respectively, were also observed in hydrated, aligned DMPC-d54 bilayers in D_2O at 295 K and 271 K (using an instrumental resolution of $\sim$1 ns), indicating a lack of dynamical coupling between water and the collective dynamics of the membrane (Rheinstädter et al. 2005). Other studies demonstrated that the lateral diffusion constants of water and those of individual lipid molecules were separated by two orders of magnitude at full hydration condition, with water diffusing at a rate of 4.7×10^{-10} m^{2}s^{-1} and lipids diffusing at a rate of 8.6×10^{-12} m^{2}s^{-1} (Gaede and Gawrisch 2003).

Experiments using NMR have shown the presence of a small, dynamically distinct, water population at the surface in a number of different PC bilayers (Volke et al. 1994). More recently, 2D-FTIR on weakly hydrated DMPC membrane fragments demonstrated a wide distribution of stretching and vibration relaxation times for water (Volkov et al. 2007). While the latter is perhaps an artifact of isolated water molecules at these weakly hydrated surfaces, it may also be typical of how geometrically restricted some of the carbonyl associated waters are, i.e., isolated from the collective bilayer relaxational mechanism. Independent simulations on fully hydrated systems have clearly illustrated how decoupled the small water population bound to the glycerol region of the head group is from the population of water molecules associated with the choline region, or the bulk (Nagata and Mukamel 2010).

A slowdown in local water dynamics near the bilayer, equivalent to a threefold increase in local viscosity, has recently been invoked to explain anomalously rigid measurements of bending moduli from neutron spin echo experiments (Yi et al. 2009). However, alternative explanations of this anomaly have cited internal bilayer friction as a possible explanation (Watson and Brown 2010; Woodka et al. 2012).

The dynamics of water in the hydration shell of lipid bilayers are an interesting counter point to what has been observed in proteins. In proteins, the solvent fluctuations drive the dynamics of the macromolecule – a phenomenon termed as 'slaving' (Fenimore et al. 2002). Lipid bilayers, despite their close contact with water, do not appear to behave in this manner. This stems from the lack of an extensive hydrogen bond network and a large solvent excluded volume in the acyl chain region of lipid bilayers. This is not meant to suggest that there is no influence of water in the protruding region of the head group (the choline region), as evidence clearly shows a role for water in this region (Ulrich and Watts 1994), and a precipitous reduction in lateral lipid diffusion rates occurs at hydration levels below 30 % (Filippov et al. 2003). (* N of E: This point is discussed by Arsov in Chap. 6) Nonetheless, further modeling of the diffusive properties of interfacial water in DPPC illustrates how the diffusivity in this population of interfacial water follows the phase behavior of the lipids (Debnath et al. 2013). Indeed, membrane proteins also seem to follow the phase behavior of the bilayer, as studies on purple membrane have shown that protein dynamics are substantially different when incorporated into the lipid bilayer (Fitter et al. 1999).

3.5 Concluding Remarks

In this chapter we have described some of the basic concepts defining the interaction of lipid bilayers and water. Fundamentally, it is the energetic cost of perturbing the equilibrium disorder of bulk water that dictates many lipid bilayer properties. Repulsive interactions between bilayers define solution stability of vesicles and multilamellar spacing. These repulsive forces are sensitive to properties such as ionic strength, and lipid chemistry. Water is also important within the bilayer, and water permeation is critical to osmotic equilibrium. Yet, the distribution of water within a bilayer is still a subject of current study, with new nanopore models being put forth, aided by liquid crystallography and NMR techniques. These methods also give important estimates of the amounts of water associated at the interface of the bilayer. Finally, we described the dynamics of water in and near a lipid bilayer, along with the primary techniques used to study this system. Taken together, structural and dynamical studies have advanced our understanding of this important biological interface, which we hope will lead us to a deeper understanding as to how the membrane/water interface defines and modulates biological functions.

Acknowledgments Support for the authors was received from the Department of Energy (DOE), Scientific User Facilities Division, Office of Basic Energy Sciences (BES) through Oak Ridge National Laboratory (ORNL), which is managed by UT-Battelle, LLC, for the U.S. DOE under contract no. DE-AC05-00OR2275. JDN was partially supported through EPSCoR grant no. DEFG02-08ER46528.

References

Ackerman DG, Heberle FA, Feigenson GW (2013) Limited perturbation of a DPPC bilayer by fluorescent lipid probes: a molecular dynamics study. J Phys Chem B 117(17):4844–4852

Al-Awqati Q (1999) One hundred years of membrane permeability: does Overton still rule? Nat Cell Biol 1(8):E201–E202

Almeida PF, Vaz WL, Thompson T (1992) Lateral diffusion in the liquid phases of dimyristoylphosphatidylcholine/cholesterol lipid bilayers: a free volume analysis. Biochemistry 31(29):6739–6747

Åman K, Lindahl E, Edholm O et al (2003) Structure and dynamics of interfacial water in an L_α phase lipid bilayer from molecular dynamics simulations. Biophys J 84(1):102–115

Armstrong BD, Han S (2009) Overhauser dynamic nuclear polarization to study local water dynamics. J Am Chem Soc 131(13):4641–4647

Auer B, Kumar R, Schmidt J et al (2007) Hydrogen bonding and Raman, IR, and 2D-IR spectroscopy of dilute HOD in liquid D2O. Proc Natl Acad Sci 104(36):14215–14220

Balasubramanian S, Pal S, Bagchi B (2002) Hydrogen-bond dynamics near a micellar surface: origin of the universal slow relaxation at complex aqueous interfaces. Phys Rev Lett 89(11):115505

Bee M (1988) Quasielastic neutron scattering: principles and applications in solid state chemistry, biology, and materials science. Adam Hilger, Bristol

Bemporad D, Essex JW, Luttmann C (2004) Permeation of small molecules through a lipid bilayer: a computer simulation study. J Phys Chem B 108(15):4875–4884

Bernal J, Fowler R (1933) A theory of water and ionic solution, with particular reference to hydrogen and hydroxyl ions. J Chem Phys 1(8):515–548
Binder H (2007) Water near lipid membranes as seen by infrared spectroscopy. Eur Biophys J 36(4–5):265–279
Böckmann RA, Hac A, Heimburg T et al (2003) Effect of sodium chloride on a lipid bilayer. Biophys J 85(3):1647–1655
Büldt G, Gally H, Seelig A et al (1978) Neutron diffraction studies on selectively deuterated phospholipid bilayers. Nature 271(5641):182–184
Cevc G (1991) Hydration force and the interfacial structure of the polar surface. J Chem Soc Faraday Trans 87(17):2733–2739
Chen S, Liao C, Huang H et al (2001) Collective dynamics in fully hydrated phospholipid bilayers studied by inelastic X-ray scattering. Phys Rev Lett 86(4):740
De Groot BL, Grubmüller H (2001) Water permeation across biological membranes: mechanism and dynamics of aquaporin-1 and GlpF. Science 294(5550):2353–2357
Deamer DW, Bramhall J (1986) Permeability of lipid bilayers to water and ionic solutes. Chem Phys Lipids 40(2–4):167–188
Debnath A, Ayappa KG, Maiti PK (2013) Simulation of influence of bilayer melting on dynamics and thermodynamics of interfacial water. Phys Rev Lett 110(1):018303
Debye P, Hückel E (1923) De la theorie des electrolytes. I. abaissement du point de congelation et phenomenes associes. Phys Z 24(9):185–206
Devaux P, Mcconnell H (1972) Lateral diffusion in spin-labeled phosphatidylcholine multilayers. J Am Chem Soc 94(13):4475–4481
Disalvo E, Lairion F, Martini F et al (2008) Structural and functional properties of hydration and confined water in membrane interfaces. Biochim Biophys Acta Biomembr 1778(12):2655–2670
Doyle DA, Cabral JM, Pfuetzner RA et al (1998) The structure of the potassium channel: molecular basis of K+ conduction and selectivity. Science 280(5360):69–77
Fecko CJ, Eaves JD, Loparo JJ et al (2003) Ultrafast hydrogen-bond dynamics in the infrared spectroscopy of water. Science 301(5640):1698–1702
Fenimore PW, Frauenfelder H, Mcmahon BH et al (2002) Slaving: solvent fluctuations dominate protein dynamics and functions. Proc Natl Acad Sci 99(25):16047–16051
Filippov A, Orädd G, Lindblom G (2003) Influence of cholesterol and water content on phospholipid lateral diffusion in bilayers. Langmuir 19(16):6397–6400
Finer EG, Darke A (1974) Phospholipid hydration studied by deuteron magnetic resonance spectroscopy. Chem Phys Lipids 12(1):1–16
Finkelstein A (1976) Water and nonelectrolyte permeability of lipid bilayer membranes. J Gen Physiol 68(2):127–135
Finkelstein A (1987) Water movement through lipid bilayers, pores, and plasma membranes: theory and reality. Wiley, New York
Fitter J, Lechner RE, Dencher NA (1999) Interactions of hydration water and biological membranes studied by neutron scattering. J Phys Chem B 103(38):8036–8050
Frank HS, Wen W-Y (1957) Ion-solvent interaction. Structural aspects of ion-solvent interaction in aqueous solutions: a suggested picture of water structure. Discuss Faraday Soc 24(0):133–140
Gaede HC, Gawrisch K (2003) Lateral diffusion rates of lipid, water, and a hydrophobic drug in a multilamellar liposome. Biophys J 85(3):1734–1740
Gawrisch K, Ruston D, Zimmerberg J et al (1992) Membrane dipole potentials, hydration forces, and the ordering of water at membrane surfaces. Biophys J 61(5):1213–1223
Ghosh A, Smits M, Bredenbeck J et al (2007) Membrane-bound water is energetically decoupled from nearby bulk water: an ultrafast surface-specific investigation. J Am Chem Soc 129(31):9608–9609
Griffith OH, Dehlinger PJ, Van SP (1974) Shape of the hydrophobic barrier of phospholipid bilayers (evidence for water penetration in biological membranes). J Membr Biol 15(1):159–192

Halle B (2004) Protein hydration dynamics in solution: a critical survey. Philos Trans R Soc Lond B Biol Sci 359(1448):1207–1224
Hanai T, Haydon D (1966) The permeability to water of bimolecular lipid membranes. J Theor Biol 11(3):370–382
Heberle FA, Petruzielo RS, Pan J et al (2013) Bilayer thickness mismatch controls domain size in model membranes. J Am Chem Soc 135(18):6853–6859
Helfrich W, Servuss R-M (1984) Undulations, steric interaction and cohesion of fluid membranes. Il Nuovo Cimento D 3(1):137–151
Ishai PB, Mamontov E, Nickels JD et al (2013) Influence of ions on water diffusion—a neutron scattering study. J Phys Chem B 117(25):7725–7729
Israelachvili JN (2011) Intermolecular and surface forces: revised third edition. Academic, London
Jansen M, Blume A (1995) A comparative study of diffusive and osmotic water permeation across bilayers composed of phospholipids with different head groups and fatty acyl chains. Biophys J 68(3):997–1008
Jorgensen WL, Chandrasekhar J, Madura JD et al (1983) Comparison of simple potential functions for simulating liquid water. J Chem Phys 79(2):926–935
Katsaras J (1995) Structure of the subgel (Lc′) and Gel (L. beta'.) phases of oriented dipalmitoylphosphatidylcholine multibilayers. J Phys Chem 99(12):4141–4147
Katsaras J (1997) Highly aligned lipid membrane systems in the physiologically relevant "excess water" condition. Biophys J 73(6):2924–2929
Katsaras J, Tristram-Nagle S, Liu Y et al (2000) Clarification of the ripple phase of lecithin bilayers using fully hydrated, aligned samples. Phys Rev E 61(5):5668–5677
Kausik R, Han S (2011) Dynamics and state of lipid bilayer-internal water unraveled with solution state 1H dynamic nuclear polarization. Phys Chem Chem Phys 13(17):7732–7746
Kedem OT, Katchalsky A (1958) Thermodynamic analysis of the permeability of biological membranes to non-electrolytes. Biochim Biophys Acta 27:229–246
Kolano C, Helbing J, Kozinski M et al (2006) Watching hydrogen-bond dynamics in a beta-turn by transient two-dimensional infrared spectroscopy. Nature 444(7118):469–472
König S, Sackmann E, Richter D et al (1994) Molecular dynamics of water in oriented DPPC multilayers studied by quasielastic neutron scattering and deuterium-nuclear magnetic resonance relaxation. J Chem Phys 100:3307
Kornyshev A, Leikin S (1989) Fluctuation theory of hydration forces: the dramatic effects of inhomogeneous boundary conditions. Phys Rev A 40(11):6431
Kučerka N, Nagle JF, Feller SE et al (2004) Models to analyze small-angle neutron scattering from unilamellar lipid vesicles. Phys Rev E 69(5):051903
Kučerka N, Tristram-Nagle S, Nagle J (2006) Structure of fully hydrated fluid phase lipid bilayers with monounsaturated chains. J Membr Biol 208(3):193–202
Kučerka N, Nagle JF, Sachs JN et al (2008a) Lipid bilayer structure determined by the simultaneous analysis of neutron and X-ray scattering data. Biophys J 95(5):2356–2367
Kučerka N, Papp-Szabo E, Nieh M-P et al (2008b) Effect of cations on the structure of bilayers formed by lipopolysaccharides isolated from pseudomonas aeruginosa PAO1. J Phys Chem B 112(27):8057–8062
Laage D, Hynes JT (2006) A molecular jump mechanism of water reorientation. Science 311(5762):832–835
Laage D, Stirnemann G, Hynes JT (2009) Why water reorientation slows without iceberg formation around hydrophobic solutes. J Phys Chem B 113(8):2428–2435
Leikin S, Parsegian VA, Rau DC et al (1993) Hydration forces. Annu Rev Phys Chem 44(1):369–395
Lelkes P, Miller I (1980) Perturbations of membrane structure by optical probes: I. Location and structural sensitivity of merocyanine 540 bound to phospholipid membranes. J Membr Biol 52(1):1–15
Leontiadou H, Mark AE, Marrink SJ (2004) Molecular dynamics simulations of hydrophilic pores in lipid bilayers. Biophys J 86(4):2156–2164

Marčelja S, Radić N (1976) Repulsion of interfaces due to boundary water. Chem Phys Lett 42(1):129–130
Marchi M, Sterpone F, Ceccarelli M (2002) Water rotational relaxation and diffusion in hydrated lysozyme. J Am Chem Soc 124(23):6787–6791
Marrink S-J, Berendsen HJ (1994) Simulation of water transport through a lipid membrane. J Phys Chem 98(15):4155–4168
Marsh D (2002) Membrane water-penetration profiles from spin labels. Eur Biophys J 31(7):559–562
Mathai JC, Tristram-Nagle S, Nagle JF et al (2008) Structural determinants of water permeability through the lipid membrane. J Gen Physiol 131(1):69–76
Mazur K, Heisler IA, Meech SR (2010) Ultrafast dynamics and hydrogen-bond structure in aqueous solutions of model peptides. J Phys Chem B 114(32):10684–10691
Mcintosh TJ, Simon SA (1986) Area per molecule and distribution of water in fully hydrated dilauroylphosphatidylethanolamine bilayers. Biochemistry 25(17):4948–4952
Mcintosh TJ, Simon SA (1994) Hydration and steric pressures between phospholipid bilayers. Annu Rev Biophys Biomol Struct 23(1):27–51
Murata K, Mitsuoka K, Hirai T et al (2000) Structural determinants of water permeation through aquaporin-1. Nature 407(6804):599–605
Murzyn K, Zhao W, Karttunen M et al (2006) Dynamics of water at membrane surfaces: effect of headgroup structure. Biointerphases 1(3):98–105
Nagata Y, Mukamel S (2010) Vibrational Sum-frequency generation spectroscopy at the water/lipid interface: molecular dynamics simulation study. J Am Chem Soc 132(18):6434–6442
Nagle JF, Tristram-Nagle S (2000) Structure of lipid bilayers. Biochim Biophys Acta Rev Biomembr 1469(3):159–195
Nagle JF, Wiener MC (1988) Structure of fully hydrated bilayer dispersions. Biochim Biophys Acta Biomembr 942(1):1–10
Nagle JF, Zhang R, Tristram-Nagle S et al (1996) X-ray structure determination of fully hydrated L alpha phase dipalmitoylphosphatidylcholine bilayers. Biophys J 70(3):1419–1431
Nagle JF, Mathai JC, Zeidel ML et al (2008) Theory of passive permeability through lipid bilayers. J Gen Physiol 131(1):77–85
Nickels JD, O'Neill H, Hong L, et al (2012) Dynamics of protein and its hydration water: neutron scattering studies on fully deuterated GFP. Biophys J 103(7):1566–1575.
Overton E (1899) Ueber die osmotischen Eigenshaften der Zelle in ihrer Bedeutung Fur die Toxikologie und Pharmakologie. Vierteljahrsschr Naturforsch Ges Zurich 44:88–135
Parsegian VA, Rand RP (1991) On molecular protrusion as the source of hydration forces. Langmuir 7(6):1299–1301
Pasenkiewicz-Gierula M, Takaoka Y, Miyagawa H et al (1997) Hydrogen bonding of water to phosphatidylcholine in the membrane as studied by a molecular dynamics simulation: location, geometry, and lipid-lipid bridging via hydrogen-bonded water. J Phys Chem A 101(20):3677–3691
Paula S, Volkov AG, Van Hoek AN et al (1996) Permeation of protons, potassium ions, and small polar molecules through phospholipid bilayers as a function of membrane thickness. Biophys J 70(1):339–348
Perticaroli S, Comez L, Paolantoni M et al (2011) Extended frequency range depolarized light scattering study of N-acetyl-leucine-methylamide–water solutions. J Am Chem Soc 133(31):12063–12068
Perticaroli S, Nakanishi M, Pashkovski E et al (2013) Dynamics of hydration water in sugars and peptides solutions. J Phys Chem B 117(25):7729–7736
Petrache HI, Feller SE, Nagle JF (1997) Determination of component volumes of lipid bilayers from simulations. Biophys J 72(5):2237–2242
Pfeiffer W, Henkel T, Sackmann E et al (1989) Local dynamics of lipid bilayers studied by incoherent quasi-elastic neutron scattering. EPL Europhys Lett 8(2):201
Raghunathan V, Katsaras J (1995) Structure of the Lc′ phase in a hydrated lipid multilamellar system. Phys Rev Lett 74(22):4456

Rand R, Parsegian V (1989) Hydration forces between phospholipid bilayers. Biochim Biophys Acta Rev Biomembr 988(3):351–376
Rheinstädter MC (2012) Lipid membrane dynamics. Dynamics of soft matter. Springer, New York, USA, pp 263–286
Rheinstädter MC, Ollinger C, Fragneto G et al (2004) Collective dynamics of lipid membranes studied by inelastic neutron scattering. Phys Rev Lett 93(10):108107
Rheinstädter MC, Seydel T, Demmel F et al (2005) Molecular motions in lipid bilayers studied by the neutron backscattering technique. Phy Rev E 71(6):061908
Sears VF (1992) Neutron scattering lengths and cross sections. Neutron News 3(3):26–37
Seelig J (1977) Deuterium magnetic resonance: theory and application to lipid membranes. Q Rev Biophys 10(3):353–418
Settles M, Doster W (1996) Anomalous diffusion of adsorbed water: a neutron scattering study of hydrated myoglobin. Faraday Discuss 103:269–279
Sirota EB, Smith GS, Safinya CR et al (1988) X-ray scattering studies of aligned, stacked surfactant membranes. Science 242(4884):1406–1409
Subczynski WK, Wisniewska A, Yin J-J et al (1994) Hydrophobic barriers of lipid bilayer membranes formed by reduction of water penetration by alkyl chain unsaturation and cholesterol. Biochemistry 33(24):7670–7681
Sui H, Han B-G, Lee JK et al (2001) Structural basis of water-specific transport through the AQP1 water channel. Nature 414(6866):872–878
Swenson J, Kargl F, Berntsen P et al (2008) Solvent and lipid dynamics of hydrated lipid bilayers by incoherent quasielastic neutron scattering. J Chem Phys 129:045101
Taschin A, Bartolini P, Eramo R et al (2013) Evidence of two distinct local structures of water from ambient to supercooled conditions. Nat Commun 4
Tepper HL, Voth GA (2005) Protons may leak through pure lipid bilayers via a concerted mechanism. Biophys J 88(5):3095–3108
Tielrooij KJ, Paparo D, Piatkowski L et al (2009) Dielectric relaxation dynamics of water in model membranes probed by terahertz spectroscopy. Biophys J 97(9):2484–2492
Tielrooij K, Garcia-Araez N, Bonn M et al (2010) Cooperativity in ion hydration. Science 328(5981):1006–1009
Torre R, Bartolini P, Righini R (2004) Structural relaxation in supercooled water by time-resolved spectroscopy. Nature 428(6980):296–299
Träuble H (1971) The movement of molecules across lipid membranes: a molecular theory. J Membr Biol 4(1):193–208
Tristram-Nagle S, Petrache HI, Nagle JF (1998) Structure and interactions of fully hydrated dioleoylphosphatidylcholine bilayers. Biophys J 75(2):917–925
Ulrich AS, Watts A (1994) Molecular response of the lipid headgroup to bilayer hydration monitored by 2H-NMR. Biophys J 66(5):1441–1449
Verwey EJW, Overbeek JTG, Van Nes K (1948) Theory of the stability of lyophobic colloids: the interaction of sol particles having an electric double layer. Elsevier, New York
Victor KG, Korb J-P, Bryant RG (2013) Translational dynamics of water at the phospholipid interface. J Phys Chem B 117(41):12475–12478
Volke F, Eisenblätter S, Galle J et al (1994) Dynamic properties of water at phosphatidylcholine lipid-bilayer surfaces as seen by deuterium and pulsed field gradient proton NMR. Chem Phys Lipids 70(2):121–131
Volkov VV, Palmer DJ, Righini R (2007) Heterogeneity of water at the phospholipid membrane interface. J Phys Chem B 111(6):1377–1383
Wassall SR (1996) Pulsed field gradient-spin echo NMR studies of water diffusion in a phospholipid model membrane. Biophys J 71(5):2724–2732
Watson MC, Brown FL (2010) Interpreting membrane scattering experiments at the mesoscale: the contribution of dissipation within the bilayer. Biophys J 98(6):L9–L11
Wood K, Plazanet M, Gabel F et al (2007) Coupling of protein and hydration-water dynamics in biological membranes. Proc Natl Acad Sci 104(46):18049–18054

Woodka AC, Butler PD, Porcar L et al (2012) Lipid bilayers and membrane dynamics: insight into thickness fluctuations. Phys Rev Lett 109(5):058102
Xiang T-X, Anderson BD (1998) Influence of chain ordering on the selectivity of dipalmitoylphosphatidylcholine bilayer membranes for permeant size and shape. Biophys J 75(6):2658–2671
Yi Z, Nagao M, Bossev DP (2009) Bending elasticity of saturated and monounsaturated phospholipid membranes studied by the neutron spin echo technique. J Phys Condens Matter 21:155104
Zaccai G, Blasie J, Schoenborn B (1975) Neutron diffraction studies on the location of water in lecithin bilayer model membranes. Proc Natl Acad Sci 72(1):376–380
Zhou F, Schulten K (1995) Molecular dynamics study of a membrane-water interface. J Phys Chem 99(7):2194–2207

Chapter 4
Hydration Forces Between Lipid Bilayers: A Theoretical Overview and a Look on Methods Exploring Dehydration

Helge Pfeiffer

Abstract Although, many biological systems fulfil their functions under the condition of excess hydration, the behaviour of bound water as well as the processes accompanying dehydration are nevertheless important to investigate. Dehydration can be a result of applied mechanical pressure, lowered humidity or cryogenic conditions. The effort required to dehydrate a lipid membrane at relatively low degree of hydration can be described by a disjoining pressure which is called hydration pressure or hydration force. This force is short-ranging (a few nm) and is usually considered to be independent of other surface forces, such as ionic or undulation forces. Different theories were developed to explain hydration forces that are usually not consistent with each other and which are also partially in conflict with experimental or numerical data.

Over the last decades it has been more and more realised that one experimental method alone is not capable of providing much new insight into the world of such hydration forces. Therefore, research requires the comparison of results obtained from the different methods. This chapter thus deals with an overview on the theory of hydration forces, ranging from polarisation theory to protrusion forces, and presents a selection of experimental techniques appropriate for their characterisation, such as X-ray diffraction, atomic force microscopy and even calorimetry.

Keywords Hydration forces • Controlled hydration • Osmotic stress • Polarization • Protrusion • Sorption • Percolation • Plasticizer

---

H. Pfeiffer (✉)
Department of Metallurgy and Materials Engineering (MTM), University of Leuven (KU Leuven), Kasteelpark Arenberg 44 Bus 2450, Leuven, Belgium
e-mail: Helge.Pfeiffer@mtm.kuleuven.be

E.A. Disalvo (ed.), *Membrane Hydration*, Subcellular Biochemistry 71,
DOI 10.1007/978-3-319-19060-0_4

4.1 Introduction

4.1.1 Fundamentals

During the last decades, it has been more and more accepted that hydration plays a complex, functional role in living cells, and in this context it becomes clear that phospholipid membranes are much more than semi-permeable barriers (Disalvo et al. 2008; Scharnagl et al. 2005).

Native membranes usually exist at excess water conditions. However, investigations on phenomena arising from partial or complete dehydration in membranes and other biological surfaces are important as well, such as for the case of cell fusion, stress on cartilage or especially for anhydrobiosis, i.e. the ability of different organisms to survive even complete dehydration. There are different potential applications for anhydrobiosis, such as tailored cryosurgery in medicine (Balasubramanian et al. 2009), biopreservation, i.e. the storage of biological material (Aksan et al. 2009; Franca et al. 2007), and also applications in space projects (Ricci et al. 2005; Rothschild and Mancinelli 2001).

The degree of hydration can on the one hand be described by the integral water concentration, and on the other hand by the fraction of water that is chemically or physically attached on their respective host molecules under excess water conditions. The quantification of such bound water mostly depends on methods and models (Jayne 1982) and one should not be confused if there appear deviating or even contradicting results in the literature; bound water can e.g. be defined by a decay constant, or by the amount of non-freezable water (Kodama et al. 2004). Furthermore, one should emphasise that in some cases, the integral water concentration or the hydration number do not necessarily give information on the actual position of the water molecules within the membrane. Some water can e.g. also be trapped in diffusionally restricted or sealed micro-volumes such as water-filled pockets ore pores (see e.g. Binder and Gawrisch 2001).

The global hydration between lipid surfaces is frequently expressed by the molar hydration number, i.e. the molar ratio of water and lipid R_W (Eq. 4.1), where n_W and n_L are the molar number of water and lipid resp.

$$R_w = \frac{n_w}{n_L} \tag{4.1}$$

or by the water layer thickness d_W (corresponds to the broad arrows in Figs. 4.9 and 4.10). The water layer thickness was frequently calculated according the model of Luzatti (Marsh 1990) but also other approaches are used that e.g. make use of the electron density profile of lipid headgroups (Schmiedel et al. 2001).

The free enthalpy of water transferred from a reference phase into the lipid bilayer is expressed by the chemical potential of water, μ_w. It has close relationships with the water activity, $a_{w'}$ or the water potential, Ψ_w, and the corresponding relationship is given by Eq. 4.2 (Moore 1972):

$$\mu_W = \mu_{W,0} + RT \ \ln \ a_w \tag{4.2}$$

where $\mu_{w,0}$ is the chemical potential of water at reference conditions, R is the gas constant and T the absolute temperature. The water potential Ψ_w is defined by Eq. 4.3:

$$\Psi_W = \frac{\mu_W - \mu_{W,0}}{V_W} \tag{4.3}$$

where V_W the molar volume of water. The water potential is in the respective literature usually composed of different terms (Adam et al. 1995) representing the osmotic potential (entropic origin), "turgid potential" (hydrostatic pressure) and matrix potential (contains contributions from the interactions with colloids, as e.g. in capillary and surface interactions).

4.1.2 *Phase Behaviour at Low Hydration*

During dehydration, intermolecular interactions and motional degrees of freedom within the membrane are affected and the loss of mobility leads in general to solidification. For almost all pure, dehydrated lipids with saturated hydrocarbon chains, a solid lamellar phase can be found at ambient temperatures, in most cases the L_β, $L_{\beta'}$ or the L_c phase (Kranenburg and Smit 2005). For lipids with unsaturated chains, such as DOPC, at room temperature a liquid-like phase can exist. For higher temperatures, also non-lamellar phases are reported for the case of low water content (see e.g. Jürgens et al. 1983) (Fig. 4.1).

Even in lipid-water systems containing only one lipid species, up to three coexisting lipid phases can be observed, at least in the range of broad phase transitions and at low hydration degrees as e.g. revealed by calorimetry, X-ray diffraction and ^{2}H and ^{31}P-NMR spectroscopy for the case of POPC (Pfeiffer et al. 2013a). This can be rationalised by the Gibbs phase rule and experimental data showing that these phases are all lamellar phases with different degrees of hydration, i.e. there is obviously an equilibrium between lipid layers with different, but distinct hydration degrees. Outside phase transitions, these separate phases cannot be observed, but this might also mean that our methods are not yet able to detect them.

The question of non-lamellar phases typically applies to mixtures, especially if lipids are mixed with other kinds of amphiphiles, such as surfactants. Here, diverse

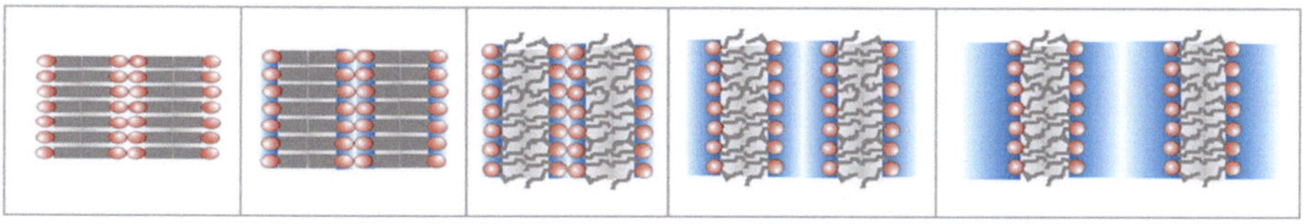

Fig. 4.1 Hydration of solid lipids, frequently inducing the liquid-crystalline phase and swelling

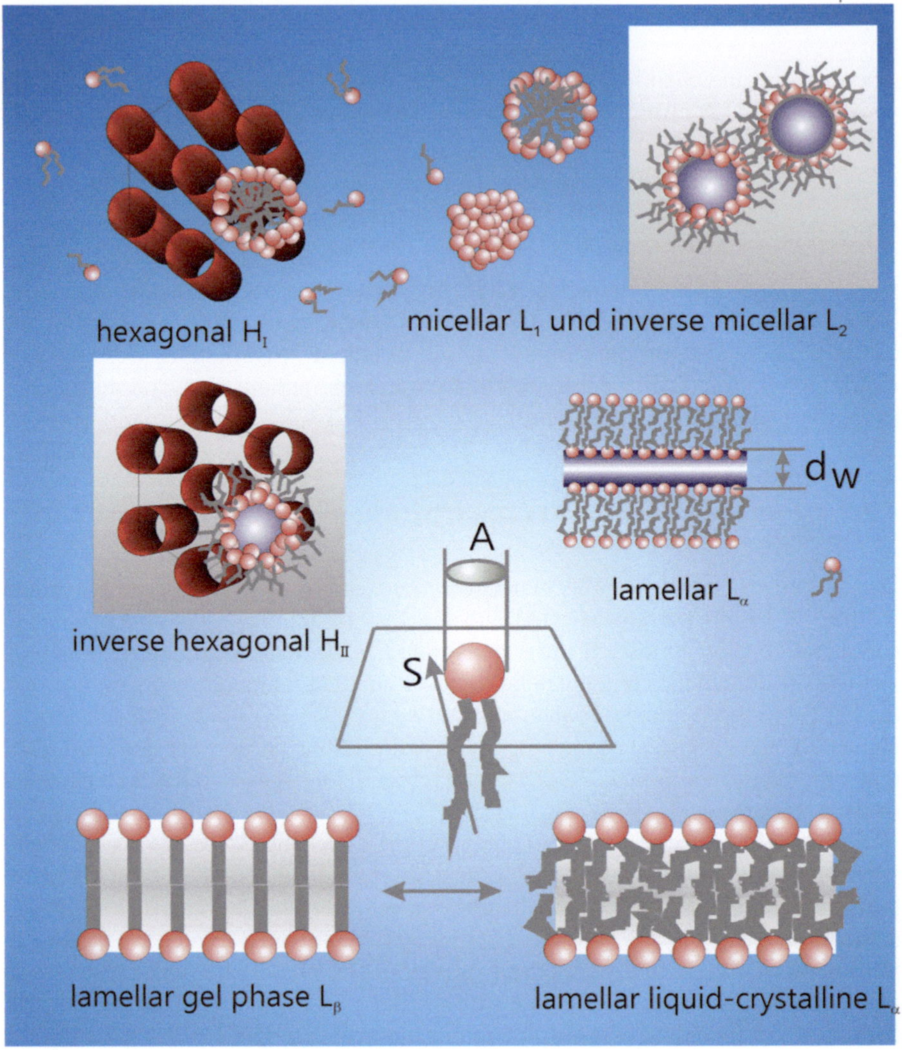

Fig. 4.2 Different hydration states of lipids induce different lyotropic phases

non-lamellar phases, such as hexagonal, inverse-hexagonal or inverse micellar phases can be observed at ambient temperatures when hydration is sufficiently reduced (Koynova and Tenchov 2001; Klose et al. 1995a; Funari et al. 1996; Pfeiffer et al. 2012). In native systems, the transition into non-lamellar phases denatures membranes as such which is considered to be a major factor of concern in cryobiology.

Table 4.1 Overview of surface forces with their principal dependence upon water layer distance, d_w

Interaction	Pressure $P(d_w)$
Electrostatic	$P_{el} \propto \exp(-d_w/C_{el})$
Van der Waals	$P_{vdw} \propto 1/d_w{}^3$
Undulation	$P_{und} \propto 1/d_w{}^3$ (Helfrich 1978)
Undulation	$P_{und} \propto 1/d_w$ (Freund 2013)
Peristaltic	$P_{per} \propto 1/d_w{}^5$ (Helfrich 1978)
Steric	$P_{ster} \propto 1/d_w{}^{(9/4)} - d_w{}^{(3/4)}$
Protrusion	$P_{pro} \propto \exp(-d_w/C_{prot})$ (Israelachvili and Wennerström 1992)
Polarisation	$P_{pol} \propto \exp(-d_w/C_{pol})$ (Cevc 1991)

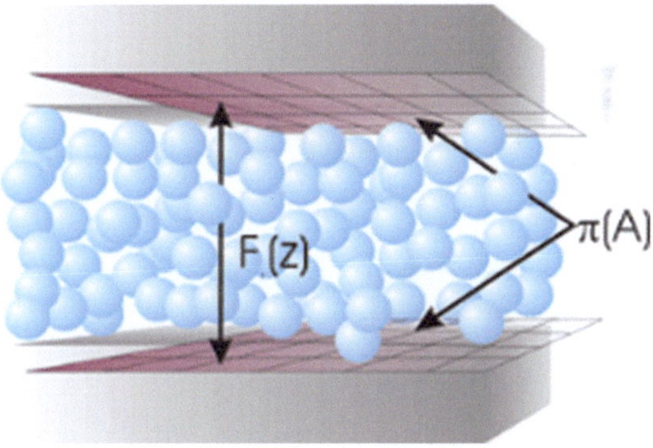

Fig. 4.3 Highly simplified representation of hydration force F(z) between polar surfaces, π(A) denotes the surface pressure

4.2 Hydration Force

4.2.1 Motivation and Possible Definitions

In physical-chemical terms, the force between hydrated surfaces is mostly defined by the hydrostatic pressure maintaining the chemical equilibrium between the hydration water and free water at reference conditions.

It is generally accepted that various interactions apply when hydrated phospholipid membranes approach, such as the Van der Waals, electrostatic (not applicable for zwitterionic lipids in pure solvents), undulatory, peristaltic and steric (=hard core) interactions (Nagle and Tristram-Nagle 2000) (Table 4.1).

There is however one interaction concerning magnitude and distance between steric repulsion and Van der Waals attraction, that is not predicted by established interaction models (Israelachvili and Wennerström 1996). It is mostly called hydration pressure or hydration force (Fig. 4.3) and it can empirically be described by an exponential function versus the water layer distance (Eq. 4.4) or the molar ratio of water and lipid, R_w (see Eq. 4.1):

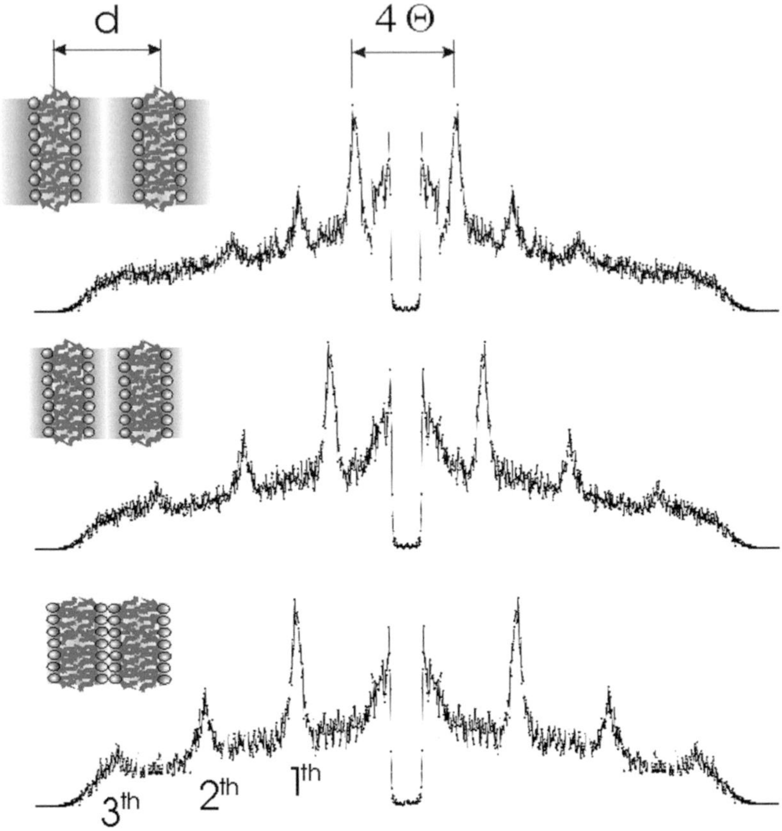

Fig. 4.4 Idealised representation of the wide-angle diffraction patterns at varying repeat distances of the corresponding lipid layers. The numbers give the diffraction orders

$$P_h = P_{h,0} \exp\left(-\frac{R_w}{R_{wo}}\right), P_h = P_{h,0} \exp\left(-\frac{d_w}{d_{wo}}\right) \tag{4.4}$$

The characteristic constants $P_{h,0}$ are the hydration pressure or hydration force at hypothetical zero-hydration and $R_{W,0}$ and $d_{W,0}$ are the corresponding decay constants (Eq. 4.4). The repeat distance is frequently determined by X-ray diffraction (Fig. 4.4).

In the range from 0.2 to 2 nm, the hydration force is considered to be the dominating interaction, therefore, in many cases the force curves obtained for that hydration range are considered as the hydration forces curves as such.

One should not be confused that there are different representations of hydration force used in the literature. The hydration force, F_h, for instance is for convenience frequently expressed in pressure units. When measuring with the surface force apparatus (SFA) however, the force is usually given as the force normalized with respect of the "effective radius" of approaching cylinders, typically with the

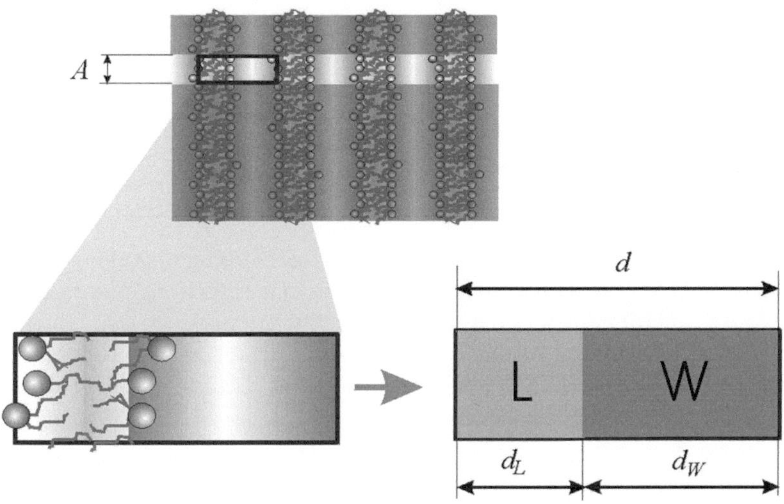

Fig. 4.5 Composition of the repeat distance according to the LUZATTI model

dimensions [F/R] = μN/m (see also Sect. 4.2.3.2). In the case of atomic force microscopy however, a force as such is provided, mostly given in "nanoscopic" dimensions, i.e. [F] = nN or pN (see also Sect. 4.2.3.3) and derived from the frequencies of an oscillating scanning tip.

For comparing hydration forces in lipids with different surface cross sections, proteins and other materials, it makes sense to use the water layer thickness or a surface related hydration number instead of a molar hydration number, otherwise it would be difficult to compare hydration force parameters. One should however keep in mind that geometrical quantities at this molecular level strongly depend on models, such as the model of Luzzati (Luzzati and Chapman 1968) (Fig. 4.5) which is presented in short below.

Thus, the repeat distance, d, is considered as the sum of the water layer thickness, d_W, and the thickness of the bilayer, d_L:

$$d = d_W + d_L \tag{4.5}$$

The volume fractions, Φ_L and Φ_W, can be expressed by the corresponding apparent specific volumes V weighted by their concentrations. Using Eq. 4.5 one can write:

$$\frac{d - d_W}{d_W} = \frac{\overline{V}_L \, c_L}{\overline{V}_W \, (1 - c_L)} \tag{4.6}$$

The weight concentration, c_L, is known from sample preparation and the apparent specific volumes are obtained from densitometry (Wilkinson and Nagle 1981). Usually, one assumes that they are independent of hydration. However, this is an assumption that is not always reasonable, especially at very low hydration. A detailed investigation even shows contradicting results, to be attributable to different methods, assumptions and used instrumentation (White et al. 1987; Scherer 1987).

The final expression for the water layer thickness is:

$$d_W = d\,\frac{1}{\left(\frac{\overline{V_L}}{\overline{V_W}}\,\frac{c_L}{1-c_L} - 1\right)} \qquad (4.7)$$

Hydration pressure, introduced by Langmuir for explaining disjoining pressure in colloids, is thus a phenomenological expression and it was introduced to explain the colloidal stability for many hydrophilic biological surfaces such as for stress on cartilage, cell surfaces, osmotic dehydration or freezing induced dehydration (Wolfe et al. 1994); it is found for DNA, proteins (Valle-Delgado et al. 2011), polyelectrolytes and polysaccharides (Rand and Parsegian 1989; Parsegian et al. 1995). Interesting options of using hydration pressure related phenomena in gels are even given by recent applications in liquid-detecting sensors using hydration triggered thresholds for percolation conductivity (Pfeiffer et al. 2011, 2014).

As mentioned above, the first concept of hydration pressure was introduced by Langmuir in 1938 (Evans and Wennerström 1994) and Le Neveu et al. (1976) applied the concept to lipid bilayers. In 1985, Marra and Israelachvili (1985) have been the first who directly measured the force curves with a surface force apparatus (SFA). Nowadays, two basic concepts try to explain the origin of this force. But these concepts are sometimes contradictory (Israelachvili and Wennerström 1996; Parsegian and Rand 1991) and the efforts to make a definitive decision for one of these theories have not been wholly successful (see also Sect. 4.3).

4.2.2 Controlling Hydration and Pressure

The exploring of hydration-dependent pressure effects requires defined thermodynamic boundary conditions. Therefore, this section will present a small overview on principal configurations to adjust hydrostatic pressure, chemical potential and/or water concentration (Fig. 4.6), see also Table 4.2.

Configuration I – Iso-compositional: This is the standard configuration to investigate high-pressure effects on closed lipid dispersions (Winter and Pilgrim 1989) which are situated in a sealed vessel preventing a change of the global solvent concentration. Changes in the hydration state of the lipid by hydrostatic pressure means amplification or weakening of solvent-solute interactions probably leading to a change of hydration numbers. Furthermore, the thermodynamic activity of all components is enhanced with respect to a reference phase because the hydrostatic pressure is increased (Moore 1972).

Configuration II – Iso-potential: This configuration represents the original hydration pressure experiment (Leneveu et al. 1976, 1977), see Sect. 4.2.3.1. The chemical potential of water is kept constant (= iso-potential) at every step of

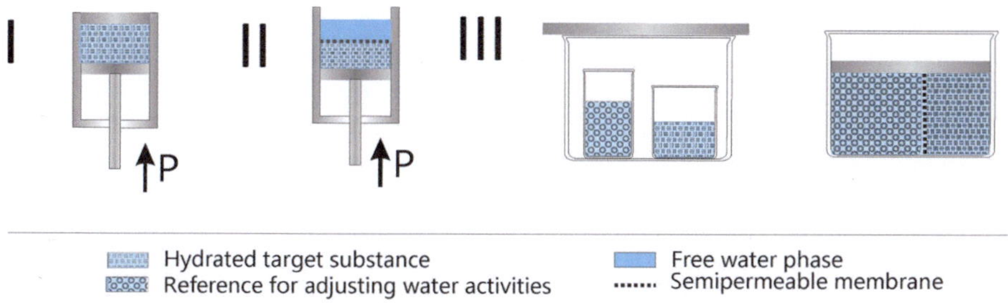

Fig. 4.6 Different configurations to adjust hydration and/or hydrostatic pressure, *left* iso-compositional, *middle*: iso-potential, *right*: isopiestic

Table 4.2 Relationship between hydration n_w, hydrostatic pressure P_{hst} and chemical potential of water μ_w at different experimental configurations

Configuration	Constant	Variable
I	n_w	P, μ_w
II	μ_w	n_w, P
III	P_{hst}	μ_w, n_w

hydration enabled by a semi-permeable membrane that mediates the chemical equilibrium of the membrane water with a water phase at standard conditions (free water). The amount of water released into the free water phase is according to the hydration force function correlated to the hydrostatic pressure.

Configuration III – Isopiestic: This configuration represents hydration under constant, ambient pressure and the chemical potential is varied by humidities or by appropriate osmotic solutions (see e.g. vapour pressure and osmotic stress method) (Parsegian et al. 1979). One obtains functional pairs of chemical potential hydration which are also known as sorption isotherms. The water activity can be adjusted by the water vapour via saturated salt solutions, by inert co-solutes, such as dextran or polyethylene-glycols (PEG) or by devices that use a calibrated gas stream composed of water vapour and other gasses, such as nitrogen (Binder et al. 1999a; Baumgart and Offenhäusser 2002).

4.2.3 *Measuring Hydration Force: A Selection of Piezotropic Methods*

4.2.3.1 The Original Method by LeNeveu and Rand

The original experiment on hydration force, published in 1976, is performed according to the iso-potential configuration (Configuration II). For achieving the osmotic pressure, the lecithin bilayers were deposited in a dextran solution which was in contact with a reference water phase via a semi-permeable membrane.

Dextran is chemically inert with respect to the lipids and the macromolecules do not penetrate into the bilayer phase (Leneveu et al. 1976, 1977). The osmosis-driven hydrostatic pressure in the dextran/lipid solution was directly determined by conventional pressure gauges. Furthermore, the repeat distances were determined by small angle X-ray diffraction (SAXD) (Fig. 4.3) and for calculating the water layer thickness, the Luzzati model was applied (Luzzati and Chapman 1968).

The Luzzati model (see above) considers the explicit assumption that water and lipid layers do not penetrate. Although, this assumption is relatively unrealistic, that model was also frequently applied in the literature, and even if the weaknesses are taken into account, valuable results were obtained. The determination of the zero approach is however also here a big challenge, such as in many other methods for determining hydration force (Butt et al. 2005). Newer models make use of the electron density profile obtained from X-ray scattering (see e.g. (Schmiedel et al. 2001)).

4.2.3.2 Surface Force Apparatus (SFA)

The surface force apparatus (Israelachvili and Adams 1978) also operates under the iso-potential configuration (Configuration II) because the pressurised water layer is also (should be) in equilibrium with an outer water phase at reference conditions. Practically, two crossed cylinders with a diameter of approximately 1 cm are moved against each other along a line perpendicular to their axes. When measuring the force between the cylinders by a deflected spring and the distance by optical interferometry using fringes, accurate surface force curves are obtained. For the cylinder, pure or coated mica surfaces are used, (Marra and Israelachvili 1985) but also silica surfaces are possible as substrate (Orozco-Alcaraz and Kuhl 2013). In contrast to the AFM method described below, relatively flat surfaces are applied with a radius that is large with respect to the thickness of the water layer. In this sense, the experiments are very close to the original experiment on hydration force (Sect. 4.2.3.1), except of the fact that with the SFA, the lipid layers are supported leading to reduced undulatory motions.

Furthermore, the force curves of several systems show oscillatory behaviour (Christenson and Horn 1983) attributed to the layer-wise removal of hydration water, also observed for other solvents. This is an interesting phenomenon because oscillatory means that even the sign of the force changes, i.e. hydration repulsion changes into hydration attraction and so forth.

When the cylinders are coated with lipids, no oscillatory but smooth curves are measured (Trokhymchuk et al. 1999). These smooth curves can be interpreted as Hilbert transforms because that mathematical operation provides envelopes connecting the maxima of an oscillation curve. In the literature, the lack of oscillatory behaviour was in physical terms explained by a dynamically "smearing out" of those oscillations (Leckband and Israelachvili 2001) (see also Sect. 4.2.4.1) which is in fact nevertheless a remarkable phenomenon; one could alternatively also assume that the positive and negative oscillatory "peaks" arithmetically average.

4.2.3.3 Atomic Force Microscopy (AFM)

Another direct method for measuring hydration forces is atomic force microscopy (AFM) (Fukuma et al. 2007; Higgins et al. 2006). Here, a ultrathin tip scans the sample perpendicular and planar along different hydration layers, comparable to the operation mode of a record player. Usually, a frequency modulation technique is applied (FM-AFM) and variations of the resonance frequency of the oscillating tip are directly correlated to local forces.

It is more difficult to categorise that experiment under the configurations mentioned under Sect. 4.2.2. This is due to the very local nature of the measurements being at molecular scale. But the set-up strongly resembles the iso-potential configuration (Configuration II) because pressurised water molecules under the tip can dynamically exchange with free water in the neighbourhood of the tip.

Already in 1995, Cleveland et al. reported "oscillatory" forces acting between water-filled mineral surfaces measured by AFM (Cleveland et al. 1995), comparable to the results obtained with surface force apparatus (see Sect. 4.2.3.2). But later on, the "oscillatory" hydration forces were also observed in the presence of lipids (Fukuma et al. 2007), in contrast to the SFA measurements where these oscillations are usually "smeared out" (Leckband and Israelachvili 2001).

Unfortunately, phospholipids obviously need to be in the solid state for in-depth analysis of force curves, otherwise the ultrathin tip breaks through the soft lipid-crystalline layer at a certain distance. The "breakthrough distance", i.e. the distance between the tip and zero-distance at membrane disruption is a well-investigated parameter seriously limiting further exploration of hydration pressure under liquid-crystalline conditions (Butt et al. 2005). Finally, it was also reported that the chemical nature of the tip itself has a strong influence on the parameters of the obtained curves (Butt et al. 2005). Accordingly, it was also tried to coat the tip with a lipid bilayer which however seems to be a quite demanding procedure.

4.2.3.4 The Piezotropic Phase Transitions Method (PPM)

Another method for obtaining hydration pressure in bulk phases was proposed a couple of years ago (Pfeiffer et al. 2003a). It uses configuration I, i.e. a hydrostatic pressure is applied on a closed lipid dispersion. In contrast to AFM or the SFA method, the pressure does not act on supported lipid bilayers, but on non-oriented bulk layers.

Using the PPM, the shift of the main phase transition pressure of lipids is measured at different hydration degrees. For that purpose, the diamond anvil cell (DAC) can be used that is filled with lipids at known water concentration and spectroscopic pressure gauges, such as quartz (Wong et al. 1985) or $BaSO_4$. The phase transition can be determined by FTIR spectroscopy making use of the pressure-tuned gauche – all-trans transition, detectable by the ν-CH_2 stretching vibrations (Dunstan and Spain 1989; Spain and Dunstan 1989). The hydration-dependent shift of the hydrostatic phase transition pressure can be considered as numerically equal to the hydration pressure (see derivation below). The approach requires almost no theoretical assumptions, the only problem is that the method requires a relatively sharp phase transition in a measurable range. In this context, it is quite difficult to determine the magnitude of the hydrostatic pressure sufficiently accurate because of spectroscopic parameters are used taken from the internal pressure gauges. The thermotropic method (Sect. 4.2.4.3) which is based on the same formalism partially establishes a solution to this problem because a scanning calorimeter can be used.

The presented piezotropic approach thus gives a simple relationship between hydration pressure and the shift of the main phase transition pressure. Let us consider a phospholipid/water dispersion existing in a two-state phase equilibrium, liquid crystalline phase, L_α – gel phase, L_β. Water and lipid are considered as the two components of the dispersion. According to the equilibrium condition, the chemical potential of the lipid in the liquid crystalline phase must be equal to the chemical potential of the gel phase (Eq. 4.8).

$$\mu_{L,liq} = \mu_{L,gel} \tag{4.8}$$

From this follows (Eq. 4.9):

$$d\mu_{L,liq} = d\mu_{L,gel} \tag{4.9}$$

The relationship between the change of the chemical potentials of lipid and water, $d\mu_L$ and $d\mu_W$ is given by the Gibbs-Duhem relation (Eq. 4.10).

$$0 = n_L\, d\mu_L + n_W\, d\mu_W \tag{4.10}$$

which is applied at both the left and the right side of Eq. 4.9 (Eq. 4.11):

$$R_{w,liq} d\mu_{L,liq} = R_{w,gel}\, d\mu_{L,gel} \tag{4.11}$$

The change of the chemical potential can be expressed by the second terms of the corresponding Taylor series (Eq. 4.12).

$$d\mu = \left(\frac{\partial\mu}{\partial p}\right)_{a_w} dp + \left(\frac{\partial\mu}{\partial a}\right)_p da \tag{4.12}$$

The derivative of the chemical potential with respect to pressure is the molar volume, V_W, and the derivative with respect to the activity can be obtained from the well-known relationship, $\mu_w = \mu_{w,0} + RT\ln a_w$, which links the activity to the chemical potential (Eq. 4.13).

$$d\mu = V_W dp + RT\, d\ln\, a_W \tag{4.13}$$

A combination with Eq. 4.11 provides Eq. 4.14:

$$\mathrm{R_{W,liq}}\left(\mathrm{V_{W,liq}dp + V_{W,liq}d\Psi_W}\right) = \mathrm{R_{W,gel}}\left(\mathrm{V_{W,gel}dp + V_{W,gel}d\Psi_W}\right) \tag{4.14}$$

where Ψ_W is the isopiestic water potential. Rearrangement leads to Eq. 4.15:

$$\left(R_{W,gel}V_{W,gel} - R_{W,liq}V_{W,liq}\right)dp = -\left(R_{W,gel}V_{W,gel} - R_{W,liq}V_{W,liq}\right)d\Psi_W \tag{4.15}$$

From this follows Eq. 4.16:

$$dp = -d\Psi_W \tag{4.16}$$

The absolute values of the differential shift of the transition pressure and the isopiestic water potential are thus numerically equal (Eq. 4.17):

$$\int_{P_{re}\ (a_w=1)}^{P_{tr}} dp = -\int_0^{\Psi_W} d\Psi_W \tag{4.17}$$

Integration of Eq. 4.17 gives the relationship between the isopiestic water potential and the shift of the piezotropic phase transition pressure (Eq. 4.18):

$$P_{tr} - P_{tr,0} = \Delta P_{tr} = -\Psi_W \tag{4.18}$$

where $P_{tr,0}$ is the transition pressure at full hydration (Eq. 4.19).

$$P_h = \Delta P_{tr} \tag{4.19}$$

Given the relationship between hydration pressure and isopiestic water potential, one can conclude that the hydration pressure is equal to the dehydration-induced shift of the main phase transition pressure, $P_h = \Delta P_{tr}$ (Eq. 4.19).

4.2.4 Measuring Hydration Force: A Selection of Isopiestic Methods

4.2.4.1 Osmotic Stress Method (OSM)

One of the most frequently applied methods for adjusting hydration force is the so-called "osmotic stress method" (OSM). It replaces the technically more demanding piezotropic experiments (configuration II) by an isopiestic set-up (configuration III), enabled by the assumed equivalence of the configurations II and III, i.e. the isopiestic water potential is considered to be numerically equal to the piezotropic hydration pressure. This is a reasonable assumption proven by consistent results (see also Sects. 4.2.3.4 and 4.2.4.3), but there also are some warning remarks on the methodological side mentioned later on.

According to the OSM, the lipids are deposited in an isopiestic arrangement according to the configuration III. The water activity, a_w, is adjusted by a water vapour or by an osmotic solution containing inert solutes. The hydration pressure can be calculated using the well-known formula (Eq. 4.20):

$$P_h = -\frac{RT}{V_w} \ln a_w \tag{4.20}$$

where R is the gas constant, T the absolute temperature and V_w is the molar volume of water (Leneveu et al. 1977). If one determines a characteristic hydration quantity, such as the water content per lipid, R_W, or/and the water layer thickness, d_W, one obtains functional pairs of pressure and hydration. It is important to mention that the curves do not show oscillatory behaviour (see Sects. 4.2.3.2 and 4.2.3.3). This is explained by an averaging, or "smearing out" of these oscillations arising from lipid domains with different hydration degrees (Leckband and Israelachvili 2001).

It was implicitly suggested (Parsegian et al. 1986) that the experiments according to the configurations II and III are equivalent in the sense that the isopiestic water potential (Eq. 4.3) is numerically equal to the hydration pressure. However, the chemical potential of water in configuration II is constant at all hydration steps, in configuration III it is not. All properties that are correlated to the chemical potential are different as well. Thus, this approach is finally based upon the silent assumption that the underlying hydrophilicity of the hydrated substances is independent of the hydrostatic pressure. However, this needs to be checked in every case. One example is the wavenumber of the ν-CH_2 stretching vibrations of POPC seen by FTIR spectroscopy (Pfeiffer et al. 2013b). With decreasing hydration, i.e. with increasing isopiestic "hydration pressure", a reduction of the wavenumber, also called red-shift, is observed, but when applying hydrostatic pressure, blue shift is measured. Another example is reported by Di Primo et al. (Diprimo et al. 1995) who have investigated the influence of "osmotic pressure" and hydrostatic pressure on the low-spin-high-spin transition for cytochrome 450. The authors found that both "pressures" induce antagonistic effects, i.e. hydrostatic pressure promotes the high spin to low spin transition and "osmotic pressure" promotes the low spin to high spin transition. But if one correlates that spin-transition with the water activity, there is indeed no antagonistic effect, because the water activity is increased at hydrostatic pressure and decreased by "osmotic pressure". The "antagonistic effect" in this view appears less mysterious. Indeed, it has been correctly suggested by Di Primo et al. that hydrostatic pressure and "osmotic pressure" probe different properties of the transition. Therefore, care has to be taken, at least when using spectroscopic quantities to derive hydration force parameters, see also Sect. 4.2.4.2.

The quantity "hydration pressure" in the configuration II (Rand and Parsegian 1989) is thus used as a synonym for the isopiestic water potential, Ψ_W. To show the difference, let us consider the basics of osmosis.

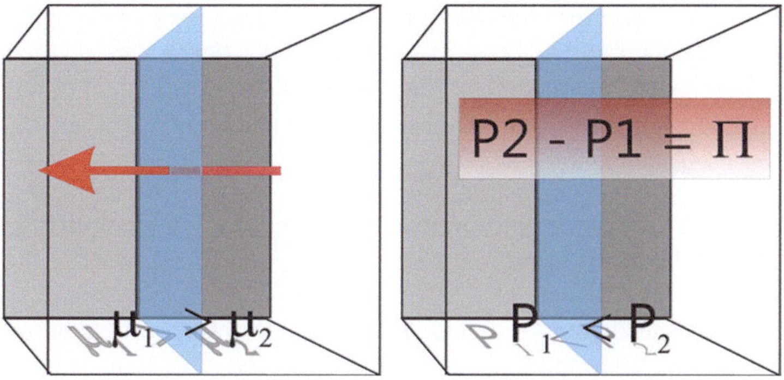

Fig. 4.7 Osmosis (*left*), a diffusion driven flux due to the gradient of the chemical potential. Osmotic pressure (*right*), the hydrostatic pressure to stop or prevent osmosis

The net flux of solvent molecules from one solution into another caused by different concentrations of solutes is called osmosis (Fig. 4.7). This flux in the direction of the solution with the higher concentration is driven by diffusion. The concentration difference is maintained by a selective diffusion barrier (semi-permeable membrane) that avoids the exchange of solute molecules. Osmotic pressure, Π, is defined as the smallest hydrostatic pressure that is required to stop or prevent osmosis (i.e. a kind of compensation pressure), i.e. the net flux of solvent. In other words, the hydrostatic excess pressure ($\Pi = P2–P1$) which must act on the higher concentrated solution (phase 2) to enable the chemical equilibrium with the less concentrated solution (phase 1 = reference phase) is called osmotic pressure. If the molar volume, V_W, of water can be considered to be constant, one can write Eq. 4.21:

$$\mu_1 = \mu_2 + V_w (P2 - P1) \tag{4.21}$$

The osmotic pressure is then defined by Eq. 4.22:

$$\Pi \underset{def}{=} P_2 - P_1 \tag{4.22}$$

The most typical feature of real osmotic pressure experiments is that these are piezotropic experiments, there is a physical hydrostatic pressure present and that the chemical potential of water is kept constant by a membrane and a reference solution.

The concept of "osmotic stress" have not only been applied to lipid hydration, but also to biochemical reactions to determine the amount of water that is exchanged at the reaction (Rand et al. 1993). This is important if the biomolecules (enzymes etc.) are situated in a solution which contains inert co-solutes.

A sometimes overlooked limitation of Eq. 4.20 regards the value of the molar volume of water. Due to change of the water structure it is unlikely that the density is a constant, i.e. the molar water volume starts deviating substantially from the bulk property (Scherer 1987). From this follows that absolute data on the hydration pressure in the range of $R_w = 1$ are most probably overestimated (Pfeiffer et al. 2003b).

4.2.4.2 Nuclear Magnetic Resonance (NMR): Quadrupolar Splitting and Relaxation Time

Also here, the isopiestic configuration was used (configuration III), but the method is probably not limited to this boundary condition.

A typical feature of the hydration force function is that it is exponentially decaying and such behaviour is also seen for other physical quantities at hydration. In this sense, one can argue that all hydration-tuned exponentially decaying parameters have a more or less direct correlation with the hydration force, or at least a correlation with its common root cause. An interesting example is the logarithm of the hydration-dependent shift of the lateral diffusion in phospholipid monolayers (Baumgart and Offenhäusser 2002) that has a linear correlation with the water activity. It was accordingly shown that a measurement of the lateral diffusion would provide at least an estimate of hydration force parameters.

An exponential behaviour is also seen for various NMR parameters, it has e.g. be shown that the quadrupolar splitting of D_2O, $\Delta\nu_0$ that is correlated to the average orientation of water molecules (Fig. 4.8), is linearly correlated with the water activity, such as reported by Volke et al. (1994a). The corresponding equation is Eq. 4.23:

$$a_W(n_W) = \frac{\Delta\upsilon_0 - A}{B} \tag{4.23}$$

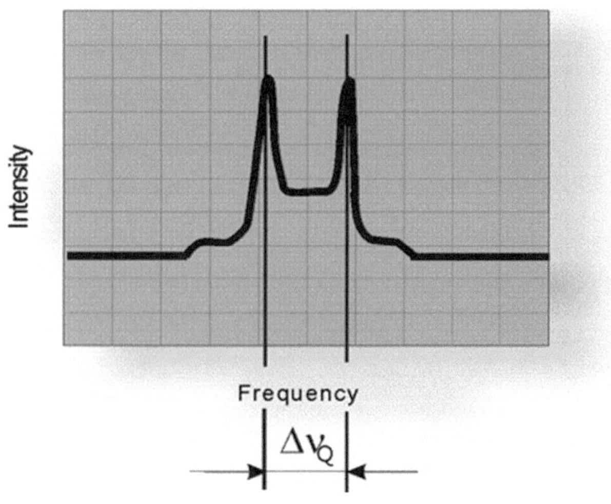

Fig. 4.8 Highly idealised representation of quadrupolar splitting spectra

The coefficients A and B are free variables which depend on the nature of the lipid. From this observation, the hydration pressure can easily be derived by Eq. 4.24 making use of Eq. 4.20.

$$P_h(n_W) = \left(-\frac{RT}{V_W}\right) \ln\left(\frac{\Delta \upsilon_0 - A_0}{B}\right) \tag{4.24}$$

The method of Volke was also applied to mixtures composed of lipids and non-ionic surfactants. Although, the linearity stated by Eq. 4.23, is slightly disturbed, most probably due to admixture effects (Pfeiffer et al. 2010), the parameters A and B are very similar to the original values published before.

An important limitation of this method should be mentioned; the quadrupolar splitting is reduced at and in the range of the main phase transition, a phenomenon which is called motional narrowing (Bryant et al. 1992). The motional narrowing proceeds over a relatively wide temperature ranges (10–20 K) and its temperature dependency can mathematically be described by Eq. 4.25 (Hawton and Doane 1987),

$$\Delta \upsilon_0 \propto \sqrt{|T - T_0|} \tag{4.25}$$

where T_0 is the main phase temperature. Motional narrowing is not yet completely explored, but one assumes that during domain formation at phase transitions, the average preferential water orientation within the lipid layer is partially lost. According to Eq. 4.25, the quadrupolar splitting can even disappear at the phase transition temperature T_0 and the water behaves apparently as free water! in the time window of the NMR experiment. However, also in the case of phase transitions, hydration forces are measured and almost no corresponding abnormal loss or uptake of water can be detected (Pfeiffer et al. 2013a). This points to the fact that this NMR parameter only correctly reflect hydration force parameters when no phase transitions are "in the neighbourhood".

An analogous formulae (Eq. 4.26) for the water activity (see Eq. 4.23) was proposed by Ulrich et al., but most interestingly, here applied to the CD_2-groups of the headgroup of lipids (Ulrich and Watts 1994a),

$$a_W(n_W) = \left(\frac{f(n_W) - f_0}{f_S - f_0}\right) \tag{4.26}$$

where f denotes the generalised NMR parameters investigated, the T_1 relaxation times of the CD_2 segments and the respective quadrupolar splitting, $\Delta \upsilon_0$. In this sense, it was consequently argued that NMR directly reflects the sorption isotherms of lipids. The equivalence of Eqs. 4.24 and 4.26 is of high importance because they show how order parameters of lipid and water are coupled, see also Ge and Freed 2003.

Also the T_1 spin-lattice relaxation time of the hydration water D_2O can directly be related to hydration phenomena (Eisenblätter et al. 1994). It was reported that the mean spin-lattice relaxation time of water in the neighbourhood of phospholipid membranes is an averaged value which is composed of the different relaxation times at different positions x, it is in the range of milliseconds. The average value can be described by the following equation, Eq. 4.27:

$$\frac{1}{T_{1,N}} = \frac{1}{N}\sum_{x=1}^{N}\frac{1}{T_1(x)} \tag{4.27}$$

The approach is based a simple two-state model for the spin lattice relaxation time T_1 assuming fast exchange between water binding sites. This model assumes two states of water, the bound state (b) and the unbound state (u). This approach resembles the assumption of a Langmuir VI isotherm where a first hydration layer is considered to be bound and the remaining one are considered to be free solvent molecules. For the spin lattice relaxation times, one finds that the relaxation of the bound water molecules are smaller than that of the unbound molecules ($1/T_{1,b} > 1/T_{1,u}$). The reason is the fixation of the water molecules at the surface of the amphiphiles which enables a faster spin lattice relaxation.

The Hamiltonian for a water molecule in the state i is given by Eq. 4.28:

$$H_i\ (N) = \left(\Delta E - \frac{E_S}{N}\right)\ \sigma_i + const. \tag{4.28}$$

where ΔE is the difference for the energy between the bound and the unbound state, and E_S the free enthalpy of hydration. The ISING-variable, σ_i, is set to be 1 for the bound state and 0 for the unbound state. From Eq. 4.28 one obtains Eq. 4.29,

$$\frac{1}{T_{1,N}} = \frac{1}{Z}\sum_{i=b,u}\frac{1}{T_{1,i}}\ \exp\left(-\frac{H_i(N)}{k_B T}\right) \tag{4.29}$$

where Z is the canonical partition function of the molecule. The final expression as published by the authors (Eisenblätter et al. 1994) is (Eq. 4.30):

$$\left(\frac{T_{1,N} - T_{1,b}}{T_{1,b}}\right) \approx \left(\frac{T_{1,F} - T_{1,b}}{T_{1,b}}\right)^{1-\frac{n_C}{N}} \tag{4.30}$$

The relaxation time of free water is given by $T_{1,F}$ and the constant n_c is defined by $n_c = E_s/\Delta E$. However, the final expression obtained from the above mentioned paper can be rewritten by using the simple relationship, Eq. 4.31:

$$a^x = e^{x\ \ln\ a} \tag{4.31}$$

In this picture, the reciprocal mean spin-lattice relaxation time $T_{1,N}$ is given by Eq. 4.32:

$$\frac{1}{T_{1,N}} = \frac{1}{T_{1,b}} \frac{1}{\exp\left(\frac{\Delta E - E_S/N}{k_B T}\right) + 1} \tag{4.32}$$

The previous equation (Eq. 4.32) shows that the relaxation time is described by an expression containing the difference of the energy states between the bound and unbound states, ΔE, as well as the averaged hydration Energy, E_S. Interestingly, the previous equation (Eq. 4.32) is formally analogous to the FERMI-DIRAC distribution. It represents the energy distribution of the fermions depending on the FERMI-energy, E_F. The original FERMI-DIRAC distribution is given by Eq. 4.33:

$$f(E) = \frac{1}{\exp\left(\frac{E - E_F}{k_B T}\right) + 1} \tag{4.33}$$

The original FERMI-energy, E_F, corresponds thus to the energy level between bound and unbound states and the energy, E, corresponds formally to the expression E_S/N in Eq. 4.32.

The analysis of hydration dependent spin-lattice relaxation times have thus been proposed as a tool for the characterisation of surface energy conditions of amphiphile/water systems (Eisenblätter et al. 1994). The approach established in Eq. 4.32 was in this way applied as a tool for the determination of the binding energy of water in phospholipids and non-ionic surfactants (Eisenblätter et al. 1994; Klose et al. 1995b). The value for POPC and $C_{12}E_4$ have been given as $E_S = 28 \pm 4$ kJ/mol and $E_S = 24 \pm 6$ kJ/mol. However, the sorption isotherm of POPC and $C_{12}E_4$ shows that the hydration behaviour is strongly different. The free enthalpy of hydration is only about $\Delta G = 4$ kJ/mol instead of 24 kJ/mol. The fitting according to Eq. 4.32 thus furnishes strongly overestimated values in the case of non-ionic surfactants. There are at least two probable reasons. The quantity E_S does not correspond to the free enthalpy of hydration, or the assumptions for Eq. 4.32 are too simplified. However, the qualitative behaviour according to the FERMI-DIRAC distribution remains an interesting fact and should be further investigated.

4.2.4.3 The Thermotropic Phase Transition Method (TPM)

This relatively direct method for obtaining hydration pressure curves was proposed in parallel with the piezotropic phase transition method, PPM, (Sect. 4.2.3.4) and it is based on the pressure-tuned variation of the main phase transition temperature in phospholipids (Pfeiffer et al. 2003b). Its advantage is that it can be performed under isopiestic conditions, such as the OSM method (Sect. 4.2.4.1).

Different approaches has been proposed in the past for using thermotropic phase transitions in lipid/water dispersions to obtain hydration pressure parameters. The approach of Ulrich et al. (Disalvo et al. 2008), demonstrated on DOPC, is based on the freezing point depression of the hydration water. Interestingly, the same approach was introduced by Bach et al. (1982) already in 1982, but in their case referred to "swelling pressure". The hydration pressure as a function of water content is given by Eq. 4.34 (Ulrich et al. 1994):

$$P_h = \frac{\Delta H_{tr,W}}{V_W}\left(1 - \frac{T\,(R_W)}{T_0}\right) \tag{4.34}$$

where $\Delta H_{tr,W}$ is the molar enthalpy change of melting ice, V_W the molar volume of liquid water, $T(R_W)$ the hydration dependent melting temperature and T_0 the melting temperature of pure water. The application of that approach provided a sound result for DOPC. However, for the hydration pressure at zero hydration one obtains $\Delta H_{W,tr}/V_W$ implying that the freezing temperature would approach zero Kelvin at dehydration, and that the pressure at dehydration is independent of the lipid species, the last outcome clearly contradicts other results from the literature.

Cevc and Marsh (Cevc and Marsh 1985) proposed an equation relating the shift of the main phase transition temperature of lipids to the degree of hydration,

$$\Delta T_m = \Delta T_m(0)\ \tanh\left(\frac{n_W V_W}{\lambda\ A}\right) \tag{4.35}$$

but experiments by Simon & McIntosh (Simon et al. 1991) could not confirm that approach experimentally, it was concluded that calorimetric investigations of the thermotropic phase transition of the lipid might only give qualitative information on hydration processes. Another equation (Eq. 4.36) based on the same formalism relates the shift in phase transition temperature to the water potential, Ψ_w (Cevc 1987),

$$\Delta T_m = \frac{A\zeta}{100\Delta S_{anh,t}}\Psi_W \tag{4.36}$$

where, $A\xi$ is a length characteristic of the water structure and $\Delta S_{anh,t}$ the transition entropy of the lipid in the dehydrated state. But, because the derivation follows the same formalism as for Eq. 4.35 the conclusions in Simon et al. 1991 also apply here.

However, when skipping intrinsic hydration force theories and applying fundamental thermodynamics, it is nevertheless possible (Pfeiffer et al. 2003b) to obtain quantitative information on hydration force from the shift of the main phase transition temperature. The underlying physics is based on the Gibbs-Duhem equation for linking hydration pressure and water potential, as well as on the Clausius-Clapeyron equation relating the pressure variations to the variations of the main phase transition temperature.

The hydration pressure parameters were determined for various lipids by differential scanning calorimetry, and the only assumption is the approximate constancy of the pressure-induced shift of the phase transition temperature ($\alpha \approx 0.2$ K/MPa), which is valid for almost all compounds containing long hydrocarbon chains (Pfeiffer et al. 2003b, c).

The derivation of the relationship between the main phase transition temperature and hydration pressure thus essentially continues the formalisms for piezotropic transitions (Sect. 4.2.3.4). The relationship between the differential shift of the piezotropic phase transitions, dP_{tr}, and the water potential, $d\psi_w$, was given by Eq. 4.37:

$$dP_{tr} = -d\Psi_W \tag{4.37}$$

After an infinitesimal shift of temperature, dT, a new equilibrium is established. The dependence of the transition pressure, P_{tr}, on the temperature is given by the equation of Clapeyron-Clausius (Eq. 4.38), which is also valid for main phase transitions of phospholipids (Winter and Pilgrim 1989):

$$\frac{dT}{dP} = \frac{\Delta V_{L,tr}}{\Delta S_{L,tr}} \tag{4.38}$$

$\Delta V_{L,tr}$ and $\Delta S_{L,tr}$ are the volume change and the entropy change of the lipid. A comparison of Eqs. 4.37 and 4.38 shows that the dependence of water potential on the temperature is simply given by Eq. 4.39:

$$\frac{dT}{d\Psi_W} = -\frac{\Delta V_{L,tr}}{\Delta S_{L,tr}} \tag{4.39}$$

From Eq. 4.39 thus follows a simple method for the determination of hydration pressure by using calorimetry. Taking the definition of the pressure-induced temperature increase, α, given in Eq. 4.40:

$$\alpha = \frac{\Delta V_{L,tr}}{\Delta S_{L,tr}} \tag{4.40}$$

one can rewrite Eq. 4.39. The crucial point is thus the knowledge of the pressure-induced temperature increase α as a function of temperature and hydration. One obtains Eq. 4.41 which gives the relationship between phase transition temperature and water potential:

$$dT_{tr} = -\alpha d\Psi_W \tag{4.41}$$

Thus, using the relationship between water potential and hydration pressure ($P_h = -\Psi_w$, see Sect. 3.3.3) one obtains Eq. 4.42:

$$dP_h = \frac{1}{\alpha} dT_{tr} \tag{4.42}$$

From Eq. 4.42 follows finally Eq. 4.43:

$$P_h = \frac{1}{\alpha} \Delta T_{tr} \tag{4.43}$$

where ΔT_{tr} is the shift of the transition temperature with respect to the fully hydrated state. Furthermore, it is important to realise that α is nearly a constant ($\alpha \approx 0.2$ K/MPa (Winter and Pilgrim 1989; Pfeiffer et al. 2003c)) for about all phospholipids, because it is mainly determined by the ratio of the volume and entropy change per CH_2-segment in the hydrocarbon chain, in this way it is also not a function of hydration itself because it is even invariant concerning the use of other solvents (Pfeiffer et al. 2003d). This finally enables an estimation of hydration pressure in phospholipids when the dehydration-induced temperature shift of the main phase transition is known, i.e. $P_h \approx 5\ \Delta T_{tr}$ with $[P_h] = \text{MPa}$ and $[\Delta T_{tr}] = \text{K}$.

Under Sect. 4.2.4.1 there were some critical remarks concerning the statement that "osmotic pressure", i.e. isopiestic water potential and the hydration pressure are numerically equal. But the consistent results offered by the piezotropic and thermotropic methods, Sects. 4.2.3.4 and 4.2.4.3 give essentially a broad confirmation that this is more or less granted.

The phase transition methods (see also Sect. 4.2.3.4) should be applied to pure lipid species. In the case of mixtures, the phase transitions becomes broad and non-ideal interactions, demixing processes occur, this is even partially a problem in one-component lipid systems, i.e. transitions become broader also in this case (Pfeiffer et al. 2013a). However, this is in fact a general problem for all methods as such. When dehydration is investigated in arbitrary lipid systems by any of these methods, admixture effects will always influence the observed force curves because also the lipid water dispersion is a highly non-ideal binary mixture, see also Sect. 4.1.2.

4.3 Hydration Force Theories

A number of theories for exploring the nature of hydration force have been proposed within the last three decades. They can roughly be systematised into two categories, i.e. some theories focus on the influence of the surface, and others see a dominant influence coming from the solvents. The final truth will be that both parts will play a role because a replacement of the solvent on the one hand, but also the change of the lipid moieties on the other hand has remarkable influence on hydration force parameters.

4.3.1 The Langmuir VI Sorption Isotherm

Usually, theories on absorptions isotherms, such as the Langmuir VI isotherm are mostly not considered as a hydration force theories. On the other hand, due to the dominating opinion that the osmotic stress method is a representation of hydration force, one must conclude that any analytical theory on absorptions is also automatically a "candidate" for a hydration force theory (Marsh 2011). In turn, also hydration force theories were applied for theories on absorption isotherms (see also (Marsh 2011; Klose et al. 1992)).

The Langmuir theory assumes a multilayer arrangement of water molecules above the absorbent and the first layer is per definition considered as bound water. Depending on the boundary conditions, the subsequent water layers belong to a free water phase, or they occupy binding sites with energies following a Boltzmann-type distribution. Many lipid isotherms fit with the Langmuir VI model, but the physical meaning of the free parameters is not always clear (König 1993).

4.3.2 *The Polarisation Theory*

The polarisation theory was originally introduced by Marčelja and Radic in 1976 (Marcelja and Radic 1976) and improved by Gruen and Marčelja in 1983 (Gruen and Marcelja 1983). The theory, based on a Landau expansion of the free energy, regards water dipoles as oriented in a confined space, showing in this way a preferential average dipole orientation (Fig. 4.9) leading to a net polarisation. With progressing surfaces approach, it is required to gradual rearrange these orientations and the entropy-based resistance against those re-orientations is supposed to be the origin of the hydration force. Refinements of that theory tried e.g. to include the explicit influence of the polarity of the surface, to modify the boundary conditions concerning the width of the polar surface and to consider non-local polarisations.

A concluding analytical expression, additionally including ideas from the Gouy-Chapman theory, was given by Cevc (1987). For the hydration pressure P_h as a function of the water layer thickness one obtains equation Eq. 4.44 (König 1993):

$$P_{pol}(x) = \frac{\chi \, \Pi_{h0}}{\xi} \exp\left(-\frac{x}{\xi}\right); P_h(d_W) = \frac{2\chi {\Pi_{h0}}^2}{\xi^2} \exp\left(-\frac{d_W}{\xi}\right) \tag{4.44}$$

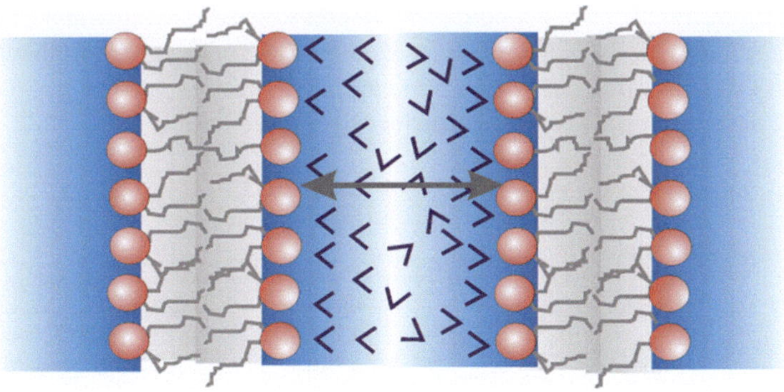

Fig. 4.9 A polar bilayer surface is the reason for a preferential dipole orientation (presented as hooks) which is supposed to be the reason of a repulsive force

The variable x is the distance between the surface and the middle of the water layer ($2x = d_w$), the water structure susceptibility is given by χ and the parameter ξ and Π_{h0} represent a length characteristics of the solvent structure and the bilayer hydration potential at surface contact.

A direct experimental support of the polarisation theory, probably overlooked so far, is the preferential orientation of water molecules with respect to the lipid layer measured by nuclear magnetic resonance (see also Sect. 4.2.4.2). The quadrupole splitting $\Delta\nu_0$ of D_2O is related to the order parameter, S by the well-known equation, $\Delta\nu_0 = ¾\ \kappa\ S$, where κ is the quadrupole coupling constant for deuterated water (Volke et al. 1994b; Gawrisch et al. 1978). The order parameter S is directly related to the angle between O-2H bond and the membrane, $S = 1/2\ (3\ cos^2\Theta - 1)$ and with it to the net polarization. One can see that there is a simple relationship between the experimentally determined water orientation and the reduced water activity, resp. enhanced hydration pressure (Eq. 4.23).

$$a_w = \frac{\frac{3}{8}\kappa\ \left(3\cos^2\Theta - 1\right) - A}{B} \tag{4.45}$$

Interestingly, exponentially decaying orientation angles with respect to the bilayer distance were obtained from computer simulations (Marrinck and Berkowitz 1995) and so it appears that the NMR results finally experimentally confirm polarisation effects. A final question is whether a measurement in the NMR time domain can represent a thermodynamic equilibrium or is it just a snapshot in the time window of the experiment.

A further interesting and puzzling aspect related to the polarisation theory was found by Binder et al. (1999b). When adding water to lipids at room temperature, an endothermic reaction is observed when carefully applying titration calorimetry. The endothermic nature of the hydration process shows that it is under these conditions essentially an entropy-driven process. Obviously, hydration enables an enhancement of the configurational space for additional degrees of freedom in water populations and most probably also in lipids (Ge and Freed 2003). The entropic nature of the interactions was also pointed out by other authors before (Marrinck and Berkowitz 1995) and does also apply to the protrusion theory, see below.

4.3.3 *The Protrusion Theory*

There is sufficient experimental evidence that there are a number of interactions arising from thermally excited motions of membrane components (Gordeliy et al. 1996), usually referred to as entropic forces. Wavelike membrane fluctuations are known as undulations, thickness variations of the bilayers are called peristaltic movements (Helfrich 1978) and thermally excited out-of-plane movements of isolated membrane components are the so-called protrusions (Table 4.1).

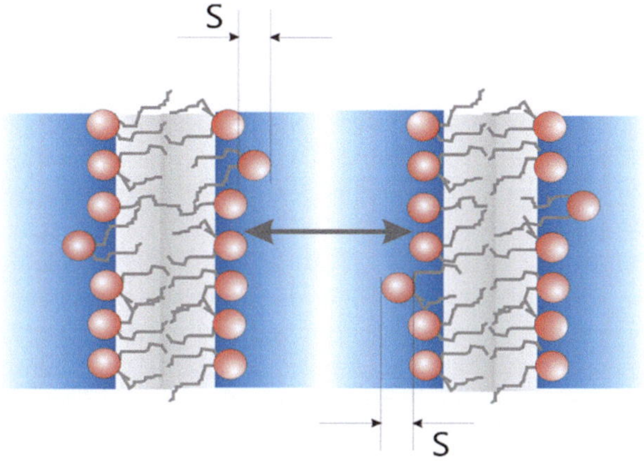

Fig. 4.10 Single-molecule displacements of membrane components are supposed to be the origin of hydration force according the protrusion theory. The structure of the water phase is neglected

The concept presented by Israelachvili and Wennerström states that protrusions are the dominant origin of hydration pressure in phospholipids. Protrusions are indicated in Fig. 4.10 by a displacement of single lipid out of a membrane by the distance S ($S = \Delta x$). It was stated that the protrusion energy $V(x)$ of one amphiphile molecule is a linear function of the displacement (Aniansson 1978) (Eq. 4.46).

$$V(x) = \alpha\, x \tag{4.46}$$

Here, the coefficient α is assumed to be the product of the average lateral dimension of an average protrusion movement and the surface energy, i.e. $\alpha = \pi\sigma\gamma$. The steric repulsion due to the "collisions" of protruding amphiphiles are thus considered as the origin of the hydration pressure and because protrusion is assumed to be governed by the Boltzmann distribution, its exponential term is finally responsible for the exponential decay of hydration pressure. The corresponding expression is given by Eq. 4.47,

$$F \approx 2,7\, \mathrm{n}\, \alpha\, \exp\left(-\frac{\alpha}{kT}\lambda_W\right) \tag{4.47}$$

where n denotes the binding sites per surface unit.

Critical comments (Parsegian and Rand 1991) referred e.g. to contradictions between the probability of protrusion movements and the solubility of the membrane components. If the protrusion theory were valid, the solubility should be larger by four orders of magnitude. Another point of criticism is the difference between the hydration pressure parameters found in membranes with solid and melted hydrocarbon chains. There is almost no such a difference, as one should expect on the basis of the protrusion theory. In the same sense, there should be a difference between the hydration pressure for membranes with one and two hydrocarbon chains, but there is not. Furthermore, addition of substances that changes the stiffness of the membrane (e.g. cholesterol) has almost no influence on hydration pressure. But hydration pressure does occur in polyelectrolytes, something which should be impossible according to Israelachvili and Wennerström. A methylation of the head group also dramatically changes the hydration pressure parameters. Furthermore, it was criticised that it was also forgotten to include the free energy created by the free volume in the hydrocarbon core during the protrusion movement.

There were also attempts to provide experimental support for the protrusion theory as well. Gordeliy (1996) proposed to use neutron and X-ray diffraction to confirm the existence of protrusion motions by relating the out-of plane movements to measurable repeat distances. From the experimental proven increase of inter-bilayer distances and the observed loss of Bragg peak intensity (Gordeliy et al. 1996) one concluded enhanced entropic motions, including protrusion and undulation. The final question is however whether these findings can prove the validity of the protrusion theory as such because it was found by Simon et al. that the short-ranging hydration repulsion does barely vary with the temperature (Simon et al. 1995), a result which would be in contradiction with the findings of Gordeliy, at least for the short-ranging part of the interaction which is assumed to be uninfluenced by undulations and peristaltic movements.

The ultimate decision on the protrusion theory was most probably given by Binder et al. (1999c). In their study, the hydration pressure parameters for polymerised and non-polymerised lipid bilayers were compared. It was found that polymerised lipids in the liquid crystalline phase show about the same hydration pressure parameters than the liquid-crystalline bilayers composed of non-polymerised lipids. In the case of the protrusion model however, the hydration pressure for non-polymerised lipids should be much stronger because the lipid molecules are not bound by covalent bonds such as in the case of the polymerised lipids.

Another study reported hydration pressure parameters for non-ionic surfactants and it was shown that the hydration pressure parameters are independent of the chain length (see parameter α in Eq. 4.46), and that they only vary with the polarity, i.e. in that case with the size of the head groups (Pfeiffer et al. 2004).

4.4 A Possible New Look on Water in Membranes

4.4.1 Recent Developments in the Literature

Finally, the debate on the validity of polarisation or protrusion models is not yet decided (Gordeliy et al. 1996; Binder et al. 1997) but it seems that the arguments for the polarisation model are at least less disputed.

It appeared that at the end of the 1990s, the big discussions on this topic were fading out, most probably because of a lack of new experimental results and inspiration. In the last years however, for instance enhanced computer power and subsequent in-silico studies enabled a renewed debate on hydration force (Zemb and Parsegian 2011). Besides support for one of the existing theories, new approaches emerge. This includes ideas that hydration repulsion only occurs in specific cases

(Schneck et al. 2012) or that hydration pressure might even be a phantom because interactions, such as Van der Waals and steric repulsion are already wrongly applied, i.e. "However if we subtract the predictions of two incorrect theories from a perfectly good measured force curve, we have in fact not a hydration force, but rather, non-sense" (Kunz et al. 2004). An argument in this direction is provided by the work of Freund who e.g. questioned the force law for undulation repulsion (see Table 4.1). Referring to this result, Sharma expressed that the debate "will hopefully encourage design of further experimental work" (Sharma 2013).

All by all, most of the knowledge on hydration force is still empirical, the satisfactory theory thus still needs to be found and there are essentially also not enough experimental results available (Parsegian and Zemb 2011).

4.5 Percolation Theory and the Decay Constants of Hydration Pressure

Jendraziak and co-workers (Jendrasiak and Smith 2004) reported studies on electrical conductivity and hydration in diverse phospholipid membranes investigated as a function of the humidity. Dehydrated lipids are essentially non-conductors, however, Jendraziak and co-workers obtained curves showing a huge increase of the electrical conductivity at relatively low humidity indicating the establishment of so-called percolation networks. This means that the first emerging water networks already enable a certain electrical conductivity that is strongly increasing with water content. Here, the continuous transfer of protons plays the major role explaining macroscopic conductance effects.

Conductance by percolation is essentially a statistical phenomenon i.e. it rises with the probability of mutual contacts between conductive domains in mixtures. For the conductive state, the percolation theory (Essam 1980) predicts a power-law behaviour for the conductance, σ, as a function of the fraction of the conductive component (Eq. 4.48) where X is the concentration of the statistically distributed conductive component with the percolation threshold X_c. The exponent is μ, a semi-universal constant that also depends on the nature of the matrix; σ_{cc} is the conductivity of the pure conductive component.

$$\sigma = \sigma_{cc}(X - X_c)^{\mu} \tag{4.48}$$

Interestingly, Jendraziak et al. never explicitly referred to the percolation theory, at least not to the best of our knowledge. A simple fit also shows that the curves obtained does not support percolation models (Eq. 4.48) established for simple two- or three- dimensional systems indicating that not only statistics is determining the network establishment and that lipid hydration also might proceed in fractal dimensions. Nevertheless, this phenomenon is a percolation process by definition and one can easily define a percolation threshold for most of the experimental curves available (Fig. 4.11, left) which can be further plotted versus the exponential decay constant (Fig. 4.11, right).

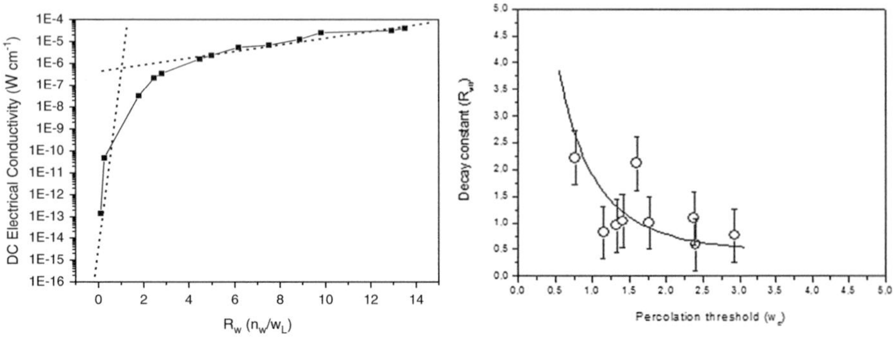

Fig. 4.11 Example of the arbitrary definition of a percolation threshold based on the data of Jendraziak et al. and relationship between decay constant as a function of the percolation threshold for a couple of different phospholipids. The *line* is just a guide to the eyes

The question is thus whether the percolation threshold has direct or indirect relationships with the hydration pressure parameters. An analysis based on the data of Jendraziak (Fig. 4.11, right) proposes that there is at least a slight trend, i.e. the decay constant of the exponential hydration force function decreases with increasing percolation threshold. This would mean that an early-established continuous water phase makes the hydration force more fare-reaching. In any case, even if the trend would be neglected, reasonable regarding the error bars (all data are situated around $R = 2 \pm 1$), one can safely state that the decay constant is in fact/aso that hydration number marking the creation of a continuous water phase.

The fact that percolation thresholds are not a constant also points to the importance of considering fractal dimensions of lipid surfaces. This easily arises from the fact that for a perfect flat surface the percolation threshold should be a constant and it should be possible to describe it by simple percolation models, which is in fact not the case.

In this "percolation-view", the decay constant would be rationalised concerning its relationship with the creation of a liquid phase out of single molecules, however, in most cases the hydration force is modelled starting from slight perturbations of water molecules within the liquid phase. Finally, relating the percolation threshold to the decay constant does not yet provide an exponential decay function. Here, further theoretical work would be required.

4.6 Water Is a Plasticizer

A plasticizer is a chemical additive that makes a material softer, more flexible and ductile. Accordingly, when water is absorbed in solid, dehydrated phospholipids, it will act as a so-called "external plasticizer", i.e. after gradual ingress of the solvent, the mobility and deformability of the membrane strongly enhances; however, which is essential for plasticizers, without challenging the global integrity of the bilayer as such. This typical behaviour is achieved by the fact that the samples change their mechanical properties when exposed to the solvent, but there is always a limiting factor preventing dissolution. This limiting factor arises e.g. from covalent polymerisation, such as in starch (Li et al. 1996; Heremans et al. 2000) or by the entropic effect ensuring lamellar stability of lipid bilayers in water (Ben-Shaul 1995). An interesting analogue is the gelation temperature in starch that shows a typical decay function on hydration, similar to the case of the main phase transition of phospholipids.

The fact that water is a plasticizer could in a certain sense be a source of inspiration for describing and understanding hydration force in lipids. There are essentially four plasticiser theories, the lubrication theory, the gel theory, the free volume theory and the mechanistic theory of plasticization (Daniels 2009). When compared to the hydration force theories, the mechanistic theory of plasticization is the closest one because it considers solvent molecules exchanging between different binding sites. Furthermore, also the free volume theory applies in a certain sense, especially when one considers that water acts like a spacer when entering the headgroup region of lipids forcing free volume in the hydrocarbon chains leading to liquid phases. Finally, it is very likely that hydration, resp. solvation force theories could in turn also apply in plastification theories.

4.7 Summary

There are arguments supporting the polarisation or protrusion theory but also reasonable arguments against them. This means that the final theory still needs to be found and/or that nature is too complex preventing a generalised solution. It might even be allowed to ask whether it is reasonable to talk about surface forces if the structural roughness of the surface itself exceeds the size of the solvent molecules and the decay length is also of the same order. Recent computer simulations seem to confirm some of these considerations leading to conclusions such as "hydration repulsion is less universal as previously assumed" (Schneck et al. 2012). It was even stated that hydration force might be a phantom because Van der Waals and steric repulsion models might be already wrongly applied, i.e. "... if we subtract the predictions of two incorrect theories from a perfectly good measured force curve, we have in fact not a hydration force, but rather, non-sense" (Kunz et al. 2004).

As a summary, one can state that what we call hydration force is observed from stiff mica surfaces up to liquid-crystalline lipid membranes indicating that the dynamics of the hydrocarbon chains has minor influence compared to water structure and surface polarity. Furthermore, the essentially endothermic nature of hydration detectable by titration calorimetry indicates an entropic origin of hydration force. Finally, the spectroscopically confirmed preferential water orientation could lead to the conclusion that hydration pressure is driven by the change of the configurational space of water molecules turning from an relatively disordered 3 dimensional bulk phase into a differently ordered intermembrane phase.

As a final summary, Table 4.3 gives a short overview to the most important facts on hydration force.

Table 4.3 Fact sheet on hydration force in phospholipid layers

Topic	Selection
Nomenclature and/or synonyms	Hydration force
	Hydration pressure
	Hydration repulsion (Schneck et al. 2012)
	Swelling pressure (Bach et al. 1982)
	Osmotic pressure
Common characteristic	Exponentially decaying surface force with a relatively uniform decay constant (0.2–0.4 nm), the pre-exponential factor can however differ considerably
	Cannot be modelled by as a superposition of Van der Waals, undulations or steric interactions
Typical range in terms of water layer thickness	$\lambda_w \approx 0.2$–2 nm
Methods and models for determining/demonstrating hydration force	Pressure gauge in solution under osmotic pressure (Leneveu et al. 1976)
	Deflected spring within the surface force apparatus (SFA) (Marra and Israelachvili 1985)
	Modulated tip oscillations in atomic force microscopy (AFM) (Fukuma et al. 2007)
	Adjusted chemical potential in osmotic stress method (OSM) (Parsegian et al. 1979)
	Main phase transition pressure of lipids (PPM) (Pfeiffer et al. 2003a)
	Main phase transition temperature of lipids (PTM) (Pfeiffer et al. 2003b)
	Melting temperature of hydration water (Ulrich et al. 1994)
	NMR Quadrupolar splitting in lipid (-CD) and/or water (D_2O) (Volke et al. 1994a)
	NMR relaxation times of lipids (Ulrich and Watts 1994b)
	Diffusion of lipids by fluorescence microscopy (Baumgart and Offenhäusser 2002)

Selection of methods for water layer thickness[a] determination	X-ray, using the Luzzati model (Luzzati and Chapman 1968)
	X-ray, analysing electron density profiles e.g. (Schmiedel et al. 2001)
	Optical fringes observed in surface force apparatus (SFA) (Marra and Israelachvili 1985)
Selected theories	Langmuir VI isotherm (statistical mechanics) (Marsh 2011)
	Polarisation (electrodynamics, statistical mechanics and thermodynamics) (Cevc 1991)
	Protrusion (statistical mechanics) (Israelachvili and Wennerström 1990)
Temperature dependence	Only a few studies, obviously negligible (Simon et al. 1995)
Computer modelling	Atomistic simulations (Schneck et al. 2012), Molecular dynamics (Leng 2012)
	Monte Carlo (Hayashi et al. 2002)?

[a]Sometimes, only a molar hydration number is given. This avoids geometrical model assumptions and is acceptable for equal or similar lipid head groups

References

Adam J, Langer P, Stark G (1995) Physikalische Chemie und Biophysik. Springer, Berlin

Aksan A, Hubel A, Bischof JC (2009) Frontiers in biotransport: water transport and hydration. J Biomech Eng-Trans ASME 131. doi:10.1115/1.3173281 074004–1

Aniansson GEA (1978) Dynamics and structure of micelles and other amphiphile structures. J Phys Chem 82:2805–2808

Bach D, Sela B, Miller CR (1982) Compositional aspects of lipid hydration. Chem Phys Lipids 31:381–394

Balasubramanian SK, Wolkers WF, Bischof JC (2009) Membrane hydration correlates to cellular biophysics during freezing in mammalian cells. Biochim Biophys Acta-Biomembr 1788:945–953

Baumgart T, Offenhäusser A (2002) Lateral diffusion in substrate-supported lipid monolayers as a function of ambient relative humidity. Biophys J 83:1489–1500

Ben-Shaul A (1995) Molecular theory of chain packing, elasticity and lipid-protein interaction in lipid bilayers. In: Lipowsky ES (eds) Elsevier Science North-Holland, 1995.

Binder H, Gawrisch K (2001) Dehydration induces lateral expansion of polyunsaturated 18: 0–22: 6 phosphatidylcholine in a new lamellar phase. Biophys J 81:969–982

Binder H, Anikin A, Kohlstrunk B, Klose G (1997) Hydration-induced gel states of the dienic lipid 1,2-bis(2,4- octadecadienoyl)-sn-glycero-3-phosphorylcholine and their characterization using infrared spectroscopy. J Phys Chem B 101:6618–6628

Binder H, Kohlstrunk B, Heerklotz HH (1999a) Hydration and lyotropic melting of amphiphilic molecules: a thermodynamic study using humidity titration calorimetry. J Colloid Interface Sci 220:235–249

Binder H, Kohlstrunk B, Heerklotz HH (1999b) A humidity titration calorimetry technique to study the thermodynamics of hydration. Chem Phys Lett 304:329–335

Binder H, Dietrich U, Schalke M, Pfeiffer H (1999c) Hydration-induced deformation of lipid aggregates before and after polymerization. Langmuir 15:4857–4866

Bryant G, Pope JM, Wolfe J (1992) Motional narrowing of the H-2 NMR-spectra near the chain melting transition of phospholipid/D2O mixtures. Eur Biophys J Biophys Lett 21:363–367

Butt H-J, Cappella B, Kappl M (2005) Force measurements with the atomic force microscope: technique, interpretation and applications. Surf Sci Rep 59:1–152

Cevc G (1987) Phospholipid bilayers. Wiley, New York

Cevc G (1991) Hydration force and the interfacial structure of the polar surface. J Chem Soc, Faraday Trans 87:2733–2739

Cevc G, Marsh D (1985) Hydration of noncharged lipid bilayer-membranes – theory and experiments with phosphatidylethanolamines. Biophys J 47:21–31

Christenson HK, Horn RG (1983) Direct measurement of the force between solid surfaces in a polar liquid. Chem Phys Lett 98:45–48

Cleveland J, Schäffer T, Hansma P (1995) Probing oscillatory hydration potentials using thermal-mechanical noise in an atomic-force microscope. Phys Rev B 52:R8692–R8695

Daniels PH (2009) A brief overview of theories of PVC plasticization and methods used to evaluate PVC-plasticizer interaction. J Vinyl Addit Techn 15:219–223

Diprimo C, Deprez E, Hoa GHB, Douzou P (1995) Antagonistic effects of hydrostatic-pressure and osmotic- pressure on cytochrome p-450(cam) spin transition. Biophys J 68:2056–2061

Disalvo EA, Lairion F, Martini F, Tymczyszyn E, Frias M, Almaleck H, Gordillo GJ (2008) Structural and functional properties of hydration and confined water in membrane interfaces. Biochim Biophys Acta-Biomembr 1778:2655–2670

Dunstan DJ, Spain IL (1989) The technology of diamond anvil high-pressure cells .1. Principles, design and construction. J Phys E-Sci Instrum 22:913–923

Eisenblätter S, Galle J, Volke F (1994) Spin–lattice relaxation of (H2O)-H-2 at amphiphile water interfaces as seen by NMR. Chem Phys Lett 228:89–93

Essam JW (1980) Percolation theory. Rep Prog Phys 43:833–912

Evans DF, Wennerström H (1994) The colloidal domain: where physics, chemistry, biology, and technology meet. Wiley, New York
Franca MB, Panek AD, Eleutherio ECA (2007) Oxidative stress and its effects during dehydration. Comp Biochem Physiol A-Mol Integr Physiol 146:621–631
Freund LB (2013) Entropic pressure between biomembranes in a periodic stack due to thermal fluctuations. Proc Natl Acad Sci U S A 110:2047–2051
Fukuma T, Higgins MJ, Jarvis SP (2007) Direct imaging of individual intrinsic hydration layers on lipid bilayers at Angstrom resolution. Biophys J 92:3603–3609
Funari SS, Mädler B, Rapp G (1996) Cubic topology in surfactant and lipid mixtures. Eur Biophys J Biophys Lett 24:293–299
Gawrisch K, Arnold K, Gottwald T, Klose G, Volke F (1978) D-2 NMR-studies of phosphate – water interaction in dipalmitoyl phosphatidylcholine – water-systems. Stud Biophys 74:13–14
Ge MT, Freed JH (2003) Hydration, structure, and molecular interactions in the headgroup region of dioleoylphosphatidylcholine bilayers: an electron spin resonance study. Biophys J 85:4023–4040
Gordeliy VI (1996) Possibility of direct experimental check up of the theory of repulsion forces between amphiphilic surfaces via neutron and X-ray diffraction. Langmuir 12:3498–3502
Gordeliy VI, Cherezov VG, Teixeira J (1996) Evidence of entropic contribution to "hydration" forces between membranes.2. Temperature dependence of the "hydration" force: a small angle neutron scattering study. J Mol Struct 383:117–124
Gruen DWR, Marcelja S (1983) Spatially varying polarization in water – a model for the electric double-layer and the hydration force. J Chem Soc, Faraday TransII 79:225–242
Hawton MH, Doane JW (1987) Pretransitional phenomena in phospholipid water multilayers. Biophys J 52:401–404
Hayashi T, Pertsin AJ, Grunze M (2002) Grand canonical Monte Carlo simulation of hydration forces between nonorienting and orienting structureless walls. J Chem Phys 117:6271–6280
Helfrich W (1978) Steric interaction of fluid membranes in multilayer systems. Z. Naturforsch, A: Phys Sci 33:305–315
Heremans K, Meersman F, Pfeiffer H, Rubens P, Smeller L (2000) Pressure effects on biopolymer structure and dynamics. High Pressure Res 19:623–630
Higgins MJ, Polcik M, Fukuma T, Sader JE, Nakayama Y, Jarvis SP (2006) Structured water layers adjacent to biological membranes. Biophys J 91:2532–2542
Israelachvili JN, Adams GE (1978) Measurement of forces between two mica surfaces in aqueous electrolyte solutions in the range 0–100 nm. J Chem Soc, Faraday Trans 1 74:975–1001
Israelachvili JN, Wennerström H (1990) Hydration or steric forces between amphiphilic surfaces. Langmuir 6:873–876
Israelachvili JN, Wennerström H (1992) Entropic forces between amphiphilic surfaces in liquids. J Phys Chem 96:520–531
Israelachvili JN, Wennerström H (1996) Role of hydration and water structure in biological and colloidal interactions. Nature 379:219–225
Jürgens E, Höhne G, Sackmann E (1983) Calorimetric study of the dipalmitoylphosphatidylcholine water phase-diagram. Ber Bunsen-Ges Phys Chem 87:95–104
Jayne J (1982) Determination of hydration numbers by near-infrared – modification of an earlier approach. J Chem Educ 59:882–884
Jendrasiak GL, Smith RL (2004) The interaction of water with the phospholipid head group and its relationship to the lipid electrical conductivity. Chem Phys Lipids 131:183–195
Klose G, König B, Paltauf F (1992) Sorption isotherms and swelling of POPC in H2O and (H2O)-H-2. Chem Phys Lipids 61:265–270
Klose G, Eisenblätter S, König B (1995a) Ternary phase-diagram of mixtures of palmitoyl-oleoyl-phosphatidylcholine, tetraoxyethylene dodecyl ether, and heavy- water as seen by p-31 and h-2 nmr. J Colloid Interface Sci 172:438–446
Klose G, Eisenblätter S, Galle J, Islamov A, Dietrich U (1995b) Hydration and structural-properties of a homologous series of nonionic alkyl oligo(ethylene oxide) surfactants. Langmuir 11:2889–2892

Kodama M, Kawasaki Y, Aoki H, Furukawa Y (2004) Components and fractions for differently bound water molecules of dipalmitoylphosphatidylcholine-water system as studied by DSC and H-2-NMR spectroscopy. Biochim Biophys Acta-Biomembr 1667:56–66
König B (1993) Untersuchungen zum Hydratationsverhalten von mit nichtionischen Tensiden modifizierten Phospholipidmembranen. University of Leipzig
Koynova R, Tenchov B (2001) Interactions of surfactants and fatty acids with lipids. Curr Opin Colloid Interface Sci 6:277–286
Kranenburg M, Smit B (2005) Phase behavior of model lipid bilayers. J Phys Chem B 109:6553
Kunz W, Lo Nostro P, Ninham BW (2004) The present state of affairs with Hofmeister effects. Curr Opin Colloid Interface Sci 9:1–18
Leckband D, Israelachvili JN (2001) Intermolecular forces in biology. Q Rev Biophys 34:105–267
Leneveu DM, Rand RP, Parsegian VA (1976) Measurement of forces between lecithin bilayers. Nature 259:601–603
Leneveu DM, Rand RP, Parsegian VA, Gingell D (1977) Measurement and modification of forces between lecithin bilayers. Biophys J 18:209–230
Leng YS (2012) Hydration Force between Mica Surfaces in Aqueous KCl Electrolyte Solution. Langmuir 28:5339–5349
Li S, Tang J, Chinachoti P (1996) Thermodynamics of starch-water systems: an analysis from solution-gel model on water sorption isotherms. J Polym Sci Part B Polym Phys 34:2579–2589
Luzzati V, Chapman D (1968) X-ray diffraction studies of lipid-water systems. In: Biological membranes, physical fact and function. Academic press, London, pp 71–124
Marcelja S, Radic N (1976) Repulsion of interfaces due to boundary water. Chem Phys Lett 42:129–130
Marra J, Israelachvili JN (1985) Direct measurements of forces between phosphatidylcholine and phosphatidylethanolamine bilayers in aqueous-electrolyte solutions. Biochemistry 24:4608–4618
Marrinck JS, Berkowitz M (1995) Water and membranes. In: Disalvo EA, Simon SA (eds) Permeability and stability of lipid bilayers. CRC Press, Boca Raton, pp 21–48
Marsh D (1990) Handbook of lipid bilayers. CRC Press, Boca Raton
Marsh D (2011) Water adsorption isotherms of lipids. Biophys J 101:2704–2712
Moore WJ (1972) Physical chemistry. Prentice Hall, Englewood Cliffs
Nagle JF, Tristram-Nagle S (2000) Structure of lipid bilayers. Biochim Biophys Acta Rev Biomembranes 1469:159–195
Orozco-Alcaraz R, Kuhl TL (2013) Interaction forces between DPPC bilayers on glass. Langmuir 29:337
Parsegian VA, Rand RP (1991) On molecular protrusion as the source of hydration forces. Langmuir 7:1299–1301
Parsegian VA, Zemb T (2011) Hydration forces: observations, explanations, expectations, questions. Curr Opin Colloid Interface Sci 16:618–624
Parsegian VA, Fuller N, Rand RP (1979) Measured work of deformation and repulsion of lecithin bilayers. Proc Natl Acad Sci U S A 76:2750–2754
Parsegian VA, Rand RP, Fuller NL, Rau DC (1986) Osmotic-stress for the direct measurement of intermolecular forces. Methods Enzymol 127:400–416
Parsegian VA, Rand RP, Rau DC (1995) Macromolecules and water: probing with osmotic stress. Methods Enzymol 259:43–94
Pfeiffer H, Binder H, Klose G, Heremans K (2003a) Hydration pressure and phase transitions of phospholipids – I. Piezotropic approach. Biochim Biophys Acta-Rev Biomembr 1609:144–147
Pfeiffer H, Binder H, Klose G, Heremans K (2003b) Hydration pressure and phase transitions of phospholipids – II. Thermotropic approach. Biochim Biophys Acta-Biomembr 1609:148–152
Pfeiffer H, Winter R, Klose G, Heremans K (2003c) Thermotropic and piezotropic phase behaviour of phospholipids in propanediols and water. Chem Phys Lett 367:370–374
Pfeiffer H, Winter R, Klose G, Heremans K (2003d) Thermotropic and piezotropic phase behaviour of phospholipids in propanediols and water. Chem Phys Lett 371:670–674

Pfeiffer H, Binder H, Klose G, Heremans K (2004) Hydration pressure of a homologous series of nonionic alkyl hydroxyoligo(ethylene oxide) surfactants. Phys Chem Chem Phys 6:614–618
Pfeiffer H, Klose G, Heremans K (2010) Thermodynamic and structural behaviour of equimolar POPC/CnE4 (n=8,12,16) mixtures by sorption gravimetry, 2H-NMR spectroscopy and X-ray diffraction. Chem Phys Lipids 163:318–328
Pfeiffer H, Heer P, Pitropakis I, Pyka G. Kerckhofs G, Patitsa M, Wevers M (2011) Liquid detection in confined aircraft structures based on lyotropic percolation thresholds. Sens Actuators, B
Pfeiffer H, Weichert H, Klose G, Heremans K (2012) Hydration behaviour of POPC/C-12-Bet mixtures investigated by sorption gravimetry, P-31 NMR spectroscopy and X-ray diffraction. Chem Phys Lipids 165:244–251
Pfeiffer H, Klose G, Heremans K (2013a) Reorientation of hydration water during the thermotropic main phase transition of 1-palmitoyl-2-oleolyl-sn-glycero-3-phosphocholine (POPC) bilayers at low degrees of hydration. Chem Phys Lett 572:120–124
Pfeiffer H, Klose G, Heremans K (2013b) FTIR spectroscopy study of the pressure-dependent behaviour of 1,2-dioleoyl-sn-glycero-3-phosphocholine (DOPC) and 1-palmitoyl-2-oleolyl-sn-glycero-3-phosphocholine (POPC) at low hydration degree - red shift and hindered correlation field splitting. Chem Phys Lipids 170–171:33–40
Pfeiffer H, Heer P, Winkelmans M, Taza W, Pitropakis I, Wevers M (2014) Leakage monitoring using percolation sensors for revealing structural damage in engineering structures. Struct Control Health 21:1030–1042
Rand RP, Parsegian VA (1989) Hydration forces between phospholipid-bilayers. Biochim Biophys Acta 988:351–376
Rand RP, Fuller NL, Butko P, Francis G, Nicholls P (1993) Measured change in protein solvation with substrate-binding and turnover. Biochemistry 32:5925–5929
Ricci C, Caprioli M, Boschetti C, Santo N (2005) Macrotrachela quadricornifera featured in a space experiment. Hydrobiologia 534:239–244
Rothschild LJ, Mancinelli RL (2001) Life in extreme environments. Nature 409:1092–1101
Scharnagl C, Reif M, Friedrich J (2005) Local compressibilities of proteins: comparison of optical experiments and simulations for horse heart cytochrome c. Biophys J 89:64–75
Scherer JR (1987) The partial molar volume of water in biological-membranes. Proc Natl Acad Sci U S A 84:7938–7942
Schmiedel H, Jorchel P, Kiselev M, Klose G (2001) Determination of structural parameters and hydration of unilamellar POPC/C12E4 vesicles at high water excess from neutron scattering curves using a novel method of evaluation. J Phys Chem B 105:111–117
Schneck E, Sedlmeier F, Netz RR (2012) Hydration repulsion between biomembranes results from an interplay of dehydration and depolarization. Proc Natl Acad Sci U S A 109:14405
Sharma P (2013) Entropic force between membranes reexamined. Proc Natl Acad Sci U S A 110:1976–1977
Simon SA, Fink CA, Kenworthy AK, McIntosh TJ (1991) The hydration pressure between lipid bilayers – comparison of measurements using x-ray-diffraction and calorimetry. Biophys J 59:538–546
Simon SA, Advani S, McIntosh TJ (1995) Temperature-dependence of the repulsive pressure between phosphatidylcholine bilayers. Biophys J 69:1473–1483
Spain IL, Dunstan DJ (1989) The technology of diamond anvil high-pressure cells.2. Operation and use. J Phys E-Sci Instrum 22:923–933
Trokhymchuk A, Henderson D, Wasan DT (1999) A molecular theory of the hydration force in an electrolyte solution. J Colloid Interface Sci 210:320–331
Ulrich AS, Watts A (1994a) Lipid headgroup hydration studied by H-2-NMR – a link between spectroscopy and thermodynamics. Biophys Chem 49:39–50
Ulrich AS, Watts A (1994b) Molecular response of the lipid headgroup to bilayer hydration monitored by H-2-Nmr. Biophys J 66:1441–1449
Ulrich AS, Sami M, Watts A (1994) Hydration of DOPC bilayers by differential scanning calorimetry. Biochim Biophys Acta-Biomembr 1191:225–230

Valle-Delgado JJ, Molina-Bolivar JA, Galisteo-Gonzalez F, Galvez-Ruiz MJ (2011) Evidence of hydration forces between proteins. Curr Opin Colloid Interface Sci 16:572–578
Volke F, Eisenblätter S, Klose G (1994a) Hydration force parameters of phosphatidylcholine lipid bilayers as determined from H-2-NMR studies of deuterated water. Biophys J 67:1882–1887
Volke F, Eisenblätter S, Galle J, Klose G (1994b) Dynamic properties of water at phosphatidylcholine lipid- bilayer surfaces as seen by deuterium and pulsed-field gradient proton nmr. Chem Phys Lipids 70:121–131
Wilkinson DA, Nagle JF (1981) Dilatometry and calorimetry of saturated phosphatidylethanolamine dispersions. Biochemistry 20:187–192
White SH, Jacobs RE, King GI (1987) Partial specific volumes of lipid and water in mixtures of egg lecithin and water. Biophys J 52:663–665
Winter R, Pilgrim WC (1989) A SANS study of high-pressure phase-transitions in model biomembranes. Ber Bunsen-Ges Phys Chem 93:708–717
Wolfe J, Yan ZJ, Pope JM (1994) Hydration forces and membrane stresses – cryobiological implications and a new technique for measurement. Biophys Chem 49:51–58
Wong PTT, Moffatt DJ, Baudais FL (1985) Crystalline quartz as an internal-pressure calibrant for high-pressure infrared-spectroscopy. Appl Spectrosc 39:733–735
Zemb T, Parsegian VA (2011) Editorial overview: hydration forces. Curr Opin Colloid Interface Sci 16:515–516

Chapter 5
Monitoring Membrane Hydration with 2-(Dimethylamino)-6-Acylnaphtalenes Fluorescent Probes

Luis A. Bagatolli

Abstract A family of polarity sensitive fluorescent probes (2-(dimethylamino)-6-acylnaphtalenes, i.e. LAURDAN, PRODAN, ACDAN) was introduced by Gregorio Weber in 1979, with the aim to monitor solvent relaxation phenomena on protein matrices. In the following years, however, PRODAN and particularly LAURDAN, were used to study membrane lateral structure and associated dynamics. Once incorporated into membranes, the (nanosecond) fluorescent decay of these probes is strongly affected by changes in the local polarity and relaxation dynamics of restricted water molecules existing at the membrane/water interface. For instance, when glycerophospholipid containing membranes undertake a solid ordered (gel) to liquid disordered phase transition the fluorescence emission maximum of these probes shift 50 nm with a significant change in their fluorescence lifetime. Furthermore, the fluorescence parameters of LAURDAN and PRODAN are exquisitely sensitive to cholesterol effects, allowing interpretations that correlate changes in membrane packing with membrane hydration. Different membrane model systems as well as innate biological membranes have been studied with this family of probes allowing interesting comparative studies. This chapter presents a short historical overview about these fluorescent reporters, discusses on different models proposed to explain their sensitivity to membrane hydration, and includes relevant examples from experiments performed in artificial and biological membranes.

Keywords Generalized polarization • Fluorescent probes • Laurdan

L.A. Bagatolli (✉)
Membrane Biophysics and Biophotonics Group/MEMPHYS-Center for Biomembrane Physics, Department of Biochemistry and Molecular Biology, University of Southern Denmark, Campusvej 55, DK-5230 Odense M, Denmark
e-mail: bagatolli@bmb.sdu.dk

E.A. Disalvo (ed.), *Membrane Hydration*, Subcellular Biochemistry 71,
DOI 10.1007/978-3-319-19060-0_5

5.1 Brief Historical Overview

Solutions composed of particular substituted aromatic molecules experience a bathochromic (or red) shift on their fluorescence emission spectrum when solvent polarity is increased. This phenomenon, which involves a large increase in the molecule's dipole moment in the electronic excited state over that of the ground state, was the object of intense scrutiny from both theory and experiment since the 1950s. In 1979, Gregorio Weber performed a rational design of environmentally sensitive fluorescent molecules based on a naphthalene structure substituted with an electron acceptor (acyl substituted carbonyl group) and donor (alkylamino group) groups in position 6 and 2 respectively, Fig. 5.1 (Weber and Farris 1979). These molecules hold a maximum distance between the electron acceptor and donor groups, resulting in a lowest electronic excited state with an important charge transfer character. Three compounds with different amphipathic

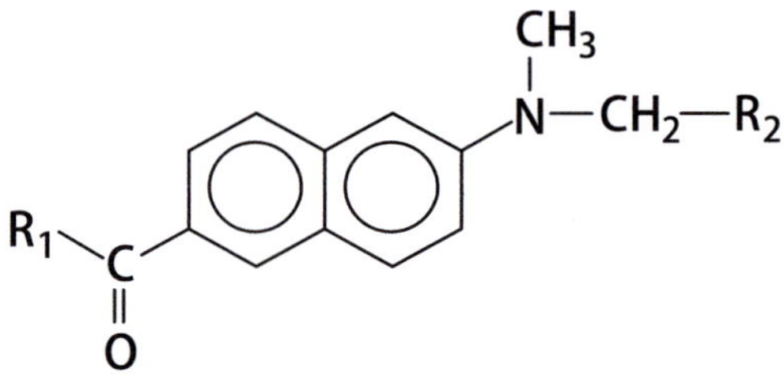

R_1	R_2	Acronymus
$-CH_3$	-H	ACDAN
CH_3-CH_2-	-H	PRODAN
$CH_3-(CH_2)_{10}-$	-H	LAURDAN
$CH_2=CH-$	-H	ACRYLODAN
$CH_3-(CH_2)_{14}-$	$-CH_2-\overset{CH_3}{\underset{CH_3}{N}}-CH_3$	PATMAN
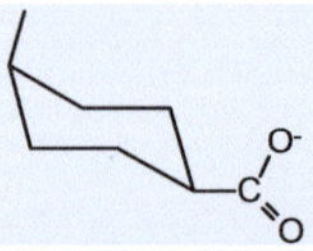	-H	DANCA
$CH_3-(CH_2)_{10}-$	$-\overset{O}{\overset{\|}{C}}-OH$	C-LAURDAN

Fig. 5.1 Chemical structures of different 2,6 substituted naphthalene derivatives

character were introduced in Weber's work, i.e. 6-acetyl-2-dimethylamino naphthalene (ACDAN), 6-propionyl-2-dimethylamino naphthalene (PRODAN), and 6-lauroyl-2-dimethylamino naphthalene (LAURDAN); see Fig. 5.1. Weber's original work also included a careful study of the absorption and fluorescence response of PRODAN to solvents of different polarity, which was interpreted using the Lippert's dipole interaction theory (Lippert 1957). For example, PRODAN exhibits a 130 nm red shift in the fluorescence emission maximum from cyclohexane to water with an important change in the magnitude of its transition dipole (Weber and Farris 1979). Additionally, a significant (larger) Stoke's shift was observed in solvents that can form hydrogen bonds with respect to aprotic solvents, showing an important effect of solvent relaxation on the characteristic nanosecond fluorescence decay of these fluorophores.

The contribution of Gregorio Weber went beyond the rational design and characterization of this family of probes. A significant contribution was their use to study biological material. Specifically, the interaction of PRODAN with bovine serum albumin was also explored in the original Weber paper, proposing the use of this molecule as "... a relaxation probe of various biological environments" (Weber and Farris 1979). In a later work, Weber's group studied nanosecond relaxation phenomenon in a protein matrix (addressed earlier in his laboratory (Lakowicz and Weber 1973)) to determine the polarity of the myoglobin haeme pocket (Macgregor and Weber 1986). Specifically he designed another 2,6 substituted naphthalene derivative, i.e. 2'-(N,N-dimethyl)amino-6-naphtophyl-4-trans-cyclohexanoic acid (DANCA) (Macgregor and Weber 1986), a probe with higher affinity than PRODAN for apo-myoglobin (Fig. 5.1).

Similar strategies were used in the early 1980s to explore solvent relaxation phenomena in membranous systems using different 2-(dimethylamino)-6-acylnaphtalene derivatives, i.e. particularly LAURDAN (Lakowicz and Sheppard 1981; Sumbilla and Lakowicz 1982) and 6-palmitoyl-2-[[[(2-trimethylammonium) ethyl]methyl]amino] naphthalene (PATMAN), another related probe introduced by Lakowicz and collaborators (Lakowicz et al. 1983), Fig. 5.1. During 1986 Parasassi and collaborators first reported a systematic study of LAURDAN sensitivity to membrane phase state in glycerophospholipid vesicles (Parasassi et al. 1986b). Specifically, these authors described a pronounced red shift on the probe's steady state and time resolved emission spectra during (and above) the solid ordered (s_o) to liquid disordered (l_d) membrane phase transition (Parasassi et al. 1986b). As will be discussed in the following sections, this phenomenon is a consequence of a drastic change in the content and dynamics of water associated to the membrane interface during the membrane phase transition. A two state model was proposed to describe LAURDAN response to the $s_o \rightarrow l_d$ phase transition (Parasassi et al. 1986a, b), providing the basis for the well-recognized *Generalized Polarization (GP) function* introduced by these authors in subsequent publications (Parasassi et al. 1990, 1991). The GP function was defined to exploit a simple steady state parameter (i.e. the probe's emission spectra) for studying structural and dynamical aspects of membrane lateral organization. The GP function, which is a generalization of the theory of fluorescence polarization, provides a more rational tool than traditional ratiometric fluorescence measurements, which are generally empirical and devoid of a defined theoretical framework.

Although LAURDAN is presently the most widely used 2-(dimethylamino)-6-acylnaphtalene derivative, PRODAN, PATMAN and ACDAN have been also employed in membrane related studies. The main difference among these probes is their location in the membrane, providing valuable information about trans-bilayer structure and associated dynamics. LAURDAN's fluorescent group was reported to be located 10 Å from the center of the bilayer (Antollini and Barrantes 1998), quite similar to PATMAN which is located 1 Å deeper (Jurkiewicz et al. 2006). PATMAN has been mainly used to study dipolar relaxation processes in membranes composed of different lipids (Hutterer et al. 1996; Jurkiewicz et al. 2006; Olzynska et al. 2007) using a solvent relaxation technique that is based on time-resolved fluorescence measurements (Jurkiewicz et al. 2005). PRODAN on the other hand, is more loosely anchored to the lipid bilayer (it has a shorter hydrophobic tail in position 6 of the naphthalene moiety), and the localization of this probe was shown to be closer to the membrane interface (Chong 1990). The partition of PRODAN into phospholipid membranes was reported to be 2 orders of magnitude lower than LAURDAN (which is virtually insoluble in water). Therefore, a fraction of PRODAN is contributing from the aqueous medium, a situation that is reflected in the characteristic shoulder at 520 nm in its emission spectrum (Zeng and Chong 1995). Experiments exploring dipolar relaxation phenomena in membranes undergoing phase transitions (Krasnowska et al. 1998, 2001) were performed with PRODAN, including effects of pressure on the structure/dynamics of phospholipid containing membranes (Chong 1988, 1990). Finally, ACDAN was also used together with LAURDAN and PRODAN to study lipid interdigitation caused by ethanol in phospholipid bilayers (Zeng and Chong 1995). ACDAN has an even shorter alkyl substitution in the position 6 of the naphthalene ring (i.e. a methyl group) showing a very low partition to the membrane interface, which is approximately three orders of magnitude lower than LAURDAN. Other derivatives were also synthesized in Weber's laboratory, i.e. 2-diisopropylamino-6-lauroylnaphthalene (LAURISAN), 2-methoxy-6-lauroylnaphthalene (LAURMEN) and 2-hydroxy-6-lauroylnaphthalene (LAURNA), and characterized further by Parasassi et al. (1998). Very recently, a new 2,6-naphthalene derivative has been introduced 6-dodecanoyl-2-[N-methyl-N-(carboxymethyl)amino]naphthalene (C-LAURDAN, see Fig. 5.1), to explore membranes mainly using fluorescence microscopy (Kim et al. 2007; Dodes Traian et al. 2012).

5.2 Models for Laurdan Relaxation in Membranes

LAURDAN is a suitable probe to study membrane related phenomena for different reasons: (i) its partition coefficient from aqueous environments to membranes is relatively high; (ii) it displays an exquisite sensitivity to distinct lipid packing in membranes; and (iii) it generally shows an even partition in membranes displaying phase coexistence (Parasassi et al. 1993b; Bagatolli and Gratton 2000a, b). The last property is not common for many of the amphiphilic fluorescent probes, which show preferential partition to one of the coexisting lipid phases. As I will discuss in the following sections, the even lateral distribution of LAURDAN plus it fluorescence response to membrane packing become a powerful experimental tool to gain information about membrane lateral heterogeneity.

As mentioned above, when LAURDAN is incorporated in membranes composed of single glycerophospholipids an emission red shift of 50 nm was observed when the $s_o \rightarrow l_d$ phase transition takes place, Fig. 5.2 (Parasassi et al. 1986b, 1990, 1991). This emission shift was reported to be independent of the nature of the glycerophospholipid polar head group and also the pH (between 4 and 10) (Parasassi et al. 1991). Since changes in the "static" dielectric constant between the two membrane phases is not sufficient to explain the observed fluorescence emission shift, a model to interpret the changes in LAURDAN's emission properties was originally provided by Parasassi et al. (1991). These authors proposed that the nanosecond relaxation process observed in the l_d phase (Parasassi et al. 1986b, 1990; Parasassi and Gratton 1992) is caused by the presence of water molecules with restricted

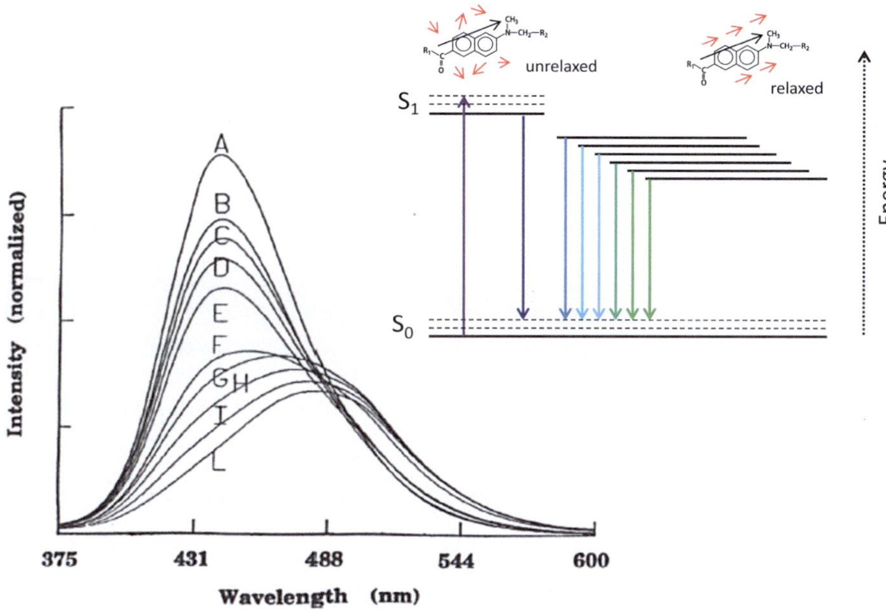

Fig. 5.2 LAURDAN emission spectra in DPPC multilamellar vesicles at different temperatures: (*A*), 30 °C; (*B*), 35 °C; (*C*), 37 °C; (*D*), 38 °C; (*E*), 39 °C; (*F*), 41 °C; (*G*), 42 °C; (*H*), 45 °C; (*I*), 50 °C; (*L*), 60 °C. The $s_o \rightarrow l_d$ main phase transition is 41.5 °C (Adapted from (Parasassi et al. 1991) with permission). The *inset* (*top right*) shows a scheme of the ground (S_0) and excited-state (S_1) energy levels in the presence of the solvent dipolar relaxation. S_1 decreases in energy as the extent of dipolar relaxation of interfacial water increases, *red* shifting the probe's emission spectrum

mobility in the region where LAURDAN is located (nearby the glycerol backbone of the glycerophospholipids) (Parasassi et al. 1991; Antollini and Barrantes 1998; Jurkiewicz et al. 2006). Part of LAURDAN's excited-state energy is utilized for the reorientation of the water dipoles diminishing the singlet excited state (S_1) energy (see Fig. 5.2, inset), consequently shifting the emission spectrum of the probe to longer wavelengths (lower energies). Importantly, the relaxation caused by these water molecules is different to the water molecules existing in bulk that have an orientational relaxation time below one picosecond (Parasassi et al. 1991). The environmental relaxation times have been measured for both the l_d and s_o phases using an equivalent expression of the classical Perrin equation for the GP (for details of the GP definition see below), assuming a two state process (Parasassi and Gratton 1992). These values were reported to be $2.5 \times 10^9\ s^{-1}$ for the l_d phase and $4 \times 10^7\ s^{-1}$ for the s_o phase. Notice that the relaxation time measured in the l_d phase is similar to the lifetime of the probe (3 ns) (Bagatolli et al. 1998).

Taking into account observations obtained by two photon excitation fluorescence microscopy, Parasassi et al. refined the model of water relaxation (Parasassi et al. 1997). This seminal article reported for the first time spatially resolved information of the LAURDAN GP function in micrometer sized (multilamellar) vesicles, showing that the GP images obtained in l_d phase membranes show a much broader GP distribution relative to the s_o phase. This observation indicates the existence of a large dynamical heterogeneity in the l_d phase, unexpected to the authors for a membrane displaying a single l_d phase. To explain these observations Parasassi et al. proposed a distribution of different sites (or cavities) in which the LAURDAN molecule can reside (Parasassi et al. 1997). These sites are characterized by a different number of dynamically restricted water molecules (Parasassi and Gratton 1995). The average number of water molecules at the location of the LAURDAN fluorescent moiety was estimated to be no more than two or three (Parasassi et al. 1997). (*For a further insight see Chaps. 2, 3, and 4). Considering a Poisson distribution of these few water molecules, the authors concluded that there is a distribution of LAURDAN environments with no, one, two, three, etc. molecules of water. For example, for an average of two molecules of water per cavity around the LAURDAN fluorescent moiety, the Poisson distribution of water molecules at the different sites is $0 \rightarrow 0.135$, $1 \rightarrow 0.270$, $2 \rightarrow 0.270$, $3 \rightarrow 0.203$, $4 \rightarrow 0.090$, $5 \rightarrow 0.031$, and more than $5 \rightarrow 0.020$, concluding that the larger the number of water molecules, the lower is the GP, and the larger is the cavity around the probe. This model has been recently supported by scanning-fluctuation correlation spectroscopy measurements of the LAURDAN GP function on the scale of few pixels in single glycerophospholipid containing membranes existing in the l_d phase (Celli and Gratton 2010).

There is considerable experimental evidence that supports the idea that structured water molecules in the vicinity of the probe are the cause of LAURDAN's emission shift in model membranes. One of the most conclusive evidences comes from comparative experiments performed in DMPC membranes prepared in D_2O and H_2O. At and above the $s_o \rightarrow l_d$ membrane phase transition the red shift observed in H_2O is more prominent than in D_2O, indicating a slower relaxation dynamics in heavy water, i.e. an expectable effect considering the solvent mass difference (Parasassi et al. 1998). Another important experimental proof supporting this model is the consistency on the extent of probe's relaxation in membranes containing different glycerophospholipid polar head groups and pH, i.e. implying that the position of emission spectrum depends on the membrane phase state (Parasassi et al. 1991). The same behavior was also observed in ether derivatives of glycerophospholipids (Bagatolli et al. 1999) and sphingolipid-containing bilayers (Bagatolli et al. 1997, 1998).

Additional experiments were performed to check if structural features of the probe itself can be the cause of the observed fluorescence changes during a membrane phase transition. For example, results obtained using the LAURDAN derivatives 2-hydroxy-6-dimethyl aminonaphthalene (LAURNA) and 2-methoxy-6-dimethyl aminonaphthalene (LAURMEN) suggest that the partial charge separation is necessary to observe the fluorescence emission shift. Also, it was observed that steric hindrance to intramolecular reorientation expected for 6-lauroyl-2-diisopropylamino naphthalene (LAURISAN) has no effect on the probe's red shift, suggesting that the molecular entity responsible for the dipolar relaxation cannot be the fluorophore itself (Parasassi et al. 1998). Additionally, information obtained from classical fluorescence polarization spectra also excludes probe reorientation along its molecular axis to be the cause of the observed emission shift (Parasassi et al. 1998).

5.2.1 Alternative Models

Based on experiments performed in alcohols at very low temperature, Viard et al. (1997) proposed that, depending on the relative orientation of the dimethylamino and carbonyl groups, two states contribute to the emission of LAURDAN: the locally excited state and the charge transfer state. This model was applied with some modifications to explore the nanosecond dynamics of membrane water interface in reverse micelles of AOT-water in isooctane (Vincent et al. 2005). Although the locally excited state can be confirmed in apolar solvents (where the lifetime and quantum yield of the probe are very low), there is no clear evidence that this state contributes to LAURDAN emission in glycerophospholipid membranes. For example, the high quantum yield observed in membranes, the clear lifetime dependence on the membrane phase (3 and 6 ns in the s_o and l_d phases, respectively (Bagatolli et al. 1998)) and the lack of spectral isosbestic point during the phase transition support that the charge transfer state is dominant (Parasassi et al. 1998), i.e. there is one state that relaxes upon the lipid main phase transition causing the emission spectrum shift, not an intramolecular relaxation of the probe. This view is also supported by the above described experiments using LAURNA, LAURMEN and LAURISAN.

The model of LAURDAN relaxation in membranes has been recently revisited by Jurkiewicz et al. (2012). Based on time dependent fluorescence shift measurements complemented with molecular dynamic simulations, these authors proposed that the nanosecond relaxation times reported by LAURDAN in membranes carry information solely on the mobility of hydrated lipid moieties at the probe's location and not the local dynamic of water molecules theirself, which interchange or even exchange with those from bulk water on a picosecond timescale. The authors' conclusion is based on the fact that at the glycerol backbone level in glycerophospholipid bilayers (i.e. where the probe is located) water molecules are sparse and fully bound to the glycerophospholipid carbonyl groups. However, the aforementioned D_2O experiments together with the fact that similar probe's emission response was observed in bilayers composed of very distinct lipid polar head groups (which coordinate different number of water molecules) undergoing s_o to l_d phase transition, e.g. DPPC and its ether derivative (Bagatolli et al. 1999), sphingomyelin, Gg_3Cer, Phrenosine, sulfatide, and galactosyl-ceramide (Bagatolli et al. 1998), reinforce the idea that the (nanosecond time scale) relaxation of associated water is the main cause of the observed probe's emission shift.

5.3 The Generalized Polarization Function

The generalized polarization (GP) function was proposed by Parasassi et al. in 1990 (Parasassi et al. 1990) as an analytical method to quantitatively determine the relative amount and temporal fluctuations of s_o and l_d phases when they coexist in a model membrane. This function was originally defined as:

$$GP = \frac{I_B - I_R}{I_B + I_R} \tag{5.1}$$

where I_B and I_R are the measured fluorescence intensities under conditions in which a wavelength (or a band of wavelengths) B and R are both observed using a given excitation wavelength. This definition corresponds to the classical fluorescence polarization definition (Jameson et al. 2003) if B and R represent two different orientations of the observation polarizers. The advantage of the GP for the analysis of the spectral properties of the 6-acyl-2-dimethylamino naphthalene probes is related to the well-known properties of the classical polarization function, which contains information on the interconversion between different "states". In the classical polarization definition the "states" correspond to different orientations of the emitting dipole with respect to the laboratory axis. In the GP function the condition of interconversion between two states *a* and *b*, i.e. unrelaxed and relaxed, has been linked respectively to the extent of solvent relaxation observed in the s_o and l_d phases in glycerophospholipid membranes respectively (Parasassi et al. 1990, 1991). Using a general formulation for the decay of *a two state system* the general expression of a steady state GP can be derived ((Parasassi et al. 1990) also reviewed in (Bagatolli 2013)) as:

$$\frac{B-R}{\langle GP\rangle}-(B+R)=\frac{2\left(b_0+\frac{k_{ab}}{k_a}\right)}{a_0\frac{k_b}{k_a}-b_0+\frac{k_{ba}}{k_a}-\frac{k_{ab}}{k_a}} \tag{5.2}$$

where GP is the generalized polarization (Eq. 5.1), B and R represent the fractional emission of the unrelaxed state in the blue and red edges of the spectrum, a_0 and b_0 are the relative absorption of the unrelaxed and relaxed states at the wavelength of excitation, k_a and k_b are the intrinsic decay rates of the unrelaxed and relaxed state and k_{ab} and k_{ba} are the forward and backward relaxation rates. It is reasonable to assume that $k_a = k_b$ (the intrinsic decay rates are equal) and that $k_{ba} = 0$, i.e. the decay rate is independent of the relaxation process and the back reaction is slow. After these assumptions Eq. 5.2 reduces to:

$$\frac{B-R}{GP}-(B+R)=\frac{2\left(b_0+\frac{k_{ab}}{k_a}\right)}{a_0-b_0-\frac{k_{ab}}{k_a}} \tag{5.3}$$

From this expression it is interesting to see that it is the ratio k_{ab}/k_a that determines the modalities of the relaxation process and the GP value. Parasassi et al. analyzed three particular limiting cases in the context of this definition (Parasassi et al. 1991):

(i) $k_{ab} >> k_a$, fast relaxation or long lifetime value of the probe. In this case the GP value depends only on the spectral emission properties of the unrelaxed state at the wavelength of excitation, i.e. on B and R and not on the relaxation process of the solvent.
(ii) $k_{ab} << k_a$, slow relaxation or short lifetime value of the probe. The GP depends also on the relative absorption of both the relaxed and unrelaxed state at the wavelength of excitation. If no relaxed state is excited ($b_0 = 0$), then there should be no dependence of the GP value on the excitation wavelength because $a_0 = 1$.
(iii) $k_{ab} = k_a$, the GP value also depends on the dynamic properties of the solvent that relaxes the probe's excited state.

Condition (ii) satisfies what is observed in membranes displaying s_o phase whereas condition (iii) reflects what is observed in a l_d phase (Parasassi et al. 1991).

Although the GP function has been applied to interpreted LAURDAN behavior in diverse type of membranous systems a modification of the GP function has been reported for PRODAN (Krasnowska et al. 1998, 2001). Because of the different lengths of their acyl residues, the partitioning of the two probes between water and membrane differs profoundly. To account for the contribution of PRODAN fluorescence arising from water, Kranowska et al. introduced a three-wavelength generalized polarization method (3wGP) that makes it possible to separate the spectral properties of PRODAN in the lipid phase from water (Krasnowska et al. 1998). This function also allows to determine the probe partition coefficient to membranes as well its distribution between s_o and l_d phases. The major conclusions of this work are that in contrast to LAURDAN, PRODAN preferentially partitions in the l_d phase with respect to the s_o phase and is sensitive to the membrane pre-transition. Additionally it was reported that PRODAN partition coefficient between the membrane and water depends on the phase state, i.e., PRODAN partitioning is higher in the l_d phase by a factor of 35 respect to the s_o phase. This last observation has been confirmed using two photon excitation fluorescence microscopy in giant unilamellar vesicles labelled with PRODAN (Bagatolli and Gratton 2000b). Last but not least, the 3wGP has been also applied to study the effect of cholesterol in phospholipid containing membranes. Particularly it was noticed that a low amount of cholesterol (3 % mol) causes higher partition of the probe into gel domains of DLPC-DPPC mixtures, reaching a maximum at 15 mol %, a phenomenon associated with the progressive formation of a liquid ordered (l_o) phase (Krasnowska et al. 2001).

In summary, the GP function applied to LAURDAN (or PRODAN) is responsive to membrane hydration, which in turn is related to membrane lateral structure. This function exhibits different values depending of the membrane lateral packing. Figure 5.3 shows a graphical example for computation of the GP function in model membranes displaying s_o and l_d phases. The function values scale from 0.6 ± 0.1, characteristic of the s_o phase, to below -0.1 for the l_d phase. In the latter case the GP function shows an additional decrease (although minor compared with the change observed at the membrane main phase transition temperature) when the temperature is increased associated with an augmentation in the extent of solvent relaxation upon increasing the temperature of the system. Finally, and importantly, membranes

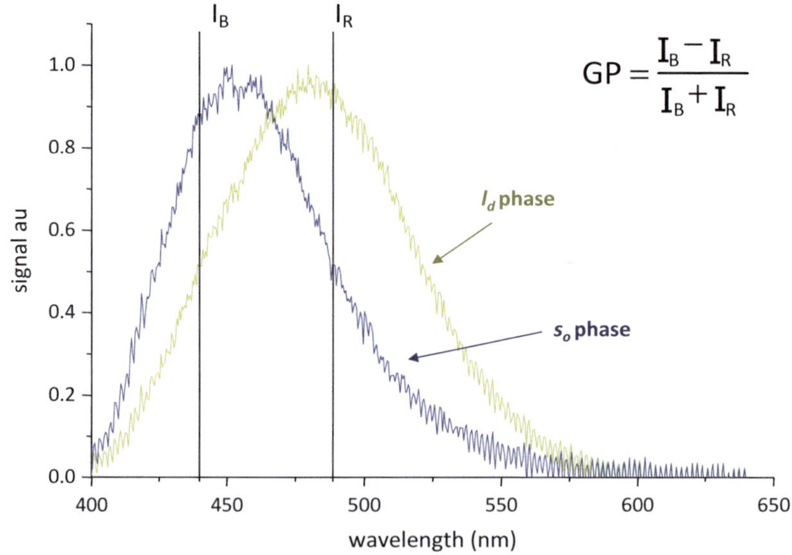

Fig. 5.3 LAURDAN emission spectra in the solid ordered (*gel; blue*) and liquid disordered (*green*) phases. The emission spectrum shift is 50 nm. The Generalized Polarization parameter, which depends on the position of the LAURDAN emission spectrum, contains information about water content and dynamics existing at the membrane interface

displaying l_o phase show intermediate GP values that depend on the amount of cholesterol and the temperature ((Parasassi et al. 1994a, b; Dietrich et al. 2001), see also Fig. 5.6 in Sect. 5.4.2). Cholesterol effects are briefly addressed in this chapter (see Sect. 5.4) as they have been widely reviewed elsewhere (Bagatolli 2013).

5.3.1 Generalized Polarization Spectra

By measurement of LAURDAN time resolved emission spectra, Parasassi et al. showed that s_o and l_d domains co-exist in multilamellar vesicles composed of DLPC/DPPC (Parasassi et al. 1993b). The authors estimated membrane domain dimensions of 20 to 50 Å, and measured domain fluctuations of 25 ns (Parasassi and Gratton 1995). By applying the additive property of the GP function (Parasassi et al. 1991, 1993b) the fractions of the coexisting phases were also obtained. These authors also concluded that the intrinsic characteristics of the s_o and l_d phases in the single lipid containing membranes is somehow modified when these phases coexist in the lipid mixture, i.e., it reflects partial lipid miscibility in the DLPC/DPPC mixture (Parasassi et al. 1993b). Based on these findings, the authors presented a simple strategy to ascertain phase coexistence in membranes using steady state "bulk" fluorescent measurements (Parasassi et al. 1991) built on the particular response of LAURDAN emission (and excitation) spectra to the phase state (s_o, l_d or s_o/l_d phase coexistence) in glycerophospholipid containing membranes. Specifically, they showed that the excitation or emission wavelength response of the GP function (i.e. called GP_{ex} and GP_{em}), provided a fingerprint to discriminate among different membrane phase state scenarios (Parasassi et al. 1991, 1994b; Parasassi and Gratton 1995). I will only discuss below the characteristics of the GP_{ex} spectrum. A complete description of the GP_{em} can be found in (Parasassi and Gratton 1995); also recently reviewed in (Bagatolli 2013).

In order to obtain a GP_{ex} spectrum the excitation spectra of LAURDAN are measured at two fixed emission wavelengths (440 and 490 nm) and the GP calculated at each excitation wavelength according to Eq. 5.1 (I_B is the intensity at 440 nm and I_R the intensity at 490 nm). In glycerophospholipid membranes in the s_o phase the GP_{ex} spectrum is independent of the excitation wavelength, since no relaxation occurs (see above condition (ii) in Sect. 5.3). However in the l_d phase the response is different. One of the characteristics of the dipolar relaxation process is that by exciting in the blue part of the excitation spectrum, molecules with an energetically unfavorable ground state (surrounded by randomly oriented dipoles) are photoselected. As a consequence, by moving the excitation toward the blue, more blue emitting molecules are excited. Since the GP_{ex} value depends on the difference between the emission intensities at 440 and 490 nm, when relaxation occurs (as it happens in the l_d phase), higher GP_{ex} values are expected at shorter excitation values, i.e. the slope of the GP_{ex} spectrum is negative, Fig. 5.4 (Parasassi et al. 1990, 1991, 1994b). When the membrane displays s_o/l_d phase coexistence the GP_{ex} spectrum shows a positive slope (Parasassi et al. 1991), since there are relaxed and unrelaxed states coexisting in the membrane. Particularly, the unrelaxed

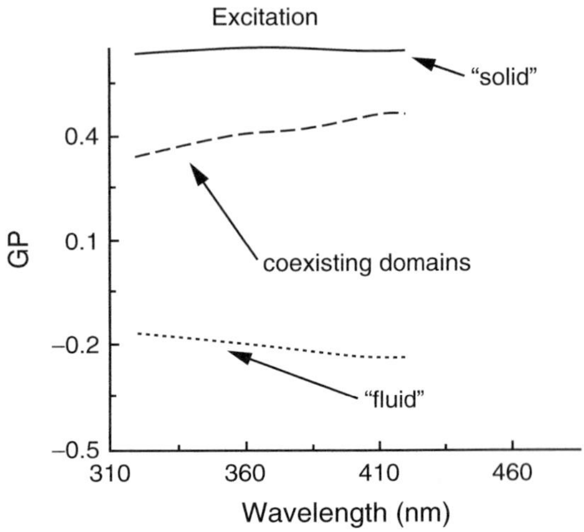

Fig. 5.4 LAURDAN GP_{ex} spectra obtained in phospholipids membranes in different phase scenarios. GP_{ex} spectra were calculated by $GP = (I_{440} - I_{490})/(I_{440} + I_{490})$, using two excitation spectra obtained at 440 and 490 nm emission wavelenghts (Adapted from (Parasassi and Gratton 1995) with permission)

state contributes in the red region of the excitation spectrum (i.e., the red band centered at 390 nm of the LAURDAN excitation spectrum is prominent in the s_o phase but not in the l_d phase (Parasassi and Gratton 1995; Bagatolli et al. 1999); also recently reviewed in (Bagatolli 2013)) and in the blue side of the emission spectrum, increasing the GP value at longer excitation wavelengths, Fig. 5.4. This strategy has been used to ascertain phase coexistence in model membranes and also in natural membranes (Parasassi et al. 1993a; Parasassi and Gratton 1995). However, it is important to remark that in membranes composed of sphingolipids, the changes observed in the LAURDAN excitation spectrum (particularly the red excitation band) between the s_o phase to l_d phase are not the same than in glycerophospholipid membranes (Bagatolli et al. 1998, 1999). This divergent response, attributed to specific interaction in the ground state of the probe (Bagatolli et al. 1999), has important consequences on the sensitivity of the GP_{ex} spectrum to s_o/l_d phase coexistence mentioned above, i.e. the GP_{ex} (and also the GP_{em}) spectrum in membranes composed of sphingolipids becomes insensitive to phase coexistence. Taking into account these facts, caution is recommended if the LAURDAN GP spectra are used to ascertain phase coexistence in membranes composed of lipids other than glycerophospholipids or membranes displaying complex compositions (e.g. natural membranes) (Bagatolli et al. 1999; Bagatolli 2013).

5.4 Relevant Examples

The GP function has been extensively applied to study different membrane related phenomena. For example, many studies have made use of the LAURDAN GP approach in a fluorometer (cuvette studies) using suspensions of liposomes (or cells) to dissect membrane lateral heterogeneity, membrane interactions with peptides, proteins, or other ligands (Parasassi et al. 1993a, 1998, Parasassi and Gratton 1995; Antollini and Barrantes 1998; Henshaw et al. 1998; Vanounou et al. 2002; Nielsen and Otzen 2010). Additionally, studies involving *spatially resolved* GP experiments (performed in a microscope) have been also reported using giant unilamellar vesicles (Bagatolli and Gratton 2001; Bagatolli 2006). Since space restrictions do not allow a review of the over 440 papers reported using these two experimental strategies, this section will focus on a few representative examples to provide a general idea of the potentialities and limitations of the technique.

5.4.1 Phase Transitions and the Effect of Membrane Hydration in Liposome Suspensions

Seminal LAURDAN GP studies on suspensions of liposomes of diverse composition were performed by Parasassi and collaborators (Parasassi et al. 1990, 1991, 1993a, b, 1994a, b, 1995; Parasassi and Gratton 1992, 1995). Typical examples are the detection of the s_o to l_d lipid main phase transition temperature in liposomes composed of single glycerophospholipids (Parasassi et al. 1990), including studies on the effect of cholesterol, which decreases the cooperativity and gradually abolishes the glycerophospholipid main phase transition (Parasassi et al. 1994b). However, instead of performing a systematic review on the extensive literature reporting on LAURDAN GP and membrane phase state/phase transitions, I will discuss the correlation of the GP function with the water content and extent of solvent relaxation in membranes composed of a series of sphingolipids and sphingolipid-glycerophospholipid mixtures (Bagatolli et al. 1997, 1998). These studies showed that the extent of solvent relaxation detected by the probe was strongly influenced by the chemical nature of the studied lipids, which in turn affects the overall structure of the membranous system. For example, for bilayer-forming sphingolipids such as Gg_4Cer, Gg_3Cer, Phrenosine, sulfatide, GalCer, and sphingomyelin, the LAURDAN GP function showed the characteristic values and sensitivity to the $s_o \rightarrow l_d$ phase transition observed in glycerophospholipid bilayers (Bagatolli et al. 1997, 1998). However, in glycosphingolipids with bulky sugar-based polar head groups (i.e. gangliosides G_{M2}, G_{M1}, G_{D1a}, G_{T1b}) the response of LAURDAN was different. Gangliosides form micellar structures that have a much higher curvature compared to bilayers. Consequently the measured LAURDAN GP values were low and relatively constant both below and above the lipid main phase transition temperatures detected by differential scanning calorimetry (Bagatolli et al. 1997, 1998). This lack of sensitivity to the micelle phase transition is consistent with the probe experiencing a highly relaxed environment, i.e. a highly curved and hydrated interface (Arnulphi et al. 1997), compared to that observed in bilayers. Remarkably, a close correlation between the GP function and the intermolecular spacing among these lipids measured at 30 mN/m in monolayers, which can be related to different hydration states, was reported for both neutral and anionic lipid species, Fig. 5.5 (Bagatolli et al. 1998). This finding strongly supports the water dipolar relaxation model discussed above for LAURDAN, i.e. LAURDAN resides in

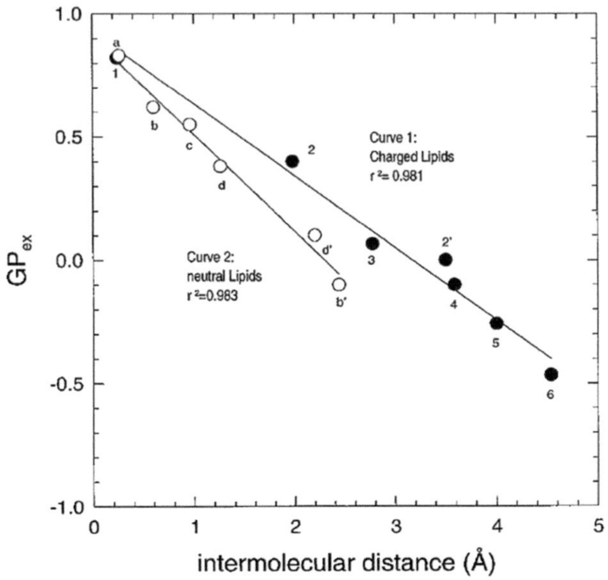

Fig. 5.5 Variation of LAURDAN GP_{ex} obtained in membranous suspensions with the calculated intermolecular distance (measured at 30 mN/m in monolayers at the air water interface) of anionic GSLs (*filled symbols*) and neutral GSLs and phospholipids (*empty symbols*). Neutral lipids: GalCer (*a*); DPPC below (*b*), and above (*b′*) the transition temperature; Gg_3Cer (*c*); Gg_4Cer below (*d*) and at (*d′*) the transition temperature. Anionic lipids: Sulf (*1*); G_{M3} below (*2*), and above (*2′*) the transition temperature, G_{M2} (*3*); G_{M1} (*4*); G_{D1a} (*5*); G_{T1b} (*6*) (Adapted from (Bagatolli et al. 1998) with permission)

cavities containing different amounts of associated water in the membrane which is regulated by curvature and the capability of the lipid polar head group to coordinate water (Bagatolli et al. 1997, 1998).

5.4.2 *Membrane Phase Coexistence Study by Fluorescence Microscopy*

Fluorescence microscopy experiments using giant unilamellar vesicles (GUVs) composed of distinct lipid mixtures further advanced the understanding of lateral phase separation in membranes. Using LAURDAN labeled GUVs and two photon excitation (TPE) fluorescence microscopy, direct information on local membrane packing can be obtained from membranes displaying coexisting domains at the level of single vesicles (Bagatolli and Gratton 2001; Bagatolli 2006). The LAURDAN GP imaging approach was shown to be superior to other confocal fluorescence microscopy strategies, which use pairs of lipophilic probes that preferentially partition to each of the coexisting membrane regions. Particularly, it has been demonstrated that the partition of those probes depends on the chemical nature of the domains, i.e. probe partition *is not* necessarily "phase" specific (Bagatolli and Gratton 2001; Bagatolli 2006; Juhasz et al. 2010). In the case of confocal fluorescence microscopy experiments, additional diffusion experiments using fluorescence correlation spectroscopy (FCS) must be performed to obtain local information on domain packing (Korlach et al. 1999). This is not needed for LAURDAN, since the probe is evenly distributed and responsive to membrane packing (Bagatolli 2006).

The effects of temperature on the lateral structure of LAURDAN labelled GUVs composed of pure phospholipids and different mixtures thereof were first reported in 1999 and 2000 (Bagatolli and Gratton 1999, 2000a, b). Micrometer-sized domains with s_o character, surrounded by membrane regions corresponding to a l_d phase, where observed at particular temperatures depending on the lipid mixture. These domains spanned the bilayer, demonstrating inter leaflet coupling in the membrane (Bagatolli and Gratton 2000a, b). Also, these experiments first demonstrated the generation of micrometer sized domains in bilayers, disproving the general assumption that nanometer-sized domains would exist in membranes composed of those lipid mixtures. This result also linked the observation of micrometer sized domains in bilayers with that previously observed in Langmuir lipid films (Nag et al. 2002; Brewer et al. 2010; Bernardino de la Serna et al. 2013). In one of these studies, for example, a correlation between domain shape and lipid miscibility was reported for binary mixtures of different phospholipids displaying s_o/l_d phase coexistence (Bagatolli and Gratton 2000a). The same strategy was implemented to explore the effect of cholesterol in lipid mixtures. Particularly, the lateral structure of canonical "raft" mixtures (DOPC/sphingomyelin/cholesterol) was first visualized in free standing bilayers using LAURDAN and TPE fluorescence microscopy (Dietrich et al. 2001). For this mixture, the coexistence of two liquid phases (l_o/l_d), characterized by the presence of round micrometer sized domains and intermediate LAURDAN GP values to those observed for the s_o/l_d case was described, Fig. 5.6. In this last case it is clear that cholesterol substantially modifies the extent of water relaxation in the membrane when l_o/l_d phases coexist. This is reflected for example in the slower water dipolar relaxation observed in the l_o phase respect to the l_d phase, but also in the different GP values observed for the l_d phase itself in presence and absence of cholesterol (Fig. 5.6). The GP function has been also applied to planar supported bilayers and Langmuir films, providing the possibility to perform texture analysis on s_o (gel) domains, to characterize membrane hydration in monolayers upon compression and also to correlate membrane packing information among the different planar membrane systems (Bernchou et al. 2009; Brewer et al. 2010; Bernardino de la Serna et al. 2013).

The information gathered using LAURDAN and TPE fluorescence microscopy has also aided in the characterization of other specialized phase separated membranes, such as those containing lipopolysaccharides, cerebrosides and their mixtures with glycerophospholipids and cholesterol (Fidorra et al. 2006, 2009; Sot et al. 2006; Kubiak et al. 2011). Additionally, the LAURDAN GP imaging approach has been also successfully applied to the study lateral heterogeneity in model systems composed of specialized innate membranes such as skin stratum

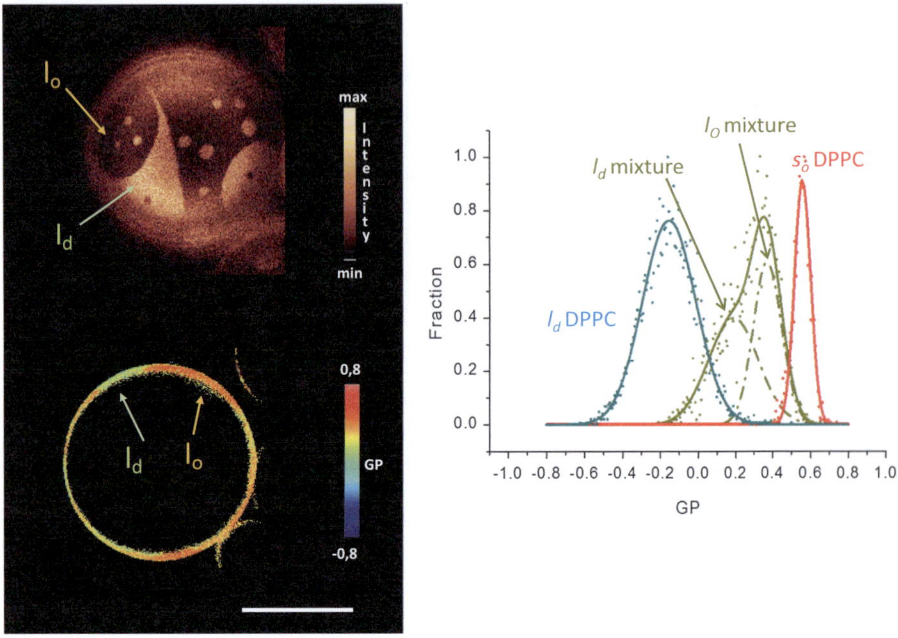

Fig. 5.6 *Left panel*: LAURDAN two photon excitation fluorescence intensity image (taken at the polar region of the vesicle) and LAURDAN GP image (taken at the equatorial region of the same vesicle) of a GUV composed of DOPC/Cholesterol/Sphingomyelin 1:1:1 mol displaying l_o/l_d phase coexistence. The *bar* corresponds to 20 μm. *Right panel*: GP distribution measured from the DOPC/Cholesterol/Sphingomyelin GP image described above. Notice that the GP distribution for this mixture is bimodal (two phases are present) and the GP center value for the l_o phase (0.4) is higher compared to the l_d phase (0.2). Additionally, the GP center values for the coexisting l_o/l_d phases are intermediate between those observed for DPPC GUVs displaying s_o and l_d phases (Adapted from (Dietrich et al. 2001) with permission)

corneum lipid membranes (Plasencia et al. 2007; Norlen et al. 2008), lung surfactant (Nag et al. 2002; Bernardino de la Serna et al. 2009, 2013) and red blood cells (Montes et al. 2007), or to study non-equilibrium phenomena in membranes upon the action of lipases (Sanchez et al. 2002; Stock et al. 2012).

5.4.3 Natural Membranes

In the pioneering work of Yu et al. and Parasassi et al. (Yu et al. 1996; Parasassi et al. 1997), LAURDAN was proposed to be a promising tool in exploring cell membranes. In those studies, domains of sizes below, in the same range and above the microscope resolution limit (0.3 μm radial) were observed in OK cells, red blood cells and brush border native membranes respectively (Yu et al. 1996; Parasassi et al. 1997). The LAURDAN GP differences observed in compositionally complex mixtures and artificial lipid ternary mixtures containing phospholipids, sphingomyelin and cholesterol (Dietrich et al. 2001) were used to interpret LAURDAN GP images in cellular membranes (Gaus et al. 2003). In this work the LAURDAN GP function was used to directly observe in living macrophages transient micron sized-high GP regions surrounded by low GP areas, supporting the presence of lateral heterogeneity in cellular membranes under *in vivo* conditions. Lately, temporal fluctuations of the GP function have been exploited using LAURDAN to ascertain membrane heterogeneity in rabbit erythrocytes and Chinese hamster ovary cells (Sanchez et al. 2012). This approach that takes into account the amplitude and diffusion relaxation time obtained from autocorrelation analysis of LAURDAN GP fluctuations constitutes a very interesting method to measure transient membrane domains with sizes below the resolution of classical fluorescence microscopy.

Very recently, it was reported that the fluorescence lifetime of LAURDAN (instead of the emission spectra) measured at two different emission wavelengths provided the ability to resolve *in vivo* cellular membranes with different properties such as water and cholesterol content (Golfetto et al. 2013). This last study provided a comprehensive analysis of cell membrane heterogeneity by isolating and analyzing the probe's emission decay during dipolar relaxation. Finally, LAURDAN GP imaging has been also applied to tissues. Sun et al. (2004) have demonstrated that both LAURDAN multiphoton polarization and GP can be combined using a TPE fluorescence microscopy to characterize the structural changes of intercellular lipids in skin tissue. This work showed how treatment with oleic acid increases water dipolar relaxation in skin stratum corneum membranes (Sun et al. 2004). A similar strategy using LAURDAN GP has been reported in order to characterize membranous structures in pig skin (Carrer et al. 2008), perform a comparative evaluation between lipid stratum corneum membranes of normal skin and acquired cholesteatoma (Bloksgaard et al. 2012b) and evaluate the effect of detergents in excised skin (Bloksgaard et al. 2014).

5.5 Conclusions

The use of different 2,6 substituted naphtalenes derivatives (e.g. LAURDAN, PRODAN, PATMAN) allow the study of important structural and dynamical aspects of membranes. The accessibility of bulk information from fluorescence "cuvette" trials together with spatially resolved information from fluorescence microscopy experiments have been important to better describe membrane lateral heterogeneity. For example using LAURDAN, the combination of these two strategies allowed to link membrane related phenomena from simple compositionally situations (observed in distinct model membranes) to those occurring in more complex compositional scenarios such as biological membranes existing in cells and tissues. Particularly, spatially resolved information has been essential to disentangle different supramolecular processes occurring in these systems. Although most of the reported applications focus on membrane lateral packing information, the exquisite sensitivity of these probes to membrane hydration should be exploited in more detail. For example, heterogeneity reported for these probes when inserted in natural membrane systems could be interpreted in terms of distinct "structured water domains" (Bloksgaard et al. 2012a; Almaleck et al. 2013), not exclusively related to lipid packing, but to the combined ability of different membrane constituents to generate areas of different water content and dynamics.(* N.of E: see Chap. 4 by Pfeiffer, Chap. 6 by Arzov and Chap. 7 by Alarcon et al. Appignanessi)

Acknowledgments This work is supported in part by a grant from the Danish Research Council (12-124751).

References

Almaleck H, Gordillo GJ, Disalvo A (2013) Water defects induced by expansion and electrical fields in DMPC and DMPE monolayers: contribution of hydration and confined water. Colloids Surf B Biointerfaces 102:871–878

Antollini SS, Barrantes FJ (1998) Disclosure of discrete sites for phospholipid and sterols at the protein-lipid interface in native acetylcholine receptor-rich membrane. Biochemistry 37:16653–16662

Arnulphi C, Levstein PR, Ramia ME, Martin CA, Fidelio GD (1997) Ganglioside hydration study by 2H-NMR: dependence on temperature and water/lipid ratio. J Lipid Res 38:1412–1420

Bagatolli LA (2006) To see or not to see: lateral organization of biological membranes and fluorescence microscopy. Biochim Biophys Acta 1758:1541–1556

Bagatolli LA (2013) LAURDAN fluorescence properties in membranes: a journey from the fluorometer to the microscope. In: Mely Y, Duportail G (eds) Fluorescent methods to study biological membranes. Springer, Heidelberg/New York, pp 3–36

Bagatolli LA, Gratton E (1999) Two-photon fluorescence microscopy observation of shape changes at the phase transition in phospholipid giant unilamellar vesicles. Biophys J 77:2090–2101

Bagatolli LA, Gratton E (2000a) A correlation between lipid domain shape and binary phospholipid mixture composition in free standing bilayers: a two-photon fluorescence microscopy study. Biophys J 79:434–447

Bagatolli LA, Gratton E (2000b) Two photon fluorescence microscopy of coexisting lipid domains in giant unilamellar vesicles of binary phospholipid mixtures. Biophys J 78:290–305

Bagatolli LA, Gratton E (2001) Direct observation of lipid domains in free-standing bilayers using two-photon excitation fluorescence microscopy. J Fluoresc 11:141–160

Bagatolli LA, Maggio B, Aguilar F, Sotomayor CP, Fidelio GD (1997) Laurdan properties in glycosphingolipid-phospholipid mixtures: a comparative fluorescence and calorimetric study. Biochim Biophys Acta 1325:80–90

Bagatolli LA, Gratton E, Fidelio GD (1998) Water dynamics in glycosphingolipid aggregates studied by LAURDAN fluorescence. Biophys J 75:331–341

Bagatolli LA, Parasassi T, Fidelio GD, Gratton E (1999) A model for the interaction of 6-lauroyl-2-(N, N-dimethylamino)naphthalene with lipid environments: implications for spectral properties. Photochem Photobiol 70:557–564

Bernardino De La Serna J, Oradd G, Bagatolli LA, Simonsen AC, Marsh D, Lindblom G, Perez-Gil J (2009) Segregated phases in pulmonary surfactant membranes do not show coexistence of lipid populations with differentiated dynamic properties. Biophys J 97:1381–1389

Bernardino De La Serna J, Hansen S, Berzina Z, Simonsen AC, Hannibal-Bach HK, Knudsen J, Ejsing CS, Bagatolli LA (2013) Compositional and structural characterization of monolayers and bilayers composed of native pulmonary surfactant from wild type mice. Biochim Biophys Acta 1828:2450–2459

Bernchou U, Brewer J, Midtiby HS, Ipsen JH, Bagatolli LA, Simonsen AC (2009) Texture of lipid bilayer domains. J Am Chem Soc 131:14130–14131
Bloksgaard M, Bek S, Marcher AB, Neess D, Brewer J, Hannibal-Bach HK, Helledie T, Fenger C, Due M, Berzina Z, Neubert R, Chemnitz J, Finsen B, Clemmensen A, Wilbertz J, Saxtorph H, Knudsen J, Bagatolli L, Mandrup S (2012a) The acyl-CoA binding protein is required for normal epidermal barrier function in mice. J Lipid Res 53:2162–2174
Bloksgaard M, Svane-Knudsen V, Sorensen JA, Bagatolli L, Brewer J (2012b) Structural characterization and lipid composition of acquired cholesteatoma: a comparative study with normal skin. Otol Neurotol 33:177–183
Bloksgaard M, Brewer J, Pashkovski E, Ananthapadmanabhan KP, Ahm Sørensen J, Bagatolli LA (2014) Effect of detergents on the physico-chemical properties of skin stratum corneum: a two-photon excitation fluorescence microscopy study. Int J Cosmet Sci 36(1):39–45
Brewer J, Bernardino De La Serna J, Wagner K, Bagatolli LA (2010) Multiphoton excitation fluorescence microscopy in planar membrane systems. Biochim Biophys Acta 1798:1301–1308
Carrer DC, Vermehren C, Bagatolli LA (2008) Pig skin structure and transdermal delivery of liposomes: a two photon microscopy study. J Control Release 132:12–20
Celli A, Gratton E (2010) Dynamics of lipid domain formation: fluctuation analysis. Biochim Biophys Acta 1798:1368–1376
Chong PL (1988) Effects of hydrostatic pressure on the location of PRODAN in lipid bilayers and cellular membranes. Biochemistry 27:399–404
Chong PL-G (1990) Interactions of LAURDAN and PRODAN with membranes at high pressure. High Pressure Res 5:761–763
Dietrich C, Bagatolli LA, Volovyk ZN, Thompson NL, Levi M, Jacobson K, Gratton E (2001) Lipid rafts reconstituted in model membranes. Biophys J 80:1417–1428
Dodes Traian MM, González Flecha FL, Levi V (2012) Imaging lipid lateral organization in membranes with C-laurdan in a confocal microscope. J Lipid Res 53(3):609–616
Fidorra M, Duelund L, Leidy C, Simonsen AC, Bagatolli LA (2006) Absence of fluid-ordered/fluid-disordered phase coexistence in ceramide/POPC mixtures containing cholesterol. Biophys J 90:4437–4451
Fidorra M, Heimburg T, Bagatolli LA (2009) Direct visualization of the lateral structure of porcine brain cerebrosides/POPC mixtures in presence and absence of cholesterol. Biophys J 97:142–154
Gaus K, Gratton E, Kable EP, Jones AS, Gelissen I, Kritharides L, Jessup W (2003) Visualizing lipid structure and raft domains in living cells with two-photon microscopy. Proc Natl Acad Sci U S A 100:15554–15559
Golfetto O, Hinde E, Gratton E (2013) Laurdan fluorescence lifetime discriminates cholesterol content from changes in fluidity in living cell membranes. Biophys J 104:1238–1247
Henshaw JB, Olsen CA, Farnbach AR, Nielson KH, Bell JD (1998) Definition of the specific roles of lysolecithin and palmitic acid in altering the susceptibility of dipalmitoylphosphatidylcholine bilayers to phospholipase A2. Biochemistry 37:10709–10721
Hutterer R, Schneider FW, Sprinz H, Hof M (1996) Binding and relaxation behaviour of prodan and patman in phospholipid vesicles: a fluorescence and 1H NMR study. Biophys Chem 61:151–160
Jameson DM, Croney JC, Moens PD (2003) Fluorescence: basic concepts, practical aspects, and some anecdotes. Methods Enzymol 360:1–43
Juhasz J, Davis JH, Sharom FJ (2010) Fluorescent probe partitioning in giant unilamellar vesicles of 'lipid raft' mixtures. Biochem J 430:415–423
Jurkiewicz P, Sykora J, Olzynska A, Humpolickova J, Hof M (2005) Solvent relaxation in phospholipid bilayers: principles and recent applications. J Fluoresc 15:883–894
Jurkiewicz P, Olzynska A, Langner M, Hof M (2006) Headgroup hydration and mobility of DOTAP/DOPC bilayers: a fluorescence solvent relaxation study. Langmuir 22:8741–8749
Jurkiewicz P, Cwiklik L, Jungwirth P, Hof M (2012) Lipid hydration and mobility: an interplay between fluorescence solvent relaxation experiments and molecular dynamics simulations. Biochimie 94:26–32

Kim HM, Choo HJ, Jung SY, Ko YG, Park WH, Jeon SJ, Kim CH, Joo T, Cho BR (2007) A two-photon fluorescent probe for lipid raft imaging: C-laurdan. Chembiochem 8:553–559

Korlach J, Schwille P, Webb WW, Feigenson GW (1999) Characterization of lipid bilayer phases by confocal microscopy and fluorescence correlation spectroscopy. Proc Natl Acad Sci U S A 96:8461–8466

Krasnowska EK, Gratton E, Parasassi T (1998) Prodan as a membrane surface fluorescence probe: partitioning between water and phospholipid phases. Biophys J 74:1984–1993

Krasnowska EK, Bagatolli LA, Gratton E, Parasassi T (2001) Surface properties of cholesterol-containing membranes detected by Prodan fluorescence. Biochim Biophys Acta 1511:330–340

Kubiak J, Brewer J, Hansen S, Bagatolli LA (2011) Lipid lateral organization on giant unilamellar vesicles containing lipopolysaccharides. Biophys J 100:978–986

Lakowicz JR, Sheppard JR (1981) Fluorescence spectroscopic studies of Huntington fibroblast membranes. Am J Hum Genet 33:155–165

Lakowicz JR, Weber G (1973) Quenching of protein fluorescence by oxygen. Detection of structural fluctuations in proteins on the nanosecond time scale. Biochemistry 12:4171–4179

Lakowicz JR, Bevan DR, Maliwal BP, Cherek H, Balter A (1983) Synthesis and characterization of a fluorescence probe of the phase transition and dynamic properties of membranes. Biochemistry 22:5714–5722

Lippert E (1957) Spektroskopische Bestimmung des Dipolmomentes aromatischer Verbindungen im ersten angeregten Singulettzustand. Z Elektrochem 61:962–975

Macgregor RB, Weber G (1986) Estimation of the polarity of the protein interior by optical spectroscopy. Nature 319:70–73

Montes LR, Alonso A, Goni FM, Bagatolli LA (2007) Giant unilamellar vesicles electroformed from native membranes and organic lipid mixtures under physiological conditions. Biophys J 93:3548–3554

Nag K, Pao JS, Harbottle RR, Possmayer F, Petersen NO, Bagatolli LA (2002) Segregation of saturated chain lipids in pulmonary surfactant films and bilayers. Biophys J 82:2041–2051

Nielsen SB, Otzen DE (2010) Impact of the antimicrobial peptide Novicidin on membrane structure and integrity. J Colloid Interface Sci 345:248–256

Norlen L, Plasencia I, Bagatolli L (2008) Stratum corneum lipid organization as observed by atomic force, confocal and two-photon excitation fluorescence microscopy. Int J Cosmet Sci 30:391–411

Olzynska A, Zan A, Jurkiewicz P, Sykora J, Grobner G, Langner M, Hof M (2007) Molecular interpretation of fluorescence solvent relaxation of Patman and 2H NMR experiments in phosphatidylcholine bilayers. Chem Phys Lipids 147:69–77

Parasassi T, Gratton E (1992) Packing of phospholipid vesicles studied by oxygen quenching of Laurdan fluorescence. J Fluoresc 2:167–174

Parasassi T, Gratton E (1995) Membrane lipid domains and dynamics as detected by LAURDAN fluorescence. J Fluoresc 5:59–69

Parasassi T, Conti F, Gratton E (1986a) Fluorophores in a polar medium: time dependence of emission spectra detected by multifrequency phase and modulation fluorometry. Cell Mol Biol 32:99–102

Parasassi T, Conti F, Gratton E (1986b) Time-resolved fluorescence emission spectra of Laurdan in phospholipid vesicles by multifrequency phase and modulation fluorometry. Cell Mol Biol 32:103–108

Parasassi T, De Stasio G, D'ubaldo A, Gratton E (1990) Phase fluctuation in phospholipid membranes revealed by Laurdan fluorescence. Biophys J 57:1179–1186

Parasassi T, De Stasio G, Ravagnan G, Rusch RM, Gratton E (1991) Quantitation of lipid phases in phospholipid vesicles by the generalized polarization of Laurdan fluorescence. Biophys J 60:179–189

Parasassi T, Loiero M, Raimondi M, Ravagnan G, Gratton E (1993a) Absence of lipid gel-phase domains in seven mammalian cell lines and in four primary cell types. Biochim Biophys Acta 1153:143–154

Parasassi T, Ravagnan G, Rusch RM, Gratton E (1993b) Modulation and dynamics of phase properties in phospholipid mixtures detected by Laurdan fluorescence. Photochem Photobiol 57:403–410
Parasassi T, Di Stefano M, Loiero M, Ravagnan G, Gratton E (1994a) Cholesterol modifies water concentration and dynamics in phospholipid bilayers: a fluorescence study using Laurdan probe. Biophys J 66:763–768
Parasassi T, Di Stefano M, Loiero M, Ravagnan G, Gratton E (1994b) Influence of cholesterol on phospholipid bilayers phase domains as detected by Laurdan fluorescence. Biophys J 66:120–132
Parasassi T, Giusti AM, Raimondi M, Gratton E (1995) Abrupt modifications of phospholipid bilayer properties at critical cholesterol concentrations. Biophys J 68:1895–1902
Parasassi T, Gratton E, Yu WM, Wilson P, Levi M (1997) Two-photon fluorescence microscopy of laurdan generalized polarization domains in model and natural membranes. Biophys J 72:2413–2429
Parasassi T, Krasnowska EK, Bagatolli LA, Gratton E (1998) LAURDAN and Prodan as polarity sensitive fluorescent membrane probes. J Fluoresc 8:365–373
Plasencia I, Norlen L, Bagatolli LA (2007) Direct visualization of lipid domains in human skin stratum corneum's lipid membranes: effect of pH and temperature. Biophys J 93:3142–3155
Sanchez SA, Bagatolli LA, Gratton E, Hazlett TL (2002) A two-photon view of an enzyme at work: crotalus atrox venom PLA2 interaction with single-lipid and mixed-lipid giant unilamellar vesicles. Biophys J 82:2232–2243
Sanchez SA, Tricerri MA, Gratton E (2012) Laurdan generalized polarization fluctuations measures membrane packing micro-heterogeneity in vivo. Proc Natl Acad Sci U S A 109:7314–7319
Sot J, Bagatolli LA, Goni FM, Alonso A (2006) Detergent-resistant, ceramide-enriched domains in sphingomyelin/ceramide bilayers. Biophys J 90:903–914
Stock RP, Brewer J, Wagner K, Ramos-Cerrillo B, Duelund L, Jernshoj KD, Olsen LF, Bagatolli LA (2012) Sphingomyelinase D activity in model membranes: structural effects of in situ generation of ceramide-1-phosphate. PLoS One 7:e36003
Sumbilla C, Lakowicz JR (1982) Fluorescence studies of red blood cell membranes from individuals with Huntington's disease. J Neurochem 38:1699–1708
Sun Y, Lo W, Lin SJ, Jee SH, Dong CY (2004) Multiphoton polarization and generalized polarization microscopy reveal oleic-acid-induced structural changes in intercellular lipid layers of the skin. Opt Lett 29:2013–2015
Vanounou S, Pines D, Pines E, Parola AH, Fishov I (2002) Coexistence of domains with distinct order and polarity in fluid bacterial membranes. Photochem Photobiol 76:1–11
Viard M, Gallay J, Vincent M, Meyer O, Robert B, Paternostre M (1997) Laurdan solvatochromism: solvent dielectric relaxation and intramolecular excited-state reaction. Biophys J 73:2221–2234
Vincent M, De Foresta B, Gallay J (2005) Nanosecond dynamics of a mimicked membrane-water interface observed by time-resolved stokes shift of LAURDAN. Biophys J 88:4337–4350
Weber G, Farris FJ (1979) Synthesis and spectral properties of a hydrophobic fluorescent probe: 6-propionyl-2-(dimethylamino)naphthalene. Biochemistry 18:3075–3078
Yu W, So PT, French T, Gratton E (1996) Fluorescence generalized polarization of cell membranes: a two-photon scanning microscopy approach. Biophys J 70:626–636
Zeng J, Chong PL (1995) Effect of ethanol-induced lipid interdigitation on the membrane solubility of Prodan, Acdan, and Laurdan. Biophys J 68:567–573

Chapter 6
Long-Range Lipid-Water Interaction as Observed by ATR-FTIR Spectroscopy

Zoran Arsov

Abstract It is commonly assumed that the structure of water at a lipid-water interface is influenced mostly in the first hydration layer. However, recent results from different experimental methods show that perturbation extends through several hydration layers. Due to its low light penetration depth, attenuated total reflection Fourier transform infrared (ATR-FTIR) spectroscopy is specifically suited to study interlamellar water structure in multibilayers. Results obtained by this technique confirm the long-range water structure disturbance. Consequently, in confined membrane environments nearly all water molecules can be perturbed. It is important to note that the behavior of confined water molecules differs significantly in samples prepared in excess water and in partially hydrated samples. We show in what manner the interlamellar water perturbation is influenced by the hydration level and how it is sequentially modified with a step-by-step dehydration of samples either by water evaporation or by osmotic pressure. Our results also indicate that besides different levels of hydration the lipid-water interaction is modulated by different lipid headgroups and different lipid phases as well. Therefore, modification of interlamellar water properties may clarify the role of water-mediated effects in biological processes.

Keywords ATR-FTIR • Lipid bilayers • Interlamellar water • Excess water • Long-range water structure effect • Water-mediated biological processes

6.1 Introduction

Hydration dynamics and lipid-water interaction strength are closely related to molecular organization and properties of lipids. Since biological membranes play a major role in many cellular processes, they depend on the level of hydration and the structural properties of water molecules at the membrane surface (Fitter et al. 1999; Tamm and Han 2000). Therefore, we have to leave behind the perception

Z. Arsov (✉)
Laboratory of Biophysics, Department of Solid State Physics,"Jozef Stefan" Institute, Jamova 39, SI-1000 Ljubljana, Slovenia
e-mail: zoran.arsov@ijs.si

E.A. Disalvo (ed.), *Membrane Hydration*, Subcellular Biochemistry 71,
DOI 10.1007/978-3-319-19060-0_6

that the influence of water on biological processes is negligible (Chaplin 2006). In recent years, numerous experimental and theoretical evidences were compiled for a substantial role of water in the regulation of biological functions.

Water confined in narrow spaces, e.g. between two opposing lipid membranes, may no longer resemble bulk water. It may be much more viscous, bound to membrane surfaces, etc. (N. of E. * see Chap. 7) During interaction between membranes, water is usually not completely expelled from the confined space, but partly remains present and is involved in the interaction (Ball 2008a). Thus, biological function depends on a delicate interplay between two previously regarded distinct entities: the biomolecule, e.g. a lipid molecule, and its water-based environment (Ball 2008b). Consequently, we have to assign a structural and functional role to the membrane hydration water (Jendrasiak 1996; Milhaud 2004; Disalvo et al. 2008).

In this respect, the hydration force is an important component of the forces operating between phospholipid membranes in water or aqueous solutions. It results from restructuring of interlamellar water with respect to bulk water (Rand and Parsegian 1989; McIntosh and Simon 1994). Accordingly, it is important to understand how lipid bilayers influence the network of water molecules in their vicinity.

Different experimental methods can be used to extract information on the structure and dynamics of water in lipid-water systems, such as nuclear magnetic resonance (NMR) spectroscopy (Finer and Darke 1974; Faure et al. 1997; Kodama et al. 2004), neutron scattering (Kiselev et al. 1999; Fitter et al. 1999), surface force measurements (Higgins et al. 2006; Fukuma et al. 2007), calorimetry (Kodama et al. 2001; Lefèvre et al. 2002; Wennerström and Sparr 2003), fluorescence methods (Parasassi et al. 1991; Beranová et al. 2012), terahertz spectroscopy (Tielrooij et al. 2009; Hishida and Tanaka 2011), and vibrational spectroscopy or their combination (Pfeiffer et al. 2013). (N.of E. * see also Chap. 4). Vibrational spectroscopy is especially suited to probe hydrogen bonding (H-bonding) characteristics of water. In this regard, the shape and position of the hydroxyl (OH) stretching band of water can be used to detect different species of water as well as to examine the water H-bonding network.

There have been several attempts to probe the effect of lipid bilayers on the structure and dynamics of water molecules by infrared (IR) spectroscopy (Grdadolnik et al. 1994; Pohle et al. 1998; Volkov et al. 2007a; Binder 2007). However, these studies were performed on partially hydrated samples, where one gets information only about water located closely to lipid molecules. On the other hand, we are also interested to know whether lipids have any effect on the structure of water distant from the membrane. It can be safely assumed that water molecules that directly interact with lipid molecules are not bulk-like, as this interaction inevitably disturbs their three-dimensional H-bonded network (Ball 2008a). What about water molecules farther away from the membrane? How far does the influence of the membrane surface propagate into the water and what are the manifestations of such propagation (Berkowitz and Vácha 2012)?

The only way to answer these questions is to conduct experiments on lipid samples that are fully hydrated, i.e. hydrated beyond the excess water point, rather than on incompletely hydrated samples. To our knowledge, we were the first to report about the effect on the shape of the OH stretching band measured by attenuated total reflection Fourier transform infrared (ATR-FTIR) spectroscopy on lipid multibilayers prepared in excess water (Arsov and Quaroni 2007; Štrancar and Arsov 2008). The OH stretching band shift to higher frequencies reflected putative weakening of H-bonding in interlamellar water compared to usual strengthening for hydration water directly interacting with lipids. These findings are supported by similar results obtained in studies on lipid membranes prepared in excess water with Raman spectroscopy (Lafleur et al. 1989) and coherent anti-Stokes Raman scattering (CARS) microscopy (Cheng et al. 2003; Wurpel and Müller 2006). To permit direct comparison of the effect on the water structure for samples prepared in excess water and for partially hydrated samples we set to design a single experiment to test the influence of different lipid hydration levels. In addition, such experiment allows studying the corresponding impact of lipid composition/headgroup, lipid phases, and lipid phase transitions.

The aim of this contribution is to show that the structure of the hydration water is modified not only for interfacial water or the first hydration water layer but even beyond. We argue that this reflects a possible long-range lipid-water interaction, which could have implications for processes that involve membranes such as adhesion, stacking, and fusion, especially when these processes are taking place in confined environments.

6.2 ATR-FTIR Approach

In this section we will introduce the ATR-FTIR method and present particular technical details important for understanding the results described in the chapter and subsequent discussion. Due to a relatively small light penetration depth this method is particularly suitable for conducting repeatable measurements of thin film samples under well controlled conditions.

6.2.1 ATR-FTIR Methodology

ATR-FTIR is one of the most powerful methods for recording infrared spectra of biological materials in general, and for biological membranes in particular. It yields a strong signal with only a few micrograms of sample, and allows to obtain information about the orientation of various parts of a molecule in an oriented system (Goormaghtigh et al. 1999). For example, it is possible to follow the orientation of water interacting with the polar headgroups of lipids and dependence of this orientation on temperature and lipid phase transition (Okamura et al. 1990).

(N.of E. * consider here the model discussed in Chap. 2).

The (aqueous) environment of lipid membranes can be modulated so that their properties can be studied as a function of temperature, pressure and pH, as well as in the presence of specific biologically active compounds (Tamm and Tatulian 1997; Goormaghtigh et al. 1999).

Due to the small penetration depth of IR light, the ATR-FTIR method is ideal for highly absorbing samples such as water suspensions. As the sample is brought into contact with an internal reflection element (IRE), absorption of the energy of the evanescent field by the sample provides ATR-FTIR spectra. The penetration depth d_p of IR light in the sample is defined by properties of the sample as well as the IRE and can be expressed as

$$d_p = \frac{\lambda}{\sqrt{2\pi\, n_{IRE} \left(\sin^2\theta - (n/n_{IRE})^2\right)}} \tag{6.1}$$

where λ is the wavelength of incident light, n_{IRE} is the real refractive index of the IRE, θ is the angle of incidence and n is the real refractive index of the sample. In our experiments we usually use a trapezoidal germanium plate IRE with an incidence angle of 45°. Therefore, the penetration depth d_p for hydrated lipid samples ($n \approx 1.4$) for a wavelength around the peak of the OH stretching band ($\lambda \approx 3.0$ μm) is estimated from Eq. 6.1 to be around 200 nm.

For such small d_p, almost always the low absorption approximation can be taken into account, which is applicable for reflection losses that do not exceed approximately 10 % (Harrick 1987). Even when the sample is pure water, the loss for single reflection is close to this limiting value. Within this approximation the effective thickness, i.e. the thickness of a sample that would be required to obtain the same absorption in a transmission measurement, represents proportionality constant between the reflection loss and absorption coefficient. The level of total absorbance further depends on the length of the IRE, which determines the number of active internal reflections, i.e. the number of reflections for which IR light is attenuated by a sample. In the examples that will be shown in this chapter, the number of active reflections was 13. At the used experimental conditions, the effective thickness for single reflection is about equal to the penetration depth (Harrick 1987). Therefore, the total effective sample thickness for 13 reflections is around 2.6 μm at $\lambda \approx 3.0$ μm. Due to a sufficiently low effective thickness, even bulk water can be measured under reproducible conditions and without distortion of the shape of water absorption bands. This enables detection of minute changes in the OH stretching band, which is the main objective of the presented studies.

6.2.2 Sample Preparation

The most typical way to prepare stacks of lipid membranes for ATR-FTIR measurements involves spreading of a particular volume of stock solution, e.g. chloroform

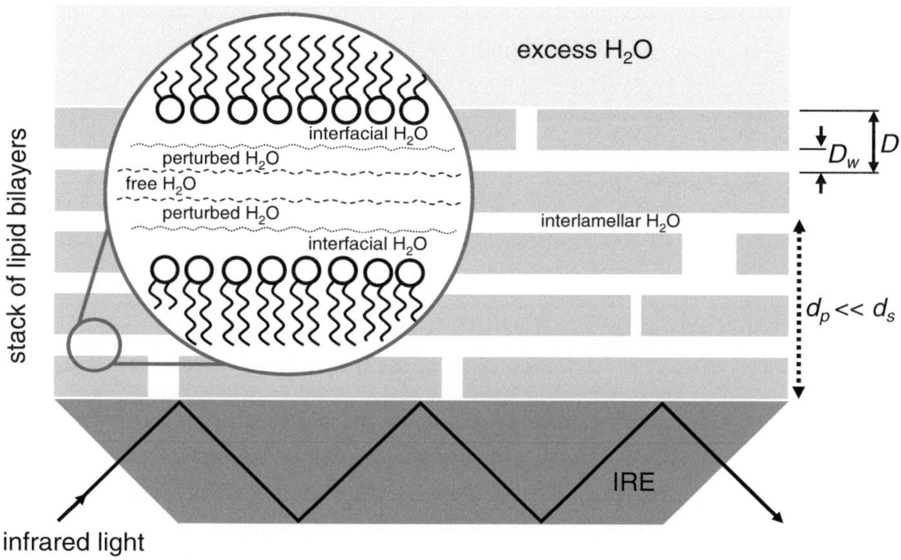

Fig. 6.1 Schematic presentation of the ATR-FTIR experimental setup. The totally reflected infrared light in the internal reflection element (IRE) is attenuated by the interaction of the sample with the evanescent field. Lipid sample in excess water is prepared to form a stack of lipid bilayers. For thick multibilayers the thickness of the stack d_s can exceed several times the penetration depth, d_p, of the evanescent field. The interlamellar water is classified as free, perturbed and interfacial water (*the zoomed inset*). The water layer thickness, D_W, and the lamellar repeat spacing, D, are also denoted. Possible defects in the lamellar structure are represented by gaps in particular bilayers

solution of lipids (Goormaghtigh et al. 1999) or water suspension of lipid vesicles (Miller and Bach 1999; Arsov and Quaroni 2007), on the properly cleaned IRE surface (Goormaghtigh et al. 1999). While the solvent is being slowly evaporated, a tip can be used to evenly spread the liquid over the surface of the IRE. This method produces a stack of highly aligned multibilayers, because capillary forces flatten the membranes during evaporation (Goormaghtigh et al. 1999). Dried lipid stacks can then be rehydrated in an atmosphere of constant relative humidity (RH) (Binder 2007) or in excess water (Arsov and Quaroni 2007).

Depending on the amount of lipids in the stock solution, thin or thick multibilayers can be formed on the IRE. In the studies that will be presented in this chapter, the thickness of the stack of lipid bilayers d_s exceeded several times the penetration depth of the evanescent wave d_p (Fig. 6.1). Figure 6.1 reveals that the water absorption in ATR-FTIR spectra for thick lipid samples, hydrated in excess water, is due to interlamellar water trapped between lipid bilayers.

It is important to note that by using multibilayers, the signal due to lipid absorption is maximized, while the signal due to water absorption is minimized. In a perfectly aligned stack without defects the water and lipid signal could be predicted from structural characteristics, such as the lamellar repeat spacing and the water layer thickness. But for samples in excess water it usually turns out that water signal is higher and lipid signal lower than expected for the ideal lamellar sample. This is due to water present in defects (Nagle and Tristram-Nagle 2000). The defect regions, illustrated by gaps in lipid film layers in Fig. 6.1, form as a consequence of irregularities in lamellar structure and because of imperfect bilayer alignment. Nevertheless, the ratio of the water and lipid signals does not change with time significantly, showing that thick stacks of membranes are relatively stable even in excess water. It is noteworthy that the way of the lipid film preparation can affect the quality of the sample by influencing the alignment of multibilayers (Pohle et al. 1998).

Based on different attributes, such as their distance from the bilayer, H-bonding characteristics, as well as dynamic and thermodynamic properties, interlamellar water molecules are usually classified into three groups (see the zoomed inset in Fig. 6.1). (N. of E. * see Chap. 7 by Appignanessi et al.).

Different notations have been used for these groups: free water (bulk-like, far water), perturbed water (intermediate, freezable interlamellar water), and interfacial water (bound, buried, neighboring, non-freezable interlamellar water) (Kiselev et al. 1999; Kodama et al. 2001; Murzyn et al. 2006; Pinnick et al. 2010; Debnath et al. 2010). Such classification has been introduced very early also for hydration of proteins and peptides (Kuntz and Kauzmann 1974). The interfacial water is represented by water molecules that directly interact with lipids or reside in the interface region, while the free water denotes putative bulk-like molecules. The perturbed water corresponds to water in the transition region whose properties are still influenced by the presence of lipid membranes.

In order to properly interpret and understand how water properties depend on its distance from the lipid membrane and how such dependence is influenced by lipid hydration level, it is useful to know the amount and properties of water molecules belonging to particular classes. There were different attempts to quantify the number of water molecules in these different classes for example by NMR (Marinov and Dufourc 1996; Faure et al. 1997), X-ray structural studies (Nagle and Tristram-Nagle 2000) and calorimetry (Kodama et al. 2001). Useful information about the amount and properties of particular water molecules can be obtained also from molecular dynamics (MD) simulations (Pasenkiewicz-Gierula et al. 1997; Pinnick et al. 2010). However, it is difficult to define exact criteria for distinguishing different water classes.

6.2.3 Calculation of the Molar Water-to-Lipid Ratio

To relate different lipid hydration levels and the corresponding properties of interlamellar water molecules, the molar water-to-lipid ratio n_W has to be determined. It is shown in the following paragraphs how this ratio is obtained from ATR-FTIR experiments on hydrated lipid multibilayers.

In the low absorption approximation, the integral absorbance A of a vibrational mode is proportional to the concentration c of the absorbing molecules within the homogeneous film. Let us now assume that we have a layered sample. For such a heterogeneous distribution of molecules along the direction perpendicular to IRE, the contribution to absorbance at different distances from IRE should be integrated. However, in our case, where samples have a periodic structure with layers much thinner than the penetration depth, the spatial distribution of absorbing molecules within layers does not significantly influence contribution of particular layer to the absorbance. That is, for thin enough layers, the electric field amplitude can be assumed constant in each of them. Thus, the integral absorbance of water (W) or lipid (L) molecules is proportional to the corresponding effective (average) concentration across the whole sample

$$A_{W,L} \propto c_{W,L}. \tag{6.2}$$

Since the molar water-to-lipid ratio is related to the effective concentrations as

$$n_W = \frac{c_W}{c_L}, \tag{6.3}$$

it follows that n_W is proportional to the quotient of integral absorbances

$$n_W \propto \frac{A_W}{A_L}. \tag{6.4}$$

One option to determine n_W from the measured absorbance values is to find the proportionality constant in Eq. 6.4 by calibrating the absorbance quotient against n_W. Another option is presented in the proceeding paragraphs.

It was found convenient (as explained below) to use the ratio between integral absorbances of a measured sample and a reference sample, i.e. the relative integral absorbance a, for determining n_W. It is suitable to use bulk water (denoted bulk) as a reference sample for relative integral absorbance of water a_W, and dried lipid stack (denoted dry) as a reference sample for relative integral absorbance of lipids a_L. Taking into account Eq. 6.2, it then holds

$$a_W = A_W/A_W^{bulk} = c_W/c_W^{bulk} \text{ and } a_L = A_L/A_L^{dry} = c_L/c_L^{dry}, \tag{6.5}$$

where the integral absorbances are experimentally determined from spectra. It is important to note that the integral absorbance of the water OH stretching band for hydrated lipid samples should be corrected for the lipid absorption due to the overlap with lipid bands in this spectral region (and vice versa). Using Eqs. 6.3 and 6.5 the ratio n_W can be written as

$$n_W = \frac{a_W c_W^{bulk}}{a_L c_L^{dry}}. \tag{6.6}$$

The reference concentrations in Eq. 6.6 can be expressed with the corresponding molecular volumes $V^0_{W,L}$. This is useful because the values of $V^0_{W,L}$ can be evaluated from the published structural data. The general expression for the effective concentrations is

$$c_{W,L} = \frac{N_{W,L}/N_A}{V} = \frac{N_{W,L}/N_A}{N_L V_L^o + N_W V_W^o}, \tag{6.7}$$

where $N_{W,L}$ is the number of water or lipid molecules in the particular volume V, and N_A is Avogadro's number. For the bulk water reference sample the number of lipid molecules is $N_L = 0$, and for the dried lipid stack the number of water molecules is assumed to be $N_W = 0$. Therefore, it holds (as can be seen also from Eq. 6.7)

$$c_W^{bulk} = 1/N_A V_W^0,\ c_L^{dry} = 1/N_A V_L^0. \tag{6.8}$$

By considering Eqs. 6.6 and 6.8 we get

$$n_W = \frac{a_W}{a_L} \frac{V_L^0}{V_W^0}. \tag{6.9}$$

Using Eqs. 6.7 and 6.8 we can rewrite Eq. 6.5 as

$$a_W = \frac{N_W V_W^0}{N_L V_L^0 + N_W V_W^0} \text{ and } a_L = \frac{N_L V_L^0}{N_L V_L^0 + N_W V_W^0}. \tag{6.10}$$

From these two relations we can quickly realize that

$$a_W + a_L = 1. \tag{6.11}$$

Finally, Eq. 6.9 can be rewritten to obtain

$$n_W = \frac{a_W}{1 - a_W} \frac{V_L^0}{V_W^0}, \tag{6.12}$$

eliminating the need to measure a_L separately (but having to know the values of $V^0_{W,L}$). This explains the convenience of using the relative integral absorbances for determining n_W. Equation 6.12 was used to calculate n_W in the experimental results presented in this chapter. If the calculated n_W is related to the n_W obtained from structural data of ideal lamellar samples with the corresponding lipid composition, it is possible to estimate the level of defects in a lipid film.

To relate our derivations above to the formerly published relations between the water-to-lipid ratio and integral absorbances (Binder et al. 2001), we will also express the above equations with the swelling factor, which has been defined as

$$\phi = \frac{N_L V_L^0 + N_W V_W^0}{N_L V_L^0} = 1 + n_W \frac{V_W^0}{V_L^0}. \tag{6.13}$$

By introducing Eqs. 6.9 and 6.11 in Eq. 6.13, we get

$$a_W = (\phi - 1)/\phi = n_W/\phi \frac{V_W^0}{V_L^0} \text{ and } a_L = 1/\phi, \tag{6.14}$$

which agrees with expressions derived previously (Binder et al. 2001)

Besides the above presented procedure, other approaches for determination of n_W according to Eq. 6.4 have been reported (Pohle et al. 1998; Miller and Bach 1999; Tielrooij et al. 2009; Gauger et al. 2010). In these studies, methyl and methylene or carbonyl lipid band absorbances are usually used to calculate lipid absorbances, while calibration is done with a method based on adding known amounts of water to a defined amount of lipids.

6.3 Structure of Water in Fully and Partially Hydrated Multibilayers

In lipid systems water content is an important variable besides temperature and composition that affects lipid phase behavior and polymorphism. Consequently, it is important to study the effect of hydration on structural and dynamical properties of lipids. This effect is reflected in different parts of IR spectrum corresponding to different lipid molecular groups such as methylene, carbonyl and phosphate groups (Grdadolnik et al. 1991; Hübner and Blume 1998; Selle and Pohle 1998; Binder 2003).

Contrary to the lipid "perspective", we can also follow a similar approach for the part of the IR spectrum, corresponding to water vibration bands. Interestingly, such studies are much less frequent. Usually, variation in the shape of the OH stretching band is followed (Grdadolnik et al. 1994; Pohle et al. 1998; Binder 2003; Disalvo et al. 2013). In the following text, we show that the shape of the OH stretching band is influenced significantly by the hydration level of multibilayers.

6.3.1 Multibilayers in Excess Water

Before we proceed to the comparison of IR spectral properties of bulk and interlamellar water, we first have to comment on the spectral behavior of pure bulk water. The shape of the OH stretching band is appreciably influenced by the temperature (Fig. 6.2). Due to the low penetration depth, ATR-FTIR is specifically suited to obtain high quality spectra of pure bulk water as explained in Sect. 6.2.1 and also highlighted previously (Maréchal 1991). Beside the drop of absorbance with higher temperature, also the shape changes significantly. As indicated by arrows in Fig. 6.2, the band peak position shifts to higher wavenumbers, the low-frequency part of the band decreases and the high-frequency part increases. This

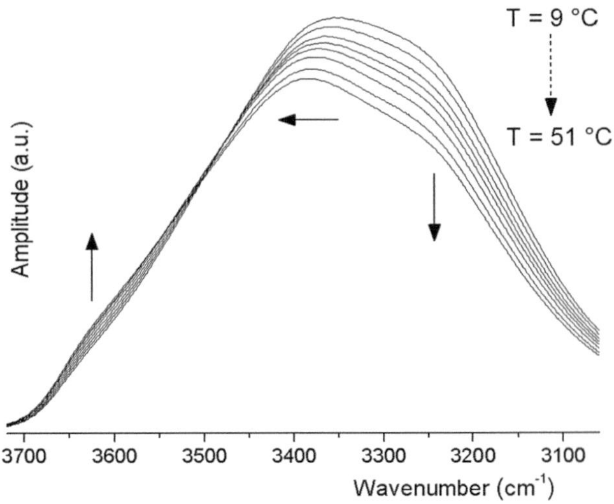

Fig. 6.2 Temperature dependence of the OH stretching band shape of pure bulk water. Spectra are shown in the interval from 9 °C to 51 °C and the temperature step is 6 °C. Beside the drop of absorbance with higher temperature also the shape changes significantly as indicated by the *solid arrows*

observation was also reported before (Brubach et al. 2005). So when we want to compare the results of the water part of IR spectrum for hydrated lipid samples at different temperatures, we have to consider also the effects of temperature on the water itself and not only the effects of lipids.

We will first present the results from multibilayers in excess water that imply the presence of interlamellar water with properties different than pure bulk water (Arsov and Quaroni 2007; Štrancar and Arsov 2008; Arsov et al. 2009). In Fig. 6.3a, the OH stretching bands for pure bulk water and interlamellar water in dimyristoylphosphatidylcholine (DMPC) multibilayers are compared. The comparison is shown for temperature above the temperature of the main phase transition for DMPC ($T_m = 24$ °C), so DMPC is observed in the liquid-crystalline or liquid-disordered phase (denoted as the Ld phase). The absorbance of water is expectedly lower in the case of the lipid sample, since beside water also lipid bilayers fill the inspected sample region.

Using Eq. 6.12 (see Sect. 6.2.3) to calculate the water-to-lipid ratio, we obtain $n_W = 52$ for the sample presented in Fig. 6.3. The values of lipid and water molecular volumes were taken from X-ray structural data (Nagle and Tristram-Nagle 2000). Let us compare this ratio to the values reported from X-ray structural data of hydrated lipid multibilayers. The values range from $n_W = 26$ for samples denoted as fully hydrated (Nagle and Tristram-Nagle 2000) to $n_W = 33$ for samples at excess water point (Costigan et al. 2000). The value calculated for our sample is significantly larger, indicating the presence of defects in our multibilayers (see Sect. 6.2.2). The value of n_W varied significantly between different samples, as it depends on the quality of the prepared multibilayers, but was always significantly higher than the n_W for ideal lamellar sample.

Despite the inferred presence of defects, it is possible to nicely discern the difference between the shape of the OH stretching band for pure bulk H_2O and

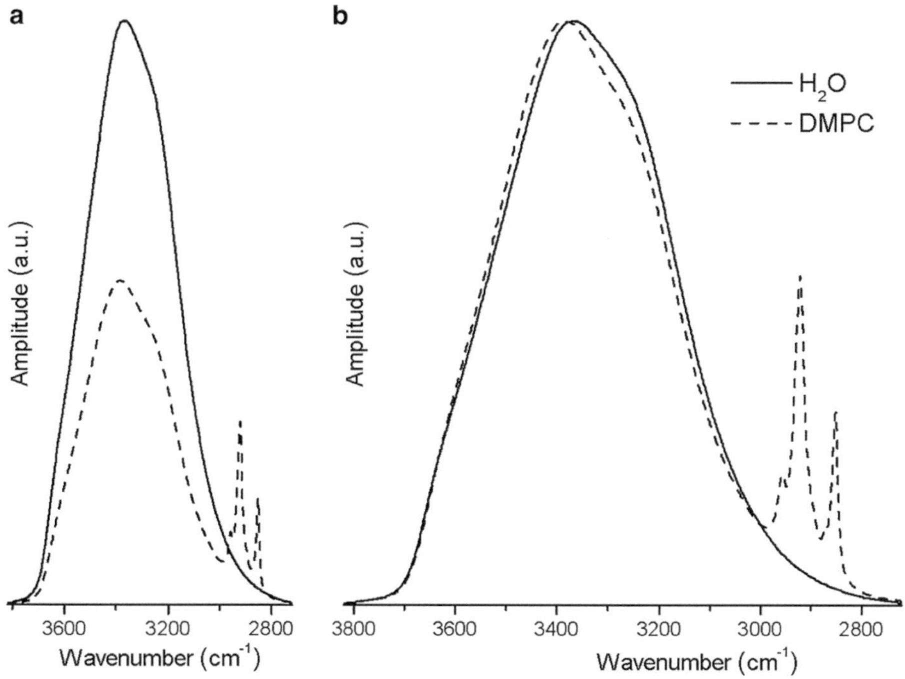

Fig. 6.3 Comparison of the OH stretching band shape for pure bulk water (H_2O) and for interlamellar water in DMPC multibilayers, prepared in excess water. Both spectra were recorded at 36 °C. (**a**) Comparison of original spectra. (**b**) Comparison of spectra after normalization to the maximum amplitude

for the interlamellar water after normalizing this part of the spectra to the maximum amplitude (Fig. 6.3b). The shift to higher frequencies could be interpreted as an overall weakening of the H-bonding among water molecules. A similar effect has been observed in the inverse micellar structures prepared from surfactants (Cringus et al. 2005), in lamellar surfactant structures (Boissière et al. 2002), and in black surfactant films (Berger et al. 2003). The observation of weakening of H-bonding is also supported by CARS data (Cheng et al. 2003; Wurpel and Müller 2006). However, for partially hydrated samples, discussed in the following section (Sect. 6.3.2), an opposite effect can be observed. This agrees with the recent statement that water properties at a weakly hydrated membrane interface are in contrast to the properties observed in inverse micelles (Volkov et al. 2007a, b).

Based on these findings, the existence of another long-range attractive contribution to the hydration force, which would compete with the fluctuation and van der Waals forces, was speculated (Arsov et al. 2009). This force should arise due to the restructuring of the interlamellar water at distances close to the equilibrium interlamellar spacing. Namely, as the water H-bonds are weakened the position of water molecules in the excess (bulk) water phase is energetically favorable, resulting in the attractive hydration force. Attractive hydration force has indeed been proposed before. It was first suggested in the case of inter-bilayer H-bonded water bridges (Rand et al. 1988). Recently, it has been introduced to explain the observations of a DNA-dendrimer complex, where long-range interactions are facilitated by the changed local water structure (Mills et al. 2013).

In order to allow a more quantitative analysis and to enable a systematic explanation of results, the broad OH stretching band can be decomposed into a few Gaussian components (Maréchal 1991; Libnau et al. 1994; Brubach et al. 2005). Each component can be interpreted as representing water molecules with particular H-bonding properties (Brubach et al. 2005), inferring an equilibrium mixture of discrete species differing according to their specific structural arrangements. Beside the mixture models, also continuum models have been proposed, assuming that distortions of the H-bonding structure result in a continuous distribution of H-bond distances, angles, and energies (Libnau et al. 1994). Since the main aim of this chapter is only a qualitative evaluation of different effects on the structure of interlamellar water, a more in-depth analysis of the shape of the OH stretching band will be omitted. We believe this topic is worth a dedicated chapter on its own.

In addition to the strength of the H-bonds, OH stretching band shape is also influenced by the intra- and intermolecular coupling, as well as Fermi coupling of different OH vibrational modes (Sokołowska and Kęcki 1986). Since in all of the above-mentioned studies water was confined between lipid layers, the influence of the confinement or lipid-water interaction on the vibrational coupling might also cause the shape difference.

To compare the extent of the change in the strength of H-bonds with the coupling effects, experiments can be conducted in dilute solution of water (H_2O) in deuterated water (D_2O), or vice versa, where HOD molecules are dilute enough so that complete decoupling of the OH vibrational modes in HOD molecules occurs (Skinner et al. 2009). Such experiment was performed on DMPC multibilayers (Arsov et al. 2009). The corresponding OH stretching band shape difference with respect to bulk HOD exhibited a similar effect on water structure as found for samples prepared in pure H_2O. The significant upshift of the peak position confirmed that H-bonds in interlamellar water became weaker. However, the observed effect was less pronounced than in the case of pure H_2O (Arsov et al. 2009). Therefore, modifications of the vibrational coupling also play an important role in the measured OH stretching band shape, which is in line with previous findings (Lafleur et al. 1989; Sovago et al. 2008; Bonn et al. 2012).

Nevertheless, also changed coupling, especially intermolecular but also intramolecular, implies the presence of different water molecular environment (Skinner et al. 2009). Therefore, in further discussion we will not deal with exact clarification of the origin of the changed OH stretching band shape. We will simply assume that different OH stretching band shapes indicate diverse interlamellar water properties.

6.3.2 Dehydration of Multibilayers by Water Evaporation and Osmotic Pressure

In contrast to the samples prepared in excess water, where the OH stretching peak shifts to higher frequencies relative to the measurements on bulk water, the corresponding peak shift towards lower frequencies for partially/weakly hydrated phosphatidylcholine (PC) membranes (Grdadolnik et al. 1994; Pohle et al. 1997; Volkov et al. 2007a; Binder 2007).

These findings were reproduced also by our measurements. In order to reduce the level of hydration, the temperature was raised to a relatively high temperature (57 °C) and water slowly evaporated. This caused deswelling of multibilayers, during which the amplitude of the OH stretching band decreased, while the amplitude of lipid bands corresponding to molecular groups involving CH bonds increased (Fig. 6.4a). To compare the shape of the OH stretching band, spectra were normalized to the same maximum amplitude at the OH stretching band peak Fig. 6.4b. The comparison with the spectrum of pure bulk H_2O nicely shows that the shape for the sample in excess water ($n_W = 47$) is opposite to the dehydrated sample ($n_W = 5$). In the latter, the high-frequency side of the OH

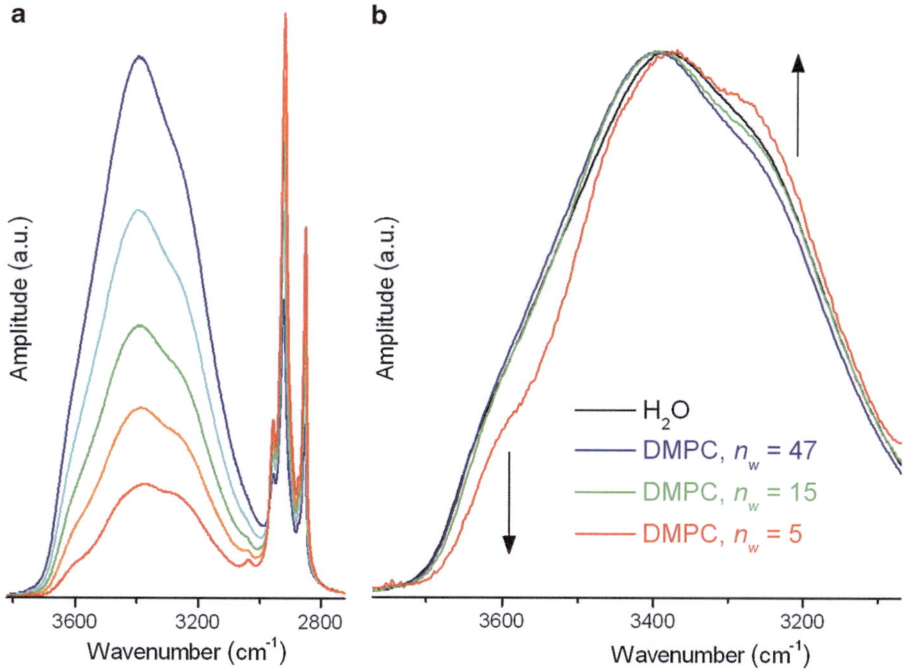

Fig. 6.4 (**a**) Deswelling of DMPC multibilayers by water evaporation (original spectra). (**b**) Comparison of the shape of the normalized OH stretching bands for pure bulk water (H_2O) and for interlamellar water in DMPC multibilayers for different n_W. *Arrows* indicate a trend in the OH stretching band shape modification upon dehydration. Spectra were recorded at 57 °C

stretching band decreases, while the low-frequency side increases relative to the spectrum of pure bulk water (as indicated by arrows in Fig. 6.4b). The level of hydration for the dehydrated sample is such that only the interfacial water remains in multibilayers (see Sect. 6.2.2 for interlamellar water classification). Therefore, the water contributing to the OH stretching band is almost entirely engaged in H-bonding directly to lipid molecular groups, e.g. phosphate and carbonyl, and not to other water molecules.

IR spectroscopy has been extensively used to characterize the hydration sites of PC showing that the primary hydration site is the phosphate group (Arrondo et al. 1984; Ter-Minassian-Saraga et al. 1988; Grdadolnik et al. 1991). Thus, the observed shift of the OH stretching band to lower frequencies is in accordance with the expected stronger hydrogen bonds between water molecules and phosphate groups relative to water-water hydrogen bonds (Bhide and Berkowitz 2005). As we will see from the results presented below (see Sect. 6.4.1), this behavior cannot be generalized to the lipids with a phosphatidylethanolamine (PE) headgroup.

It is evident that the properties of weakly hydrated lipid multibilayers significantly differ from lipid samples in excess water. The question is, how various properties are influenced for only slightly dehydrated samples. For example, lipid phase transition temperatures are not affected by water concentrations slightly below excess water point, while structural parameters such as lamellar repeat spacing are altered (Katsaras 1997; Pohle et al. 2001). As seen in Fig. 6.4b, the OH stretching band shape for DMPC already significantly changes when curves for $n_W = 47$ and $n_W = 15$ are compared, although it was reported that the thermotropic properties of DMPC are retained close to the latter value of n_W (Faure et al. 1997). Similar observations were attained by NMR measurements (Arnold et al. 1983). Thus, we have to be cautious when comparing data for samples in excess water and for the so-called "fully" hydrated samples. For example, in the experiments with IR spectroscopy where hydration level was varied through RH, the n_W values of around 12 were determined for DOPC and POPC lipids close to RH = 100 % (Pohle et al. 1998; Binder 2007). These values are much lower than expected excess water points and can probably be ascribed to the so-called vapor pressure paradox (Rand and Parsegian 1989). It was shown that the origin of this paradox is connected to a very high sensitivity of the lipid lamellar structures on small deviations of RH from 100 % (Nagle and Katsaras 1999). These deviations bring about osmotic pressure that can strongly influence the lamellar repeat spacing and consequently the value of n_W (Kucerka et al. 2005).

Another interesting conclusion can be drawn from results presented in Fig. 6.4b. Since the value $n_W = 15$ is higher than the number of interfacial water molecules per DMPC molecule of about 12 (Pasenkiewicz-Gierula et al. 1997), it can be deduced that for this sample the water removed during the evaporation came either from the population of the free water and/or from the population of the perturbed water. From the drop in the intensity of the high-frequency side of the OH stretching band, observed for DMPC as the water content was decreased from $n_W = 47$ to $n_W = 15$ (Fig. 6.4b), we can reason that the water extracted from the system cannot be classified as the bulk-like free water. Namely, if bulk-like water, whose OH stretching band is shifted to lower frequencies with respect to the sample in excess water (Fig. 6.4b), was removed, the high-frequency side of the OH stretching band would increase. Since this is not the case, it seems that all the water removed belongs to the class of perturbed water. Consequently, we can conclude that the water structure is perturbed throughout the whole interlamellar space with thickness of about 1.8 nm in DMPC (Nagle and Tristram-Nagle 2000).

We have checked how these results compare for different dehydration procedures. We performed a similar experiment where the level of hydration was reduced by osmotic pressure through the addition of a high-molecular-weight polymer polyvinylpyrolidone (PVP). Compared to the previous dehydration procedure where higher temperature (57 °C) was required to increase the rate of water evaporation, the present procedure had no such requirements and could be conducted at lower temperature (36 °C). The addition of PVP on top of DMPC multibilayers triggered deswelling and subsequent extraction of interlamellar water as judged from the decreased OH stretching band amplitude (Fig. 6.5a). Since the thickness of the prepared multibilayers exceeds many times the penetration depth, there is no

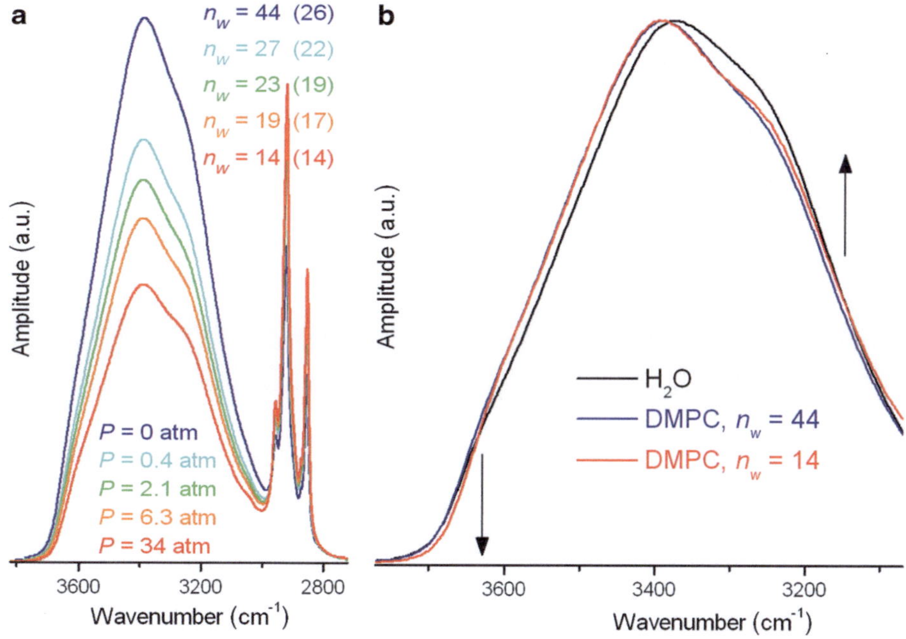

Fig. 6.5 (**a**) Deswelling of DMPC multibilayers by osmotic pressure exerted by PVP (original spectra). The marked values of osmotic pressure P were estimated from the weight concentration of PVP. The indicated experimentally derived values of n_W are compared to the values in *parentheses* obtained from X-ray structural studies (see text for details). (**b**) Comparison of the shape of the normalized OH stretching bands for pure bulk water (H_2O) and for interlamellar water in DMPC multibilayers for different n_W. *Arrows* indicate a trend in the OH stretching band shape modification by dehydration. Spectra were recorded at 36 °C

spectral contribution from PVP to the ATR-FTIR spectra. The values of osmotic pressure P presented in Fig. 6.5a were calculated from the amounts of added PVP. Calculation was based on interpolating function found by fitting the published data (McIntosh and Simon 1986) by a variant of functional form of dependence of osmotic pressure on PVP weight concentration (Parsegian et al. 1986).

The difference in the OH stretching band shape between DMPC in excess water ($n_W = 44$) and partially hydrated DMPC ($n_W = 14$) discerned in Fig. 6.5b was qualitatively similar to the difference observed when water was extracted by evaporation (compare to the difference between samples with $n_W = 47$ and $n_W = 15$ in Fig. 6.4b). This finding confirms that the majority of interlamellar water (beside the interfacial water) is perturbed. Similar results for the two dehydration methods are not unexpected. Recent NMR experiments indicate that the use of dehydration pressure or osmotic pressure should lead to similar effects on membrane. Namely, these two stresses are thermodynamically equivalent, since the change in chemical potential when transferring water from the interlamellar space to the bulk water phase corresponds to the induced pressure (Mallikarjunaiah et al. 2011).

The determined values of n_W for low P are higher than the values expected from X-ray structural studies (Petrache et al. 1998a, b) presented in parentheses in Fig. 6.5a. This is due to the presence of defects, i.e. irregularities in lamellar structure and imperfections in bilayer alignment. Instead of deswelling and subsequent reduction of the interlamellar repeat spacing, the osmotic pressure first reduces the number of defects, as observed before (Mennicke and Salditt 2002). Hence, only by appreciably increasing the osmotic pressure and by allowing enough time for equilibration, the values of n_W come close to the expected values (Fig. 6.5a).

Let us summarize the main conclusions drawn from Sects. 6.3.1 and 6.3.2. Firstly, we have to be careful when comparing properties of lipid bilayers in excess water and in the so-called fully hydrated samples or (even more so) in partially hydrated samples. Secondly, it seems that beside interfacial water, practically all the remaining interlamellar water can be regarded as perturbed.

6.3.3 Water Isotope and Salt Effect on the Structure of Water

The experiments presented in this chapter were conducted in pure H_2O. In contrast, to decrease the spectral overlap of particular lipid bands, e.g. carbonyl, amide or even methylene, with water bands, often D_2O is used as a solvent. Furthermore, when we want to mimic physiological conditions, membranes are usually prepared in a buffer or salt solution. It is important to be aware of the possible effects of deuterated water or salt on the structure of water.

Although the properties of liquid H_2O and D_2O are closely similar, there are small but definite differences in the magnitudes of several physical properties. For example, the degree of hydrogen bonding is higher in the case of D_2O (Némethy and Scheraga 1964). One experimental indication that D_2O has a slightly different influence on phospholipid bilayers relative to ordinary H_2O, was offered by comparison of the carbonyl band peak position in the two solvents (Arsov and Quaroni 2007). Several computational studies confirmed these findings. One work showed that isotope substitution affects the local arrangement of the hydrogen-bonded network (Bergmann et al. 2007), while another study offered a molecular concept for this observation (Róg et al. 2009). Theoretical predictions of the latter study were also verified by fluorescence measurements on lipid membranes, where the water isotope effect on headgroup hydration and mobility, lateral lipid diffusion and lipid backbone packing was determined (Beranová et al. 2012). Substituting H_2O for D_2O can pose another problem, as H/D exchange of exchangeable sample protons can occur (Lewis and McElhaney 2007). This results in loss of the absorption band of the protonated species and its replacement with the absorption band of the deuterated species at lower frequencies. Consequently, depending on the band concerned, spectroscopic observation may not be convenient.

The knowledge of the ion-specific hydration effect at the interface between the phospholipid bilayer and aqueous solution has greatly improved lately. But despite substantial progress, many issues remain unresolved (Parsegian and Zemb 2011). Diverse efforts have been undertaken in this respect. It was studied by ATR-FTIR how different aqueous solutions of salts at diverse concentrations modify the OH stretching band (Riemenschneider et al. 2008). The effect on the structure of water upon addition of salts was examined also by Raman spectroscopy (Cavaille et al. 1996). Furthermore, a surface-sensitive vibrational sum frequency generation method was used to examine how the water structure at the lipid-water interface is affected by the presence of ions (Chen et al. 2010). Structural properties of pure water and ionic solutions were also compared through measurement of the orientational-correlation time of water molecules by femtosecond pump–probe spectroscopy (Omta et al. 2003). Moreover, it has been suggested recently that the swelling of neutral lipid bilayers upon addition of a salt cannot be explained only by the screening of the van der Waals interactions, but that also increase in the hydration force has to be taken into account (Manciu and Ruckenstein 2007).

6.4 Influence of Lipid Composition/Phase on the Structure of Interlamellar Water

Lipid composition and lipid phase of multibilayers influence the excess water point and structural characteristics such as lamellar repeat spacing and interlamellar water layer thickness. Consequently, the force balance between bilayers changes with temperature and lipid composition of membranes. Therefore, it can be anticipated that also interlamellar water properties as well as the corresponding hydration force contribution might be modified by these characteristics, as will be outlined throughout this section.

6.4.1 Influence of Lipid Composition

In order to study the influence of lipid composition and to compare the results to those obtained on DMPC multibilayers, shown above, measurements with samples prepared from palmitoyloleoylphosphatidylethanolamine (POPE) were conducted. This lipid was chosen because it differs with regards to DMPC in its headgroup, affecting the interfacial water, and because it has much smaller interlamellar repeat spacing, implying a different balance of forces between bilayers (McIntosh and Simon 1996).

From the inset to Fig. 6.6a, we can nicely appreciate that the amplitude of DMPC spectrum is higher than that of POPE spectrum in the OH stretching band region. This observation is expected, since the lamellar repeat spacing is much smaller in the case of POPE, which explains much smaller water-to-lipid ratio n_W. In order to better compare the OH stretching band shapes, Fig. 6.6a shows the normalized spectra against the corresponding spectrum of pure bulk H_2O. Similar observations follow for POPE as mentioned previously in the case of DMPC. The increase in the high-frequency part is even more pronounced than for DMPC, while the low-

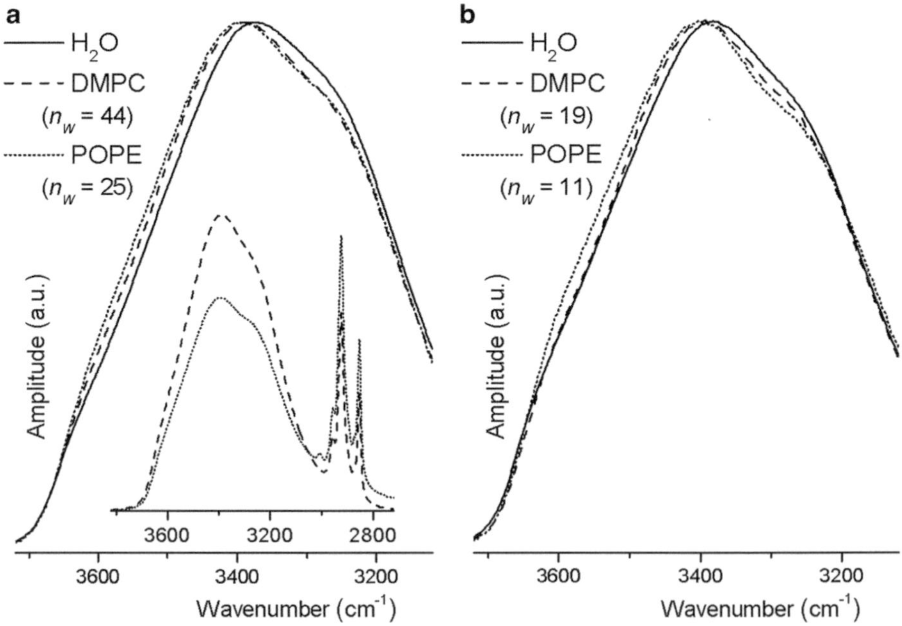

Fig. 6.6 Comparison of the OH stretching band shape for pure bulk water (H_2O) and for interlamellar water in DMPC and POPE multibilayers. (**a**) Comparison before (*inset*) and after normalization to the maximum amplitude (main plot) for samples in excess water. Spectra were recorded at 36 °C. (**b**) Comparison with normalization to the maximum amplitude for partially dehydrated samples. Spectra were recorded at 57 °C. Two representative spectra are compared to the spectrum for pure bulk water

frequency part of the spectra seems comparable. There are two possible explanations for this difference. Either the average H-bonding is even more weakened than in DMPC and/or higher confinement of water in the case of POPE more strongly influences the coupling of water vibrational modes (see Sect. 6.3.1). In the case of POPE, there is no free water present between bilayers because of a very small interlamellar repeat spacing of around 0.6 nm (Rappolt et al. 2003). So, interlamellar water is represented only by the interfacial and the perturbed water. Therefore, the increase in the intensity of the high-frequency side of the OH stretching band for POPE with respect to pure bulk H_2O is due to the structure of the perturbed water (Fig. 6.6).

Similar to DMPC, the determined value of $n_W = 25$ for POPE in excess water (Fig. 6.6a) is larger than expected from structural or NMR data that yield n_W of around 15 (Marinov and Dufourc 1996; Rappolt et al. 2003). Thus, also our POPE multibilayers contain defects and are irregular to some extent. In order to exclude a possibility that the difference in the spectral shapes between DMPC and POPE arose mainly because of a relatively larger amount of interlamellar water due to the defects for DMPC, the comparison between samples partially dehydrated by water evaporation was done (Fig. 6.6b). The value of n_W in this case is below the fully hydrated level but above the level where only the interfacial water is present, which is around $n_W = 12$ for DMPC (Pasenkiewicz-Gierula et al. 1997) and around $n_W = 8$ for POPE (Marinov and Dufourc 1996), respectively. As we saw above in the experiment with osmotic pressure, we can expect that the dehydration reduces the number of defects in the lamellar structure. The difference between DMPC and POPE remains qualitatively the same (compare Fig. 6.6a, b), which suggests that it originates from particular lipid-water interaction properties and not from the difference in the quality of sample preparation.

Further spectra of POPE multibilayers before and after sequential dehydration by water evaporation are presented in Fig. 6.7. Deswelling of POPE multibilayers is nicely discerned (Fig. 6.7a). Next, the OH stretching band shape was compared for different n_W. The situation for partially hydrated POPE samples (Fig. 6.7b) is markedly different from what is seen in DMPC (Fig. 6.4b). Here even when only interfacial water molecules remain in the system ($n_W = 5$, Fig. 6.7b), the shape of the OH stretching band and the peak position do not significantly differ from the sample in excess water ($n_W = 23$, Fig. 6.7b). This observation is especially trustful for frequencies above 3400 cm^{-1}. At lower frequencies, between roughly 3400 cm^{-1} and 3000 cm^{-1}, the POPE absorption due to the stretching vibrations of ethanolamine group (Fringeli and Günthard 1981) shown in Fig. 6.8a, disturbs the comparison. In order to diminish this disturbance, properly weighted spectrum of dry POPE was subtracted from the original spectra (Fig. 6.7b).

As seen in Fig. 6.7b, the OH stretching band for weakly hydrated POPE is not shifted to lower frequencies with respect to pure bulk H_2O, Therefore, the corresponding shift observed for partially hydrated PC membranes, is strongly correlated to the properties of the PC headgroup and not to some intrinsic lipid bilayer effect on the water structure. This specific interaction is also reflected in the fact that it is possible to remove almost all the bound water from PE membranes by

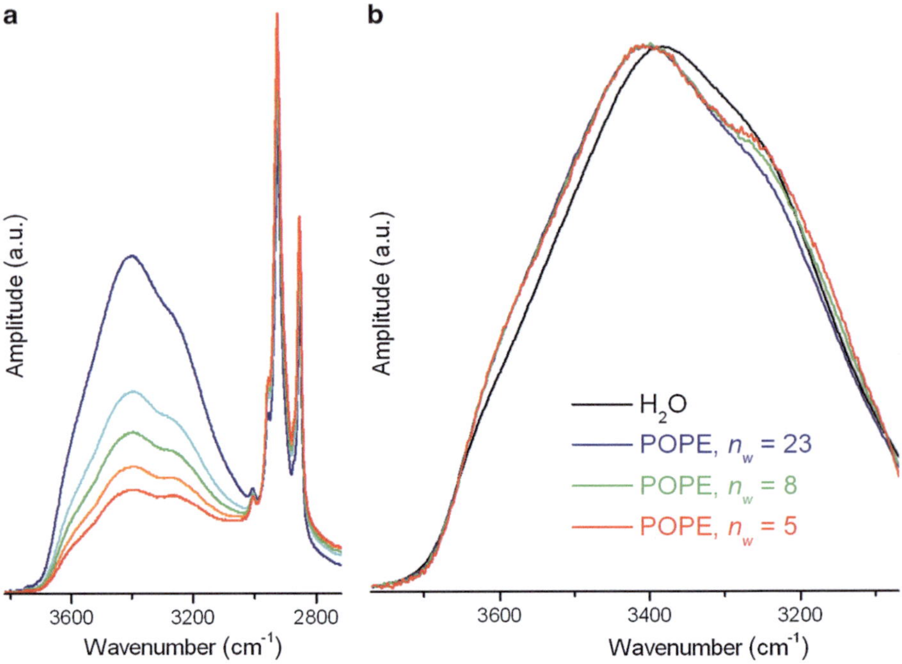

Fig. 6.7 (**a**) Deswelling of POPE multibilayers by water evaporation (original spectra). (**b**) Comparison of the shape of the normalized OH stretching bands for pure bulk water (H_2O) and for interlamellar water in POPE multibilayers for different n_W. Properly weighted spectrum of dry POPE was subtracted from the original spectra to reduce the contribution of lipid absorption in the inspected spectral range (for details see text). Spectra were recorded at 57 °C

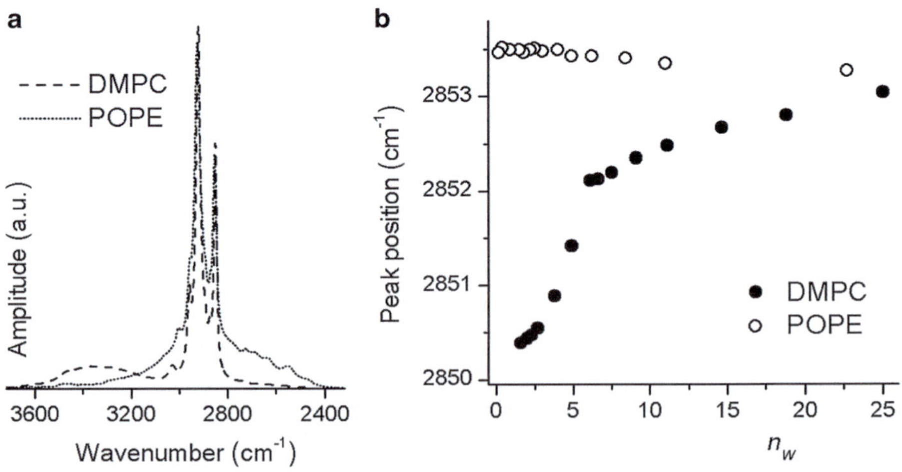

Fig. 6.8 (**a**) Comparison of spectra for DMPC and POPE multibilayers gently dried by evaporation and recorded at 57 °C. (**b**) Dependence of the symmetric methylene (CH_2) stretching band peak position on the molar water-to-lipid ratio n_W

drying the stack of lipid bilayers under gentle conditions (Pohle and Selle 1996). In contrast, some interfacial water, presumably tightly bound to phosphate groups and represented by a broad absorption band from approximately 3700 cm^{-1} to 3100 cm^{-1}, remains trapped in the case of PC (Fig. 6.8a).

Moreover, dependence of lipid phases on the water content in the system, i.e. lyotropic properties, changes as well. As can be seen from Fig. 6.8b, the lyotropic main phase transition of lipids can be detected from dependence of symmetric methylene (CH_2) stretching band peak position on hydration level of DMPC as expected from the literature (Markova et al. 2000; Binder 2003). This transition is absent for POPE, in accordance with previously published results for PE lipids (Selle et al. 1999; Binder 2003).

There are other evidences for the difference between properties of water amongst layers of PC and PE membranes. For example, significant variations exist in the portions of water molecules belonging to different classes of interlamellar water as determined by calorimetric approaches (Kodama et al. 1997, 2001). Furthermore, an FTIR study revealed that the intermolecular H-bonding interaction between amine and phosphate groups in POPE considerably modifies the hydration properties of PE lipids (Bouchet et al. 2009). In addition, computational studies showed that the ethanolamine group of PE affects water structure and motion differently than the equivalent choline group of PC (Damodaran and Merz 1993; Murzyn et al. 2006).

From the discussion above it follows that the lipid composition can affect interlamellar water structure not only in the interfacial water population but also farther away from the membrane. Therefore, the long-range lipid-water interaction could be modulated by the lipid composition.

6.4.2 *Influence of Lipid Phase*

Lipids in excess water can exist in different phases depending on the temperature. For example, at low temperatures phosphatidylcholine membranes are found in the gel or solid-ordered phase (S phase), just below the main phase transition temperature in the ripple phase, and above this temperature in the liquid-crystalline or liquid-disordered phase (Ld phase) (Janiak et al. 1979). High concentrations of added cholesterol (Chol) can lead to the liquid-ordered phase (Lo phase) (Ipsen et al. 1987). Similar phase behavior is observed also for phosphatidylethanolamines, but additionally the so-called inverted hexagonal phase (H_{II} phase) appears at higher temperatures (Mantsch 1984; Paré and Lafleur 1998).

IR spectroscopy has long been a valuable tool for studying phase transitions as well as the properties of different lipid phases. In this respect, it is especially worthwhile to follow the lipid part of an IR spectrum, i.e. the temperature dependence of the peak position of the methylene (CH_2) stretching bands or of the carbonyl ($C=O$) stretching band (Casal and Mantsch 1984; Mantsch and McElhaney 1991; Lewis and McElhaney 2007; Arsov and Quaroni 2007). Because of the interaction between lipid and water molecules, it would be interesting to examine whether lipid thermotropic characteristics are reflected also in the water part of IR spectrum.

Recently, there have been several attempts to check the influence of different lipid phases on the OH stretching band shape in samples prepared in excess water. It was reported that the difference between the S phase and the Ld phase can be appreciated for DMPC by correlating the temperature behavior of the carbonyl band with the evolution of the OH stretching band shape (Disalvo and Frias 2013). This finding is supported by MD simulations, which have shown that local hydration of carbonyl groups differ between the S and the Ld phase (Stepniewski et al. 2010).

We have shown that the difference in the water part of IR spectrum, although relatively small, can be appreciated also when we compare the S phase and the Lo phase for DMPC and DMPC samples containing 40 mol% of cholesterol (DMPC/Chol 0.4), respectively (Štrancar and Arsov 2008). We analyzed for possible changes also samples above the temperature of the main phase transition, but we observed no significant differences between samples in the Ld and the Lo phase. We know from structural measurements that the interlamellar water layer thickness decreases and bilayer thickness increases with higher concentration of cholesterol (Hodzic et al. 2008; Gallová et al. 2011). This could have two possible implications. The straightforward one is that the amplitude of the OH stretching band should decrease in the samples with cholesterol. Unfortunately, in our excess water experiments this amplitude depends significantly on the quality of the sample preparation, i.e. on the number and size of defects in multibilayers. It therefore seems that adding cholesterol enhances formation of defects. The second effect could be an alteration in the shape of the OH stretching band. Since an influence of the Lo phase was not detected, as noted above, it is possible that the influence is reflected mainly in the population of interfacial water. Consequently, due to a relatively small contribution of interfacial water to the overall water signal, the expected difference between DMPC and DMPC/Chol multibilayers might be obscured.

Similar to the expectation for the Ld and the Lo phase, an abrupt change in the structural parameters at the main phase transition between the S and the Ld phase is anticipated to affect the amplitude of the OH stretching band. In this case, contrary to the Ld/Lo comparison, the change in the amplitude is nicely discerned around the temperature of the main S-to-Ld phase transition in DMPC (Fig. 6.9a). Even though the spectra are taken at temperatures that are apart for 3 °C, this difference is large enough not to be a consequence of the intrinsic pure bulk water OH stretching band temperature dependence (Fig. 6.2). However, the observed decrease in the amplitude is opposite to what we would expect from structural data, which show that the amount of interlamellar water increases from the S phase (Tristram-Nagle et al. 2002) to the Ld phase (Kucerka et al. 2005). Possibly, the effect of additional swelling in the Ld phase competes with reduction of the number of defects due to the presence of a more adaptable liquid lipid phase.

The altered amplitude of the OH stretching band can be appreciated also at the Ld-to-H_{II} phase transition for POPE/Chol 0.2 multibilayers in excess water (Fig. 6.9b). Here the change is much more dramatic. The temperature of the amplitude jump coincides with the expected temperature for the Ld-to-H_{II} phase

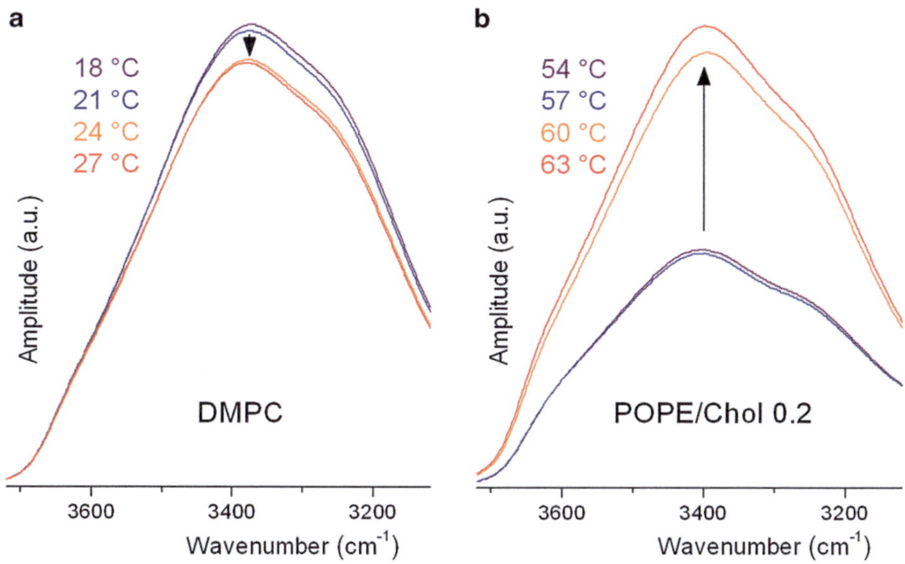

Fig. 6.9 Evolution of the shape of the OH stretching band for interlamellar water around the phase transition temperature in (**a**) DMPC and (**b**) POPE/Chol 0.2 multibilayers prepared in excess water. The temperature at which particular spectrum was recorded is indicated. In the case of DMPC, the S-to-Ld phase transition is discerned from the sudden change in the OH stretching band amplitude as denoted by the *arrow*. Similarly, for POPE/Chol 0.2 the Ld-to-H_{II} phase transition is detected

transition of around 55 °C (Paré and Lafleur 1998). The amplitude rise also agrees with the expected increase of n_W from the Ld phase to the H_{II} phase (Rand and Fuller 1994; Marinov and Dufourc 1996). However, the change is much larger than expected from this increase. It seems that in this case the phase transition triggers the formation of defects in lipid samples.

We have additionally confirmed that this really represents the Ld-to-H_{II} phase transition by measuring the temperature dependence of the peak position of the methylene symmetric stretching band for POPE and POPE/Chol 0.2 samples (Fig. 6.10a). Our measurements nicely reproduce previously published results (Paré and Lafleur 1998). The arrow in Fig. 6.10a also demonstrates that the temperature at which the jump in the peak position of the methylene symmetric stretching band for POPE/Chol 0.2 sample occurs matches the temperature at which the jump in the OH stretching band amplitude is seen (Fig. 6.9b).

To finally check the effect of the phase transition on the structure of interlamellar water, we also compared the OH stretching band shape for POPE/Chol 0.2 sample just below and above the Ld-to-H_{II} phase transition (Fig. 6.10b). As we saw in Fig. 6.9b, the lamellar structure strongly swells on account of defects that arise at the Ld-to-H_{II} phase transition. However, from the discussion in Sect. 6.4.1 and from the results presented in Fig. 6.6, where we showed that the spectral changes between DMPC and POPE did not arise from a relatively larger amount of interlamellar

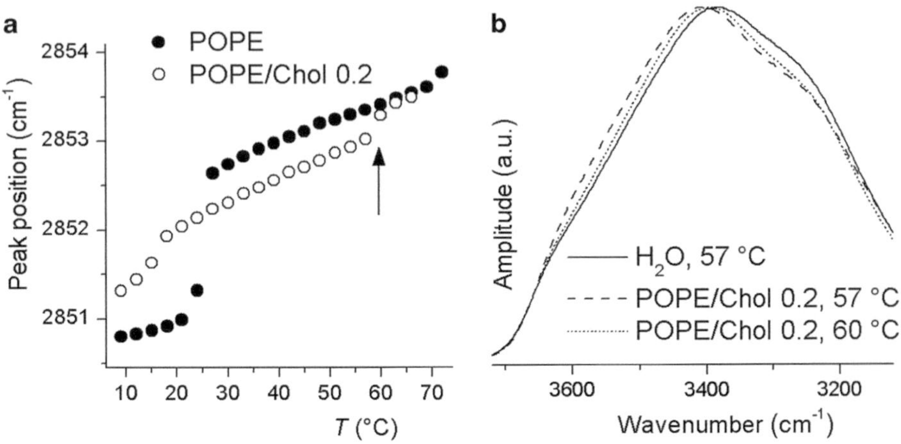

Fig. 6.10 (**a**) Temperature dependence of the symmetric methylene (CH_2) stretching band peak position for POPE and POPE/Chol 0.2. The main phase transition for POPE at around 25 °C and the Ld-to-H_{II} phase transition for POPE/Chol 0.2 (indicated by *arrow*) can nicely be recognized. (**b**) Comparison of the shape of the OH stretching band for interlamellar water in POPE/Chol 0.2 multibilayers prepared in excess water for temperature below and above the Ld-to-H_{II} phase transition. The temperature at which each particular spectrum was recorded is indicated

water due to the defects for DMPC, it seems reasonable to conclude that the OH stretching band shape in the case of POPE/Chol 0.2 changes because of the lipid phase transition and not because of the reduced quality of the sample. Therefore, in this case the influence of the lipid phase is reflected also in the interlamellar water structure.

6.5 The Long-Range Lipid-Water Interaction

Experimental evidence presented in the previous sections points to the presence of a long-range lipid-water interaction in lipid multibilayers. As emphasized above, by "long-range" it is meant that a large part or even all interlamellar water, beside interfacial water, can be classified as perturbed. Hence, contrary to the more general view, there seem to be more than one or two hydration water layers perturbed. Taking into account the typical dimension of a water molecule of around 0.3 nm, and the typical interlamellar water layer thickness in PC multibilayers of around 1.8 nm, there could be roughly three hydration layers on each side of a PC lipid bilayer expanding about 0.9 nm from the bilayer surface. Therefore, each water property in this range of distance from bilayer surface that is different from bulk water indicates a long-range effect.

Results of other experimental methods also offer support of the long-range lipid-water interaction, as discussed further in this section. This finding could influence the understanding of water-mediated processes in confined regions between membranes. Nevertheless, it is still questionable whether this long-range interaction is strong enough to influence any physical or chemical process and how this strength is modified by different lipid composition or lipid phase. However, a small perturbation of many molecules can add up to significant forces, strong enough to deform the force balance. For example, moving two bilayers together by the size of a water molecule amounts to a displacement of a large number of water molecules due to relatively large surface area of the bilayers, requiring significant work input (Parsegian and Zemb 2011).

6.5.1 The Long-Range Interaction as Observed by Different Methods

It was observed quite early by MD simulations that the water density profile along the normal to the membrane changes much farther away than for one water hydration layer (Pandit et al. 2003; Berkowitz et al. 2006; Debnath et al. 2010). In another MD study it was noticed that some properties do not change beyond the commonly established excess water point, but water diffusion coefficient still increases upon further hydration (Pinnick et al. 2010). Similarly, it was shown that diffusion rates settle to their bulk water values only as far as 1.0 nm away from the lipid surface (Hansen et al. 2012), what can be regarded as a long-range effect based on the criteria mentioned in the introduction to this section. Another example revealed that orientational relaxation of hydration water is still slower than in the bulk water, although the hydration level was above the full hydration level of the studied bilayer (Zhang and Berkowitz 2009).

As can be seen from the discussion above, both structural and dynamical properties of water can be influenced far away, i.e. around 1.0 nm, from the water-membrane interface. Therefore, we can conclude that when two opposing bilayers come close, almost none of the molecules will have bulk-like water properties.

Not only computational but also experimental techniques confirm the notion of the long-range lipid-water interaction. Here structural and dynamical properties can be characterized. It has to be noted, with regard to the experimental emphasis of this chapter, that ATR-FTIR does not offer information on the dynamics of water molecules. It has been shown that substantial differences in dynamics can exist between samples, even when IR absorption spectra are similar. By augmenting conventional spectra with time-resolved studies, e.g. IR pump-probe spectroscopy, ambiguous interpretations of the relative structure and dynamical behavior of water in different environments can be avoided (Piletic et al. 2006).

Work with frequency modulation atomic force microscopy (AFM) revealed an oscillatory hydration force with distance from the lipid surface, which suggested that water is organized in at least two or more structured hydration layers (Higgins et al. 2006; Fukuma et al. 2007). Interestingly, the force profile is influenced by the lipid phase (Higgins et al. 2006).

Fluorescence time-resolved measurements showed that three hydration water layers, including the interfacial water, are perturbed in a confined environment. These findings were based on measuring relaxation dynamics profile of confined water in aqueous nanochannels of the lipidic cubic phase mapped out with femtosecond resolution across the nanochannel (Kim et al. 2006).

(* N. of E. for details in fluorescent methodologies and interfacial water see Chap. 5)

The hydration state of phospholipid bilayers can also be precisely studied by terahertz spectroscopy. By this method, water perturbed by a lipid membrane is detected from the observation of the relaxation dynamics of water molecules in the subpicosecond time scale. Combined with X-ray observation of the lamellar structure of the lipids, a long-range hydration effect on up to four to five layers of water was confirmed (Hishida and Tanaka 2011).

Although no unique picture exists, there are many different indications about the perturbation of water properties beyond the so-called interfacial water or the first hydration water layer. This supports the view that in confined membrane environments water structure can be different than in bulk. This might have an important impact on water-mediated biological processes. (* N. of E. Further discussion on this topic can be found in Chap. 7)

6.5.2 Possible Implications of the Long-Range Interaction in Confined Environments

Coupling phenomena between membranes, such as adhesion, stacking, and fusion, could be influenced by water-mediated effects. Experimental results also demonstrate that hydration layers are stable enough to present energy barriers to approaching nanoscale objects, such as proteins and solvated ions, and are therefore expected to affect membrane permeability and transport (Fukuma et al. 2007). Therefore, an important open question is how exactly the water structure is affected by a nanometer-scale confinement (Levinger 2002), because water confined to narrow spaces may no longer resemble that in the bulk (Guo et al. 2005). This holds true for lipid model systems as well as for cellular environments. For example, organelle-specific water structure was resolved with Raman microspectroscopy (Tiwari et al. 2013). Similarly to membrane-based processes, water is expected to have an essential role also in adapting protein conformational properties in confined environments. In the following, some examples of implication of long-range lipid-water interaction will be presented.

First example is the phase-governed stacking of lipid domains in phase-separated multibilayers. It was speculated that differences in water networks for the Lo and Ld phases could produce distinct lipid-water interfaces for the coexisting lipid domains. The mismatch of the water network introduces an energetically unfavorable penalty at the interface where the two networks join. Lipid domain alignment across the lipid layers reduces such interfaces and thus lowers the overall free energy of the system, suggesting a plausible mechanism for the observed interlamellar domain ordering (Tayebi et al. 2012). The mentioned assumptions are supported by our as well as several other independent observations of lipid phase influence on hydration water properties. Ultrahigh resolution frequency modulation AFM imaging suggested possible different hydration structures around particular membrane raft-like domains that could present significant energy barriers to interacting biomolecules (Sheikh and Jarvis 2011). Although raft hydration layers would not exclude other membrane-associated proteins or external biomolecules from interacting with raft-associated proteins, they could modulate the spatial location and kinetics of these interactions.

Another example of a possible water structure-mediated interaction was brought by a computational study of vesicle fusion. In this survey, results show not only that the dynamics of water between two membranes is altered, but also that the conformational state of this water can control the fusion reaction between the two membranes. In this case, water helps the vesicles stick to each other by slowing lipid rearrangements at the interface that are necessary for fusion (Kasson et al. 2011).

Analogous to lipid bilayers, aqueous solvent and hydrophobic interaction play also a major role for the polypeptide chain folding in globular proteins and for the conformational stability and flexibility of proteins. Specifically for proteins, the dynamics of water–protein interactions govern various activities, including the facilitation of protein folding, maintenance of structural integrity, mediation of molecular recognition, and acceleration of enzymatic catalysis (Zhong et al. 2011). The coupling between dynamics of water structure and protein dynamics is considered to play an important role in protein folding. The experimental data obtained by terahertz spectroscopy suggests a long-range influence on the correlated motion of water molecules in the protein hydration water network (Ebbinghaus et al. 2007). It was also shown by MD simulation that confining both protein and solvent gives rise to a solvent-mediated effect that destabilizes the native structure of the protein. Thus, it was demonstrated that the confinement of solvent has a significant impact on protein kinetics and thermodynamics (Lucent et al. 2007). MD simulations were also used to explore the effects of unfolding on the dynamical behavior of water present in the hydration layers of different segments of the protein (Chakraborty and Bandyopadhyay 2008).

6.6 Concluding Remarks

The application of ATR-FTIR to study the properties of interlamellar water in lipid multibilayers, prepared in excess water, enabled us to show the presence of the long-range lipid-water interaction. In addition, there are many indications that lipid composition and phase properties modify this interaction. Consequently, it is likely that biological processes taking place in confined environments between membranes are influenced by this long-range interaction at appreciable distances, even before the membranes are in direct contact. This can have important implications for understanding the role of water in membrane-mediated processes. Further studies are needed to precisely elucidate the unique significance of structural and dynamical properties of water for different processes. The interest in further research in this area is rapidly growing and will in the future undoubtedly fit water into the wider picture of its biological relevance.

Acknowledgments The financial support from the state budget by the Slovenian Research Agency (program No. P1-0060) is acknowledged. The author appreciates collaboration or helpful discussions with Luca Quaroni, Michael Rappolt, Joze Grdadolnik, Primoz Ziherl and Rudolf Podgornik. The author also thanks Iztok Urbancic for carefully reading this manuscript.

References

Arnold K, Pratsch L, Gawrisch K (1983) Effect of poly(ethylene glycol) on phospholipid hydration and polarity of the external phase. Biochim Biophys Acta 728:121–128

Arrondo J, Goni F, Macarulla J (1984) Infrared spectroscopy of phosphatidylcholines in aqueous suspension a study of the phosphate group vibrations. Biochim Biophys Acta 794:165–168

Arsov Z, Quaroni L (2007) Direct interaction between cholesterol and phosphatidylcholines in hydrated membranes revealed by ATR-FTIR spectroscopy. Chem Phys Lipids 150:35–48

Arsov Z, Rappolt M, Grdadolnik J (2009) Weakened hydrogen bonds in water confined between lipid bilayers: the existence of a long-range attractive hydration force. Chemphyschem 10:1438–1441

Ball P (2008a) Water as a biomolecule. Chemphyschem 9:2677–2685

Ball P (2008b) Water as an active constituent in cell biology. Chem Rev 108:74–108

Beranová L, Humpolíčková J, Sýkora J et al (2012) Effect of heavy water on phospholipid membranes: experimental confirmation of molecular dynamics simulations. Phys Chem Chem Phys 14:14516–14522

Berger C, Desbat B, Kellay H et al (2003) Water confinement effects in black soap films. Langmuir 19:1–5

Bergmann U, Nordlund D, Wernet P et al (2007) Isotope effects in liquid water probed by x-ray Raman spectroscopy. Phys Rev B 76:024202

Berkowitz ML, Vácha R (2012) Aqueous solutions at the interface with phospholipid bilayers. Acc Chem Res 45:74–82

Berkowitz ML, Bostick DL, Pandit S (2006) Aqueous solutions next to phospholipid membrane surfaces: insights from simulations. Chem Rev 106:1527–1539

Bhide SY, Berkowitz ML (2005) Structure and dynamics of water at the interface with phospholipid bilayers. J Chem Phys 123:224702

Binder H (2003) The molecular architecture of lipid membranes—new insights from hydration-tuning infrared linear dichroism spectroscopy. Appl Spectrosc Rev 38:15–69

Binder H (2007) Water near lipid membranes as seen by infrared spectroscopy. Eur Biophys J 36:265–279

Binder H, Arnold K, Ulrich AS, Zschörnig O (2001) Interaction of Zn2+ with phospholipid membranes. Biophys Chem 90:57–74

Boissière C, Brubach JB, Mermet A et al (2002) Water confined in lamellar structures of AOT surfactants: an infrared investigation. J Phys Chem B 106:1032–1035

Bonn M, Bakker HJ, Tong Y, Backus EHG (2012) No ice-like water at aqueous biological interfaces. Biointerphases 7:20

Bouchet AM, Frías MA, Lairion F et al (2009) Structural and dynamical surface properties of phosphatidylethanolamine containing membranes. Biochim Biophys Acta 1788:918–925
Brubach J-B, Mermet A, Filabozzi A et al (2005) Signatures of the hydrogen bonding in the infrared bands of water. J Chem Phys 122:184509
Casal H, Mantsch H (1984) Polymorphic phase behaviour of phospholipid membranes studied by infrared spectroscopy. Biochim Biophys Acta Rev Biomembr 779:381–401
Cavaille D, Combes D, Zwick A (1996) Effect of high hydrostatic pressure and additives on the dynamics of water: a Raman spectroscopy study. J Raman Spectrosc 27:853–857
Chakraborty S, Bandyopadhyay S (2008) Dynamics of water in the hydration layer of a partially unfolded structure of the protein HP-36. J Phys Chem B 112:6500–6507
Chaplin M (2006) Do we underestimate the importance of water in cell biology? Nat Rev Mol Cell Biol 7:861–866
Chen X, Hua W, Huang Z, Allen HC (2010) Interfacial water structure associated with phospholipid membranes studied by phase-sensitive vibrational sum frequency generation spectroscopy. J Am Chem Soc 132:11336–11342
Cheng J-X, Pautot S, Weitz DA, Xie XS (2003) Ordering of water molecules between phospholipid bilayers visualized by coherent anti-Stokes Raman scattering microscopy. Proc Natl Acad Sci U S A 100:9826–9830
Costigan SC, Booth PJ, Templer RH (2000) Estimations of lipid bilayer geometry in fluid lamellar phases. Biochim Biophys Acta 1468:41–54
Cringus D, Lindner J, Milder MTW et al (2005) Femtosecond water dynamics in reverse-micellar nanodroplets. Chem Phys Lett 408:162–168
Damodaran K, Merz K (1993) Head group-water interactions in lipid bilayers: a comparison between DMPC-and DLPE-based lipid bilayers. Langmuir 9:1179–1183
Debnath A, Mukherjee B, Ayappa KG et al (2010) Entropy and dynamics of water in hydration layers of a bilayer. J Chem Phys 133:174704
Disalvo EA, Frias MA (2013) Water state and carbonyl distribution populations in confined regions of lipid bilayers observed by FTIR spectroscopy. Langmuir 29:6969–6974
Disalvo EA, Lairion F, Martini F et al (2008) Structural and functional properties of hydration and confined water in membrane interfaces. Biochim Biophys Acta 1778:2655–2670
Disalvo EA, Bouchet AM, Frias MA (2013) Connected and isolated CH2 populations in acyl chains and its relation to pockets of confined water in lipid membranes as observed by FTIR spectrometry. Biochim Biophys Acta 1828:1683–1689
Ebbinghaus S, Kim SJ, Heyden M et al (2007) An extended dynamical hydration shell around proteins. Proc Natl Acad Sci U S A 104:20749–20752
Faure C, Bonakdar L, Dufourc EJ (1997) Determination of DMPC hydration in the $L\alpha$ and $L\beta'$ phases by ^2H solid state NMR of D2O. FEBS Lett 405:263–266
Finer E, Darke A (1974) Phospholipid hydration studied by deuteron magnetic resonance spectroscopy. Chem Phys Lipids 12:1–16
Fitter J, Lechner RE, Dencher NA (1999) Interactions of hydration water and biological membranes studied by neutron scattering. J Phys Chem B 103:8036–8050
Fringeli UP, Günthard HH (1981) Infrared membrane spectroscopy. In: Grell E (ed) Membrane spectroscopy. Springer, Berlin, pp 270–332
Fukuma T, Higgins MJ, Jarvis SP (2007) Direct imaging of individual intrinsic hydration layers on lipid bilayers at Angstrom resolution. Biophys J 92:3603–3609
Gallová J, Uhríková D, Kučerka N et al (2011) The effects of cholesterol and β-sitosterol on the structure of saturated diacylphosphatidylcholine bilayers. Eur Biophys J 40:153–163
Gauger DR, Andrushchenko VV, Bour P, Pohle W (2010) A spectroscopic method to estimate the binding potency of amphiphile assemblies. Anal Bioanal Chem 398:1109–1123
Goormaghtigh E, Raussens V, Ruysschaert JM (1999) Attenuated total reflection infrared spectroscopy of proteins and lipids in biological membranes. Biochim Biophys Acta 1422:105–185
Grdadolnik J, Kidrič J, Hadži D (1991) Hydration of phosphatidylcholine reverse micelles and multilayers—an infrared spectroscopic study. Chem Phys Lipids 59:57–68

Grdadolnik J, Kidrič J, Hadži D (1994) An FT-IR study of water hydrating dipalmitoylphosphatidylcholine multibilayers and reversed micelles. J Mol Struct 322:93–102
Guo Y, Yui H, Minamikawa H et al (2005) FT-IR study of the interlamellar water confined in glycolipid nanotube walls. Langmuir 21:4610–4614
Hansen FY, Peters GH, Taub H, Miskowiec A (2012) Diffusion of water and selected atoms in DMPC lipid bilayer membranes. J Chem Phys 137:204910
Harrick NJ (1987) Internal reflection spectroscopy, 3rd edn. Harrick Scientific Corporation, New York
Higgins MJ, Polcik M, Fukuma T et al (2006) Structured water layers adjacent to biological membranes. Biophys J 91:2532–2542
Hishida M, Tanaka K (2011) Long-range hydration effect of lipid membrane studied by terahertz time-domain spectroscopy. Phys Rev Lett 106:158102
Hodzic A, Rappolt M, Amenitsch H et al (2008) Differential modulation of membrane structure and fluctuations by plant sterols and cholesterol. Biophys J 94:3935–3944
Hübner W, Blume A (1998) Interactions at the lipid–water interface. Chem Phys Lipids 96:99–123
Ipsen JH, Karlström G, Mourtisen OG et al (1987) Phase equilibria in the phosphatidylcholine-cholesterol system. Biochim Biophys Acta Biomembr 905:162–172
Janiak M, Small D, Shipley G (1979) Temperature and compositional dependence of the structure of hydrated dimyristoyl lecithin. J Biol Chem 254:6068–6078
Jendrasiak G (1996) The hydration of phospholipids and its biological significance. J Nutr Biochem 7:599–609
Kasson PM, Lindahl E, Pande VS (2011) Water ordering at membrane interfaces controls fusion dynamics. J Am Chem Soc 133:3812–3815
Katsaras J (1997) Highly aligned lipid membrane systems in the physiologically relevant "excess water" condition. Biophys J 73:2924–2929
Kim J, Lu W, Qiu W et al (2006) Ultrafast hydration dynamics in the lipidic cubic phase: discrete water structures in nanochannels. J Phys Chem B 110:21994–22000
Kiselev M, Lesieur P, Kisselev A et al (1999) DMSO-induced dehydration of DPPC membranes studied by X-ray diffraction, small-angle neutron scattering, and calorimetry. J Alloys Compd 286:195–202
Kodama M, Aoki H, Takahashi H, Hatta I (1997) Interlamellar waters in dimyristoylphosphatidylethanolamine-water system as studied by calorimetry and X-ray diffraction. Biochim Biophys Acta 1329:61–73
Kodama M, Kato H, Aoki H (2001) Comparison of differently bound molecules in the gel and subgel phases of a phospholipid bilayer system. J Therm Anal Calorim 64:219–230
Kodama M, Kawasaki Y, Aoki H, Furukawa Y (2004) Components and fractions for differently bound water molecules of dipalmitoylphosphatidylcholine-water system as studied by DSC and ^{2}H-NMR spectroscopy. Biochim Biophys Acta 1667:56–66
Kucerka N, Liu Y, Chu N et al (2005) Structure of fully hydrated fluid phase DMPC and DLPC lipid bilayers using X-ray scattering from oriented multilamellar arrays and from unilamellar vesicles. Biophys J 88:2626–2637
Kuntz ID, Kauzmann W (1974) Hydration of proteins and polypeptides. Adv Protein Chem 28:239–345
Lafleur M, Pigeon M, Pezolet M, Caille J-P (1989) Raman spectrum of interstitial water in biological systems. J Phys Chem 93:1522–1526
Lefèvre T, Toscani S, Picquart M, Dugué J (2002) Crystallization of water in multilamellar vesicles. Eur Biophys J 31:126–135
Levinger N (2002) Water in confinement. Science 298:1722–1723
Lewis RNAH, McElhaney RN (2007) Fourier transform infrared spectroscopy in the study of lipid phase transitions in model and biological membranes: practical considerations. Methods Mol Biol 400:207–226
Libnau FO, Toft J, Christy AA, Kvalheim OM (1994) Structure of liquid water determined from infrared temperature profiling and evolutionary curve resolution. J Am Chem Soc 116:8311–8316

Lucent D, Vishal V, Pande VS (2007) Protein folding under confinement: a role for solvent. Proc Natl Acad Sci U S A 104:10430–10434

Mallikarjunaiah KJ, Leftin A, Kinnun JJ et al (2011) Solid-state ^{2}H NMR shows equivalence of dehydration and osmotic pressures in lipid membrane deformation. Biophys J 100:98–107

Manciu M, Ruckenstein E (2007) On possible microscopic origins of the swelling of neutral lipid bilayers induced by simple salts. J Colloid Interface Sci 309:56–67

Mantsch H (1984) Biological applications of Fourier transform infrared spectroscopy: a study of phase transitions in biomembranes. J Mol Struct 113:201–212

Mantsch HH, McElhaney RN (1991) Phospholipid phase transitions in model and biological membranes as studied by infrared spectroscopy. Chem Phys Lipids 57:213–226

Maréchal Y (1991) Infrared spectra of water. I. Effect of temperature and of H/D isotopic dilution. J Chem Phys 95:5565–5573

Marinov R, Dufourc E (1996) Thermotropism and hydration properties of POPE and POPE-cholesterol systems as revealed by solid state ^{2}H and 31P-NMR. Eur Biophys J 24:423–431

Markova N, Sparr E, Wadsö L, Wennerström H (2000) A calorimetric study of phospholipid hydration. Simultaneous monitoring of enthalpy and free energy. J Phys Chem B 104:8053–8060

McIntosh TJ, Simon SA (1986) Hydration force and bilayer deformation: a reevaluation. Biochemistry 25:4058–4066

McIntosh T, Simon S (1994) Hydration and steric pressures between phospholipid bilayers. Annu Rev Biophys Biomol Struct 23:27–51

McIntosh TJ, Simon SA (1996) Adhesion between phosphatidylethanolamine bilayers. Langmuir 12:1622–1630

Mennicke U, Salditt T (2002) Preparation of solid-supported lipid bilayers by spin-coating. Langmuir 18:8172–8177

Milhaud J (2004) New insights into water-phospholipid model membrane interactions. Biochim Biophys Acta 1663:19–51

Miller IR, Bach D (1999) Hydration of phosphatidyl serine multilayers and its modulation by conformational change induced by correlated electrostatic interaction. Bioelectrochem Bioenerg 48:361–367

Mills M, Orr BG, Banaszak Holl MM, Andricioaei I (2013) Attractive hydration forces in DNA-dendrimer interactions on the nanometer scale. J Phys Chem B 117:973–981

Murzyn K, Zhao W, Karttunen M et al (2006) Dynamics of water at membrane surfaces: effect of headgroup structure. Biointerphases 1:98–105

Nagle J, Katsaras J (1999) Absence of a vestigial vapor pressure paradox. Phys Rev E 59:7018–7024

Nagle JF, Tristram-Nagle S (2000) Structure of lipid bilayers. Biochim Biophys Acta Rev Biomembr 1469:159–195

Némethy G, Scheraga HA (1964) Structure of water and hydrophobic bonding in proteins. IV. The thermodynamic properties of liquid deuterium oxide. J Chem Phys 41:680–689

Okamura E, Umemura J, Takenaka T (1990) Orientation studies of hydrated dipalmitoylphosphatidylcholine multibilayers by polarized FTIR-ATR spectroscopy. Biochim Biophys Acta 1025:94–98

Omta AW, Kropman MF, Woutersen S, Bakker HJ (2003) Influence of ions on the hydrogen-bond structure in liquid water. J Chem Phys 119:12457–12461

Pandit SA, Bostick D, Berkowitz ML (2003) An algorithm to describe molecular scale rugged surfaces and its application to the study of a water/lipid bilayer interface. J Chem Phys 119:2199–2205

Parasassi T, De Stasio G, Ravagnan G et al (1991) Quantitation of lipid phases in phospholipid vesicles by the generalized polarization of Laurdan fluorescence. Biophys J 60:179–189

Paré C, Lafleur M (1998) Polymorphism of POPE/cholesterol system: a ^{2}H nuclear magnetic resonance and infrared spectroscopic investigation. Biophys J 74:899–909

Parsegian VA, Zemb T (2011) Hydration forces: observations, explanations, expectations, questions. Curr Opin Colloid Interface Sci 16:618–624

Parsegian V, Rand R, Fuller N, Rau D (1986) Osmotic stress for the direct measurement of intermolecular forces. Methods Enzymol 127:400–416

Pasenkiewicz-Gierula M, Takaoka Y, Miyagawa H et al (1997) Hydrogen bonding of water to phosphatidylcholine in the membrane as studied by a molecular dynamics simulation: location, geometry, and lipid-lipid bridging via hydrogen-bonded water. J Phys Chem A 101:3677–3691

Petrache H, Gouliaev N, Tristram-Nagle S et al (1998a) Interbilayer interactions from high-resolution x-ray scattering. Phys Rev E 57:7014–7024

Petrache HI, Tristram-Nagle S, Nagle JF (1998b) Fluid phase structure of EPC and DMPC bilayers. Chem Phys Lipids 95:83–94

Pfeiffer H, Klose G, Heremans K (2013) Reorientation of hydration water during the thermotropic main phase transition of 1-palmitoyl-2-oleolyl-sn-glycero-3-phosphocholine (POPC) bilayers at low degrees of hydration. Chem Phys Lett 572:120–124

Piletic IR, Moilanen DE, Levinger NE, Fayer MD (2006) What nonlinear-IR experiments can tell you about water that the IR spectrum cannot. J Am Chem Soc 128:10366–10367

Pinnick ER, Erramilli S, Wang F (2010) Computational investigation of lipid hydration water of L α 1-palmitoyl-2-oleoyl- sn -glycero-3-phosphocholine at three hydration levels. Mol Phys 108:2027–2036

Pohle W, Selle C (1996) Fourier-transform infrared spectroscopic evidence for a novel lyotropic phase transition occurring in dioleoylphosphatidylethanolamine. Chem Phys Lipids 82:191–198

Pohle W, Selle C, Fritzsche H, Bohl M (1997) Comparative FTIR spectroscopic study upon the hydration of lecithins and cephalins. J Mol Struct 408–409:273–277

Pohle W, Selle C, Fritzsche H, Binder H (1998) Fourier transform infrared spectroscopy as a probe for the study of the hydration of lipid self assemblies. I. Methodology and general phenomena. Biospectroscopy 4:267–280

Pohle W, Selle C, Gauger DR, Brandenburg K (2001) Lyotropic phase transitions in phospholipids as evidenced by small-angle synchrotron X-ray scattering. J Biomol Struct Dyn 19:351–364

Rand RP, Fuller NL (1994) Structural dimensions and their changes in a reentrant hexagonal-lamellar transition of phospholipids. Biophys J 66:2127–2138

Rand RP, Parsegian VA (1989) Hydration forces between phospholipid bilayers. Biochim Biophys Acta Rev Biomembr 988:351–376

Rand RP, Fuller N, Parsegian VA, Rau DC (1988) Variation in hydration forces between neutral phospholipid bilayers: evidence for hydration attraction. Biochemistry 27:7711–7722

Rappolt M, Hickel A, Bringezu F, Lohner K (2003) Mechanism of the lamellar/inverse hexagonal phase transition examined by high resolution x-ray diffraction. Biophys J 84:3111–3122

Riemenschneider J, Holzmann J, Ludwig R (2008) Salt effects on the structure of water probed by attenuated total reflection infrared spectroscopy and molecular dynamics simulations. Chemphyschem 9:2731–2736

Róg T, Murzyn K, Milhaud J et al (2009) Water isotope effect on the phosphatidylcholine bilayer properties: a molecular dynamics simulation study. J Phys Chem B 113:2378–2387

Selle C, Pohle W (1998) Fourier transform infrared spectroscopy as a probe for the study of the hydration of lipid self-assemblies. II. Water binding versus phase transitions. Biospectroscopy 4:281–294

Selle C, Pohle W, Fritzsche H (1999) FTIR spectroscopic features of lyotropically induced phase transitions in phospholipid model membranes. J Mol Struct 480–481:401–405

Sheikh KH, Jarvis SP (2011) Crystalline hydration structure at the membrane-fluid interface of model lipid rafts indicates a highly reactive boundary region. J Am Chem Soc 133:18296–18303

Skinner JL, Auer BM, Lin YS (2009) Vibrational line shapes, spectral diffusion, and hydrogen bonding in liquid water. Adv Chem Phys 142:59–103

Sokołowska A, Kęcki Z (1986) Inter- and intra-molecular coupling and Fermi resonance in the Raman spectra of liquid water. J Raman Spectrosc 17:29–33

Sovago M, Campen RK, Wurpel GWH et al (2008) Vibrational response of hydrogen-bonded interfacial water is dominated by intramolecular coupling. Phys Rev Lett 100:173901

Stepniewski M, Bunker A, Pasenkiewicz-Gierula M et al (2010) Effects of the lipid bilayer phase state on the water membrane interface. J Phys Chem B 114:11784–11792
Štrancar J, Arsov Z (2008) Application of spin-labeling EPR and ATR-FTIR spectroscopies to the study of membrane heterogeneity. In: Leitmannova Liu A (ed) Advances in planar lipid bilayers and liposomes, vol 6. Elsevier, Amsterdam, pp 16–139
Tamm LK, Han X (2000) Viral fusion peptides: a tool set to disrupt and connect biological membranes. Biosci Rep 20:501–518
Tamm LK, Tatulian SA (1997) Infrared spectroscopy of proteins and peptides in lipid bilayers. Q Rev Biophys 30:365–429
Tayebi L, Ma Y, Vashaee D et al (2012) Long-range interlayer alignment of intralayer domains in stacked lipid bilayers. Nat Mater 11:1074–1080
Ter-Minassian-Saraga L, Okamura E, Umemura J, Takenaka T (1988) Fourier transform infrared-attenuated total reflection spectroscopy of hydration of dimyristoylphosphatidyl-choline multi-bilayers. Biochim Biophys Acta 946:417–423
Tielrooij KJ, Paparo D, Piatkowski L et al (2009) Dielectric relaxation dynamics of water in model membranes probed by terahertz spectroscopy. Biophys J 97:2484–2492
Tiwari S, Ando M, Hamaguchi H (2013) Investigation of organelle-specific intracellular water structures with Raman microspectroscopy. J Raman Spectrosc 44:167–169
Tristram-Nagle S, Liu Y, Legleiter J, Nagle JF (2002) Structure of gel phase DMPC determined by X-ray diffraction. Biophys J 83:3324–3335
Volkov VV, Palmer DJ, Righini R (2007a) Heterogeneity of water at the phospholipid membrane interface. J Phys Chem B 111:1377–1383
Volkov VV, Palmer DJ, Righini R (2007b) Distinct water species confined at the interface of a phospholipid membrane. Phys Rev Lett 99:078302
Wennerström H, Sparr E (2003) Thermodynamics of membrane lipid hydration. Pure Appl Chem 75:905–912
Wurpel GWH, Müller M (2006) Water confined by lipid bilayers: a multiplex CARS study. Chem Phys Lett 425:336–341
Zhang Z, Berkowitz ML (2009) Orientational dynamics of water in phospholipid bilayers with different hydration levels. J Phys Chem B 113:7676–7680
Zhong D, Pal SK, Zewail AH (2011) Biological water: a critique. Chem Phys Lett 503:1–11

Chapter 7
Hydration and Nanoconfined Water: Insights from Computer Simulations

Laureano M. Alarcón, J.A. Rodríguez Fris, Marcela A. Morini, M. Belén Sierra, S.A. Accordino, J.M. Montes de Oca, Viviana I. Pedroni, and Gustavo A. Appignanesi

Abstract The comprehension of the structure and behavior of water at interfaces and under nanoconfinement represents an issue of major concern in several central research areas like hydration, reaction dynamics and biology. From one side, water is known to play a dominant role in the structuring, the dynamics and the functionality of biological molecules, governing main processes like protein folding, protein binding and biological function. In turn, the same principles that rule biological organization at the molecular level are also operative for materials science processes that take place within a water environment, being responsible for the self-assembly of molecular structures to create synthetic supramolecular nanometrically-sized materials. Thus, the understanding of the principles of water hydration, including the development of a theory of hydrophobicity at the nanoscale, is imperative both from a fundamental and an applied standpoint. In this work we present some molecular dynamics studies of the structure and dynamics of water at different interfaces or confinement conditions, ranging from simple model hydrophobic interfaces with different geometrical constraints (in order to single out curvature effects), to self-assembled monolayers, proteins and phospholipid membranes. The tendency of the water molecules to sacrifice the lowest hydrogen bond (HB) coordination as possible at extended interfaces is revealed. This fact makes the first hydration layers to be highly oriented, in some situations even resembling the structure of hexagonal ice. A similar trend to maximize the number of HBs is shown to hold in cavity filling, with small subnanometric hydrophobic cavities remaining empty while larger cavities display an alternation of filled and dry states with a significant inner HB network. We also study interfaces with complex chemical and geometrical nature in order to determine how different conditions affect the local hydration properties. Thus, we show some results for protein hydration and, particularly, some preliminary

L.M. Alarcón • J.A. Rodríguez Fris • M.A. Morini • M.B. Sierra • S.A. Accordino • J.M. Montes de Oca • V.I. Pedroni • G.A. Appignanesi (✉)
Departamento de Química and INQUISUR-UNS-CONICET, Universidad Nacional del Sur, Av. Alem 1253, 8000 Bahía Blanca, Argentina
e-mail: lalarcon@uns.edu.ar; mamorini@criba.edu.ar; mbsierra@uns.edu.ar; pedroni@criba.edu.ar; appignan@criba.edu.ar

E.A. Disalvo (ed.), *Membrane Hydration*, Subcellular Biochemistry 71,
DOI 10.1007/978-3-319-19060-0_7

studies on membrane hydration. Finally, calculations of a local hydrophobicity measure of relevance for binding and self-assembly are also presented. We then conclude with a few words of further emphasis on the relevance of this kind of knowledge to biology and to the design of new materials by highlighting the context-dependent and non-additive nature of different non-covalent interactions in an aqueous nanoenvironment, an issue that is usually greatly overlooked.

Keywords Hydration water • Confined water • Geometry • Computer simulation • H-bonds • Self assembled monolayers • Hydrophobicity • Density fluctuations

7.1 Introduction

The behavior of water at interfaces and under nanoconfinement is quite different from that at bulk conditions, as many experimental and theoretical studies have demonstrated (Huang and Chandler 2000; Huang et al. 2003; Bizzarri and Cannistraro 2002; Vitkup et al. 2000; Choudhury and Montgomery Pettitt 2005; Stanley et al. 2007; Giovambattista et al. 2008; Rasaiah et al. 2008; Berne et al. 2009; Malaspina et al. 2010; Alarcón et al. 2011; Gelman Constantin et al. 2011; Accordino et al. 2012a, c; Schulz et al. 2011). However, the picture of (nano) confined water is still far from being complete (Giovambattista et al. 2012; Rasaiah et al. 2008; Berne et al. 2009; Schulz et al. 2011; Alarcón et al. 2014). This issue is not only crucial from an intrinsic, fundamental level, but also from its far reaching implications (Fernández 2010; Qvist et al. 2008; Berne et al. 2009; Young et al. 2007; Wang et al. 2011; Kulp III et al. 2011; Accordino et al. 2011a, 2012a, b, c, 2013; Schulz et al. 2011; Sierra et al. 2013; Alarcón et al. 2014; Bogan and Thorn 1998; Li and Liu 2009). For example, being water the matrix of life, a full understanding of its behavior in the nano and mesoscales would be essential to understand biology at the molecular level. In biological organization processes, water acts as a mediator between complex surfaces which tend to associate by means of interactions of non-covalent type (Giovambattista et al. 2012), thus generating nanoconfined environments for which our knowledge of the behavior of bulk water might be of little value. Such new settings claim for the development of a novel intuition. One such example is the nanoconfinement produced upon association of hydrophobic surfaces, which affects the thermodynamic properties of the hydration water in order to produce the "drying" process promoter of hydrophobic collapse. Another example is the binding of a ligand to a protein, which is expected to displace easily removable hydration water (Fernández and Scheraga 2003; Fernández and Scott 2003; Fernández 2010; Qvist et al. 2008; Berne et al. 2009; Young et al. 2007; Wang et al. 2011; Kulp III et al. 2011; Accordino et al. 2012a, b, c, 2013; Sierra et al. 2013; Alarcón et al. 2014; Bogan and Thorn 1998; Li and Liu 2009). Thus, the knowledge of how different nanoconfinement conditions (both of chemical and geometrical nature) affect hydrophobicity is therefore not only of great significance to understand and to predict the behavior of these systems, but also to guide the efforts to emulate them in bioengineering. Notwithstanding this, a complete theory of hydrophobicity at the nanoscale level is still lacking (Giovambattista et al. 2012).

In turn, the principles that govern biological organization at the molecular level are the same as the ones acting in the self-assembly of molecular structures to create synthetic supramolecular nanometrically-sized materials (Giovambattista et al. 2012). The solvent therefore plays a key role in the design of new materials (Giovambattista et al. 2012; Faul and Antonietti 2003; Rehm and Schmuck 2010). Water is the natural solvent in bioengineering and drug design, but is also becoming increasingly important in supramolecular chemistry. But despite the growing importance of green chemistry, it is interesting to note that chemists (whose synthesis is predominantly organic and non-aqueous) are finding it difficult to imitate the kind of processes that Nature performs all the time (Faul and Antonietti 2003; Rehm and Schmuck 2010) (obviously, Nature has much to teach us in this regard). Also, chemical research has tried to emulate certain rules of Biology in generating nanoscale structures by working with building blocks to attain materials of greater complexity and functionality, thus distributing a complex structural and functional problem at different levels of integration (Faul and Antonietti 2003; Rehm and Schmuck 2010) (coding three-dimensional structure in chemical sequence). However, for this to be effective we should fully understand the principles that govern the interactions, generally of non-covalent nature, between such modular units operating within nanoconfined environments in which the solvent is obviously discrete and not a continuous medium. Additionally, the way in which non-covalent interactions depend on their local environment is not fully understood (Fernández and Scheraga 2003; Fernández and Scott 2003; Fernández 2010; Accordino et al. 2012a, b, c, 2013; Sierra et al. 2013). Very often concepts derived in bulk conditions or *in vacuo*, not necessarily valid under nanoconfinement conditions, are applied. Non-covalent interactions of use in supramolecular chemistry (Faul and Antonietti 2003; Rehm and Schmuck 2010), such as hydrogen bonds and ionic interactions are extremely weak and even irrelevant in bulk water given solvation and screening. However, such interactions might be strong in nanoconfined environments, where the local "effective dielectric" is diminished and coulombic interactions are strengthened (Fernández and Scheraga 2003; Fernández and Scott 2003; Fernández 2010; Accordino et al. 2012a, b, c, 2013; Sierra et al. 2013). The glaring weakness of certain non-covalent interactions in water has led to use the so-called "Gulliver principle": to add interactions of equal or different nature, as if a giant were to be tied through many tiny ropes. However, for this strategy not to still remain conceptually "Lilliputian" it should be considered that the effects of these interactions are not necessarily simply additive, while the way they depend on the local environment should be understood. This same ignorance is also observed in other contexts such as bioengineering and drug design, so the scarcity of rational design elements and the overwhelming dominance of "trial and error" attempts (expensive and suboptimal) are not surprising. Thus, it is not enough to add-up interactions but to combine them in a suitable topology. For example, the role of the local hydrophobic content around hydrogen bonds in protein binding and drug design has been turned into a design concept (Fernández and Scheraga 2003; Fernández and Scott 2003; Fernández 2010; Accordino et al. 2012a, b, c, 2013; Sierra et al. 2013).

The above-expounded scenario is obviously extremely vast. In this chapter we do not mean to provide a comprehensive all-embracing treatment, but to focus on certain issues arbitrarily chosen by our particular interest. Thus, while we shall mainly review some recent computational work of our group, we will refer it to a generic context of wide-ranging interest. A useful introduction to this field is the investigation of the hydration properties of model systems, like simple hydrophobic-like surfaces. Such studies where geometry plays a main role can also be taken as simple first steps towards more complex situations where chemistry is also at play, like the situation encountered for the so-called biological water, that is, the hydration water around proteins and other biological molecules. Extensive hydrophobic-like surfaces disrupt the hydrogen bond network of water. In the absence of significant attractive interactions with the surface, the water molecules close to a hydrophobic surface would lack a hydrogen bond in such direction, thus diminishing their local connectivity from preferentially four to preferentially three first neighbors. In the present work we shall study by means of molecular dynamics (MD) simulations the structure and also the orientation of the water molecules at graphene surfaces, single walled carbon nanotubes of different radii and fullerenes. We shall also study the hydration of model alkane-like self-assembled monolayers (SAMs) and the filling/drying propensity of cavities and tunnels carved in the SAMs. Additionally, we shall study the hydration layers of the protein lysozyme and we shall also present some preliminary studies on hydration and water penetration in phospholipid membranes. Thus, we shall investigate both the role of curvature and chemistry on the properties of hydration water. Improved structuring as compared to the bulk will be made evident at surfaces together with clear preferential orientational ordering. The study of the orientation of the water molecules is interesting since it can be directly contrasted with experimental measurements (surface sum-frequency vibrational spectroscopy (Shen and Ostroverkhov 2006), when available. Our studies will also show the possibility of geometrically-based subnanoscale dewetting. Finally, by means of the study of water density fluctuations, we shall estimate the local hydrophobicity properties both for model surfaces and for the different groups of the lipid bilayer. These studies might be relevant for the comprehension of the behavior of binding sites in different contexts which, as already indicated, are expected to present easily removable hydration water which might be replaced by the ligand upon binding (Fernández and Scheraga 2003; Fernández and Scott 2003; Fernández 2010; Qvist et al. 2008; Berne et al. 2009; Young et al. 2007; Wang et al. 2011; Kulp III et al. 2011; Accordino et al. 2012a, b, c, 2013; Sierra et al. 2013; Alarcón et al. 2014; Bogan and Thorn 1998; Li and Liu 2009). We thus conclude with a few words of further emphasis on the relevance of this kind of knowledge to biology and to the design of new materials by highlighting the context-dependent and non-additive nature of different con-covalent interactions in an aqueous nanoenvironment, an issue that is usually greatly overlooked.

7.2 Hydration and Geometry

The hydration properties of a solute are not only defined by the chemical nature of the solute but also by its geometry. For example, a (very) small nonpolar solute can be clathrated by water while larger nonpolar solutes aggregate by means of the hydrophobic effect. Water molecules in the clathrate basically retain the fourfold hydrogen bond (HB) coordination typical of bulk water while extended nonpolar surfaces disrupt the HB network and thus the water molecules at the surface are engaged in less than four HBs. In turn, it has been shown that subnanometric nonpolar cavities (spherical pores) remain empty, whereas water penetrates nanometric size ones (Rasaiah et al. 2008; Schulz et al. 2011). This behavior is due to the reluctance of the water molecules to resign hydrogen bond coordination with other water molecules (only nanometric size cavities allow penetration retaining the coordination typical at surfaces). For example, the interiors of the spherical C_{180} and C_{140} fullerenes (the size of the last one being barely larger than 10 Å) have been shown to present stability for filled states with small clusters of water molecules connected by strong hydrogen bonds (Rasaiah et al. 2008 and references therein). Also, the filling and the conduction of water in carbon nanotubes and related systems have been extensively studied (Rasaiah et al. 2008 and references therein) and many different water phases have been discovered (from 1D trains of hydrogen-bonded water molecules at low nanotube radius to complex layered structures within larger nanotubes). It has also been shown that the hydrophobicity of the material is important since a small reduction of the van der Waals attraction between water and the carbon atoms induces the drying of previously filled nanotubes (Rasaiah et al. 2008). In addition, the behavior of water confined in cylindrical pores or tunnels is also important both from the basic and the applied viewpoints. For instance, this behavior is relevant for water flow in aquaporins and proton flow in proton pumps and enzymes (Rasaiah et al. 2008). Additionally, while it is generally believed that small protein cavities are empty (Qvist et al. 2008), there is no general consensus on whether large protein cavities are filled or empty and certain experimental results on the subject are contradictory (even when certain large cavities seem to indeed present small clusters of confined water molecules) (Rasaiah et al. 2008; Qvist et al. 2008). Thus, since in complex realistic systems both geometry and chemistry affect its hydration properties, it becomes convenient to first separate them by focusing in simple systems to study the role of geometry to then consider more realistic contexts like, for example, proteins and phospholipid membranes.

7.2.1 *Water at Graphitic-Like Surfaces: Graphene, Carbon Nanotubes and Fullerenes. The Role of Curvature and Water Orientational Ordering*

In this section we review our computational work on model hydrophobic surfaces: a graphene sheet, single-walled carbon nanotubes of different radii and C_{60} and

C_{20} fullerenes (Malaspina et al. 2010; Alarcón et al. 2011; Accordino et al. 2011a, 2012a, c; Gelman Constantin et al. 2011; Schulz et al. 2011). Given the chemical similarity between the three kinds of systems under study, the hydration of the graphene sheet should constitute the limit of infinite curvature radius for both the nanotubes and the fullerenes.

The water molecules were modeled by the TIP3P. The equilibration was carried out with a canonical NVE ensemble with a Berendsen thermostat. All simulations were done using the AMBER10 molecular simulation suite with a 1 fs time step (we used the GAFF and FF99SB force fields for carbon and water, respectively). All calculations were performed in the NpT ensemble with a Langevin thermostat.

The model graphene surface was a perfect honeycomb graphite-like sheet, consisting of a layer of 5 by 5 benzenic rings (approximately 100 Å^2) with terminations in hydrogen atoms solvated with 4,365 TIP3P water molecules in an orthogonal cubic box with periodic boundaries. The surface was centered in the middle of the box and parallel to the XY plane. To discard the presence of significant finite size effects we also employed a second layer of 10 by 10 benzenic rings (approximately 400 Å^2), solvated with 10,788 TIP3P water molecules.

In order to avoid molecules close to borders of the graphene sheet, we considered a circular tube of radius 4.8 Å (8.4 Å for the larger system) placed at the center of the graphene layer and normal to such plane. We then divided the tube in six consecutive regions or cylinders (from region or cylinder 1 to region or cylinder 6) of height 3.5 Å so that the first cylinder extended from the surface to a distance of 3.5 Å, the second one went from 3.5 to 7 Å, and so on. This distance implies that the first region contained water molecules exclusively from the first peak of the water density distribution function normal to the surface. Since the system is equivalent in the two directions normal to the graphene plane, we gathered statistical data from both sides of the sheet (that is, we also constructed a tube normal to the graphene sheet but extending to the other direction and divided it into six cylinders). Thus, from now on when we speak of a given region or cylinder number it means that we are gathering data from the water molecules of the corresponding cylinders at the two sides of the graphene layer.

In turn, the single-walled carbon nanotubes employed had radii 2.0, 3.9, 4.75, 8.15, 12.25, 20.4 and 67.5 Å. Each of the nanotubes was solvated in a box of TIP3P water molecules that extended more than 20 Å away from the nanotube borders in each direction. The axis of the nanotube coincided with the x axis and the length of the nanotubes was 14 Å in all cases and thus, this dimension was always extensive. We studied the water molecules located within six concentric annular regions of 3.5 Å thickness so that the first region contained the water molecules external to the nanotube whose distance to the nanotube surface within the (y, z) plane at its corresponding x value was lower than 3.5 Å, the second region involved the molecules whose distance to the surface lied between 3.5 and 7 Å, and so on. In other words, we divided the region external to the nanotube into six concentric tubes (the first tube radius was 3.5 Å larger than that of the nanotube, the second one's radius was 7 Å larger than the nanotube, and so on) and the region defined by any given tube excluded the molecules within all the previous ones. In all cases, the regions studied were more than 2 Å apart from the nanotube x-endings in order to avoid border effects.

Finally, the fullerenes studied were C_{60} and C_{20} and were centered at the center of coordinates. Both of them were also solvated in a box of TIP3P water molecules that extended more than 20 Å away from the fullerene surface. We also studied water molecules within six regions. The spherically concentric regions studied were such that the region 1 contained the water molecules external to the fullerene whose distance to its surface was less than 3.5 Å, the second one involved water molecules whose distance to the fullerene lied between 3.5 and 7 Å, and so on.

In all cases we carried simulations at a temperature of 300 K, mean pressure of 1 bar and average density around 1.0 kg/dm^3. We used a time step of 1 fs and a long range interaction cutoff of 8 Å with particle mesh Ewald.

Equilibration was tested by monitoring the behavior of thermodynamical properties like temperature, pressure and energy oscillations, and by dynamical properties like oscillations in the mean squared displacement of the water molecules; the equilibration times were in each case much larger than the structural relaxation time.

To characterize water structure we used the local structural index I(i, t), proposed by Shiratani and Sasai (see Shiratani and Sasai 1996, 1998; Malaspina et al. 2009, 2010; Alarcón et al. 2011; Accordino et al. 2011a, b; Appignanesi et al. 2009, for details):

For each molecule i one orders the rest of the molecules depending on the radial distance r_j between the oxygen of the molecule i and the oxygen of molecule j: $r_1 < r_2 < r_j < r_{j+1} < \ldots < r_{n(i,t)} < 3.7\ \text{Å} < r_{n(i,t)+1}$ and calculates I(i, t) as:

$$I(i,t) = \frac{1}{n(i,t)} \sum_{j=1}^{n(i,t)} \left[\Delta(j,i,t) - \overline{\Delta}(i,t)\right]^2$$

where the index i identifies a water molecule at a given time t, $\Delta(j, i, t) = r_{j+1} - r_j$, and the bar indicates that the quantity is averaged over all water molecules.

Such index is apt to be used at interfaces, at variance of other structural indices like the (tetrahedral) orientational order parameter, which we have demonstrated to be invalid at surfaces in its original form (Accordino et al. 2011a). The key observation is the existence of certain molecules which show an unoccupied gap between 3.2 and 3.8 Å in their radial-neighbor distribution for certain periods of time. Such low-density molecules are well structured and coordinated in a highly tetrahedral manner with four other water molecules. Occupancy of such gap increases the local density and distorts the tetrahedral order of the central molecule. A high value of the index I(i, t) implies that molecule i at time t has a good tetrahedral local order and low local density (and thus, a low local potential energy since it is able bind to its first four neighbors by geometrically well-shaped hydrogen bonds), while on the contrary, low values of I(i, t) indicate a molecule with defective tetrahedral order and high-local density (and thus, high local potential energy), even allowing for a fifth neighbor within the coordination shell. We have

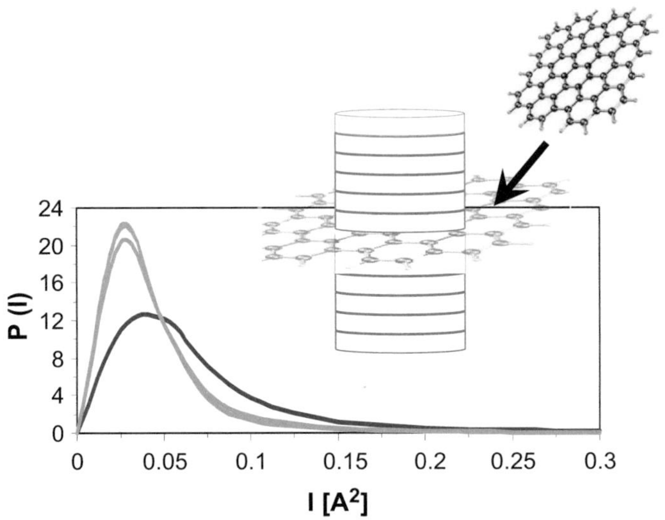

Fig. 7.1 Scheme of the graphene sheet with the different cylinders studied and the simulation box, and the distribution of the local structure index, I, for the different cylinders studied. The *black curve* corresponds to cylinder 1. All the other cylinders display a similar behavior (only cylinder 2 shows a slightly different distribution but still far form the curve for cylinder 1)

shown that when this index is calculated for minimized water configurations (called inherent structures) it is possible to obtain clear bimodal distributions, thus speaking of the existence of low and high local density water molecules (Appignanesi et al. 2009; Malaspina et al. 2009; Accordino et al. 2011b). However, in the present work we shall directly use the real dynamics configurations without minimization.

In Fig. 7.1 we show the results for the graphene sheet. We can clearly see that the distribution of the local structure index, I, for the first cylinder is clearly displaced to the right as compared to the rest of the cylinders (the water molecules in the outermost cylinders in fact show a bulk-like behavior). This fact speaks of the better structuring of the water molecules close to the surface and that this structuring effect is only local and is quickly lost when we move away from the surface.

We obtained similar results for the exterior faces of carbon nanotubes and fullerenes: Only the water molecules close to the surface of these objects showed a better structuring than bulk water. However, we were interested in determining the effect of curvature on the structure of hydration water. To that end, we studied (as indicated above) nanotubes of different radii. We found that the structuring of the first hydration layer indeed depends on the local curvature since the structuring gets poorer as the radius of the nanotube drops. The first cylinder of the smallest nanotube and the first spherical region of the small fullerenes showed less structuring (I index distribution displaced to the left) than the first cylinder of the largest nanotube which displayed a behavior very similar to that of the first cylinder of the graphene sheet. These results are displayed in Fig. 7.2.

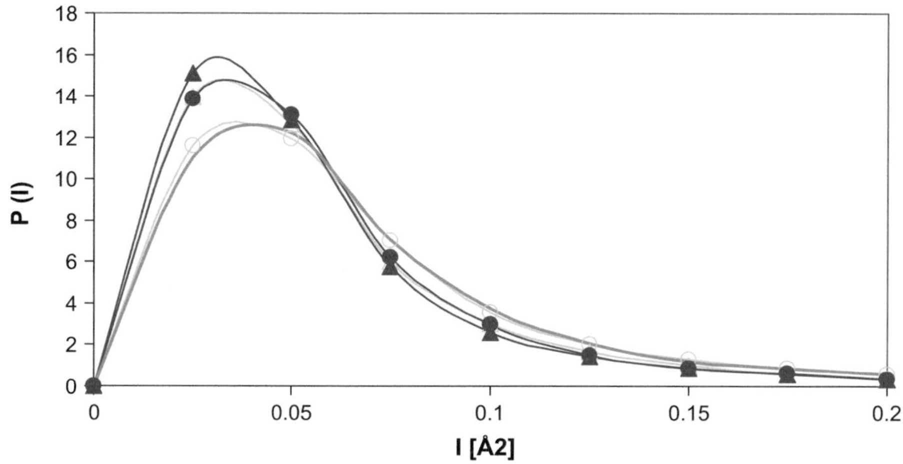

Fig. 7.2 Distribution of the index I for the first hydration layers of the smallest and largest nanotubes, the fullerenes and the graphene sheet. *Black curves* for fullerenes (*triangles* for C_{20} and *circles* for C_{60}), *light gray* curves for carbon nanotubes (*triangles* for small radius nanotubes and *circles* for large ones) and *dark gray curve* for the graphene sheet

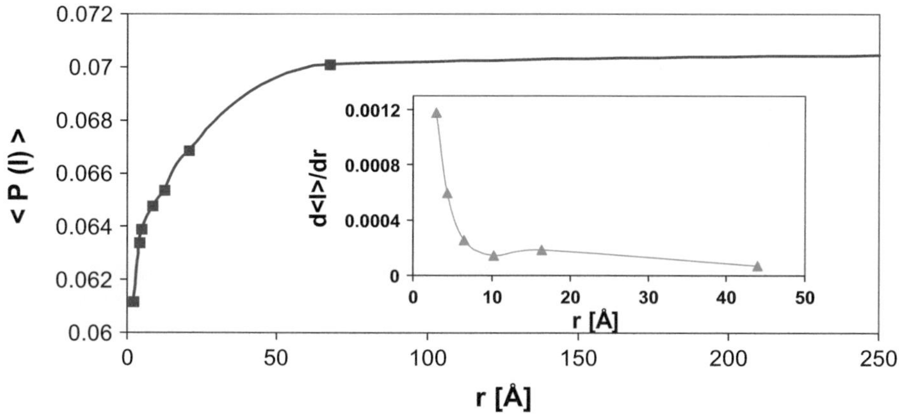

Fig. 7.3 Mean value of the local structure index I for the water molecules in the first cylinder around carbon nanotubes as a function of the nanotube radius. *Inset*: Derivative of the former quantity

In turn, in Fig. 7.3 we show the mean value of the index I for the first hydration layer of the carbon nanotubes studied as a function of nanotube radius. From such figure we can learn on the great loss in water structure around the subnanometric regime. That is, when the diameter of the nanotube gets below the nanometer (10 Å), the structuring of the water molecules of the first hydration layer is quickly lost.

The previous results speak of an increased structural ordering of the water molecules close to graphitic-like surfaces as compared to bulk water. Now we would like to know which kind of order is being developed close to the surfaces. It is expected that the water molecules close to the surface loss a hydrogen bond in the direction towards the surface plane and would be preferentially coordinated with other three water molecules via a hydrogen bond (as compared to the 4-fould coordination typical in the bulk). It is important to notice, however, that the index I is sensitive to the quality of the first coordination sphere irrespective of the number of first neighbors. Then, we studied the distribution of water molecules with three or less hydrogen bonds (HB) and with four HBs, as a function of the distance to the surface. In so doing, we verified that the molecules close to the surface (those which most contribute to cylinder 1) are indeed hydrogen bonded to less than four other water molecules (we used a geometric criterion for HBs: O...O distance lower than 3.5 Å and O-H...O angle larger than 140°). Thus, such molecules are expected to loss a HB in the direction of the surface. We found that the distribution for molecules with three or less HBs display a peak close to 3 Å from the surface. However, the distribution of the molecules with four HBs displays a peak close to 4.5 Å from the surface. Both curves intersect at roughly 3.5 Å. Thus, for convenience, from now on we shall divide the water molecules inside the first peak of the density plot in two layers: layer 1, L_1, with molecules closer than 3.5 Å from the surface and which form preferentially three HBs, and layer 2, L_2, with molecules that extend from 3.5 Å to 4.25 Å from the surface and which tend to participate in 4 HBs.

To get a better idea of the structuring of the water molecules we now study the orientational ordering of molecules in L_1 and L_2. To that end, we calculate the angle θ between the normal to the surface and the OH vector, as schematically indicated in Fig. 7.4. In such figure we can see that molecules in L_1 and L_2 show clear orientational preferences, unlike the situation in bulk layers. In particular, L_1 shows a peak close to 70° while L_2 tends to peak around 0° and 110°. In turn, the bulk layer clearly shows no orientational preference, as expected. Such orientational ordering for the first two layers is typical of ice Ih (hexagonal ice). This speaks of the fact that the molecules close to the surface tend to loss the lower HB coordination as possible. Thus, they orient a vertex of the tetrahedron, in this case a lone pair, towards the surface and keep the other three HBs unperturbed. These molecules of L_1 hydrogen-bond to molecules in L_2 which preferentially form 4 HBs. Such arrangement is compatible with a local ice-like structure that extends for the first hydration layers. In Fig. 7.5 we show a scheme of a perfect hexagonal ice arrangement in contact with a graphene surface, indicating both L_1 and L_2 molecules.

The result of the ordering of the water molecules resembling the structure of hexagonal ice is similar to the situation at the water-vacuum or water-air interface (Shen and Ostroverkhov 2006; Fan et al. 2009) and to the situation in small water clusters (Gelman Constantin et al. 2011). However the orientation is on the other direction of the basal plane. In such cases the superficial water molecules orient an H atom towards the surface. In this case, molecules in the first layer (L_1, that is, the molecules closest to the interface) would tend to display an orientation of $\theta = 180°$

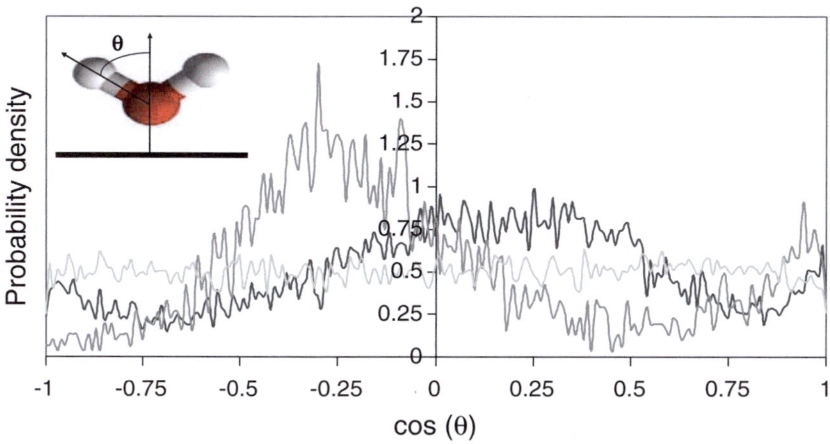

Fig. 7.4 Distribution of the orientational angle, θ, for layer L_1 (*black curve*), L_2 (*dark gray curve*) and a bulk layer (*light gray*) for the water molecules around a graphene sheet. We have lowered temperature to $T = 240$ K but similar results (with less developed peaks) are seen at $T = 300$ K. Results are for TIP3P water, but similar results are obtained with TIP5P and SPC/E

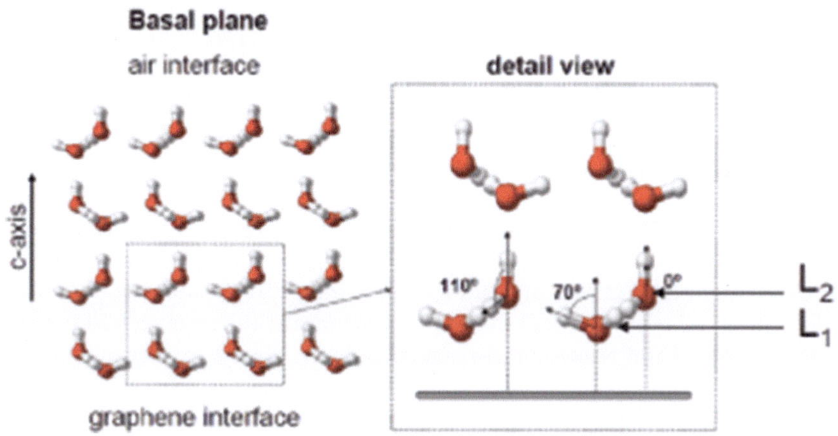

Fig. 7.5 Scheme of a perfect hexagonal ice structure indicating the orientation the molecules tend to adopt (albeit in a more disordered fashion) over a graphene sheet. The graphene sheet would be at the bottom. At the top we place an air interface, at which the water molecules would tend to orient in the other direction along the c axis of the basal plane

and $\theta = 70°$, while L_2 molecules would prefer angles of 0° and 110°. Again this situation implies a minimization of HB loss since only a vertex of the tetrahedron is lost.

Similar results are obtained for carbon nanotubes and fullerenes, with the orientational ordering getting worse as the curvature of the graphitic-like object increases.

7.2.2 Water at Self-Assembled Monolayers. Filling of Cavities and Tunnels

In order to include geometric complexity in a controlled way (as a first step towards more realistic cases like proteins and biological membranes) we study the hydration of self-assembled monolayers (SAMs) and the filling propensity of model hydrophobic pores and tunnels carved in such SAMs (for full details see Schulz et al. 2011; Alarcón et al. 2014). We note that these systems are rigid so as to preserve the desired geometry they model. Later on we shall also present some studies for flexible SAMs where water is able to penetrate and we shall also study flexible phospholipid bilayers. The SAMs studied consisted in monolayers of 81 or 144 chains (in a 9×9 or 12×12 arrangements) of n-heptadecane, CH_3-$(CH_2)_{15}$-CH_3 (in some cases we also used decane chains) aligned in a parallel fashion so as to generate a cube. This arrangement mimics the monolayer structure of stearic acid chains adopted at a water interface but replaces the acid group (COOH) by a H so that the chain ends with a methyl, i.e. it becomes a n-heptadecane chain. The original chain separation was 4.53 Å, the typical distance in a fatty acid monolayer. The monolayer was solvated with water molecules modeled by the TIP3P model with a canonical NVE ensemble with a Berendsen thermostat. All simulations were done using the AMBER10 molecular simulation suite with a 2 fs time step (we used the GAFF and FF99SB force fields). All calculations were performed in the NpT ensemble with a Langevin thermostat at $T = 300$ K. The monolayer was solvated with TIP3P water molecules in an orthogonal cubic box with periodic boundary conditions. The size of the box was such that it extended more than 20 Å away from all the monolayer faces. The surface monolayer was centered in the middle of the box. For the cavities, we carved holes in the monolayer surface parallel to the (x, y) plane in the z direction (the positions of the hydrogens at the top of the monolayer was $z = 42.5$ Å while the corresponding carbon atoms were placed at $z = 41.6$ Å. In all cases (perfect monolayer or with the different carved holes) the AMBER equilibration reduced the distance between the n-heptadecane molecules (the chain separation) to 4.2 Å.

To study the effect of concavity on hydrophobicity, we carved several cavities on the hydrophobic SAMs. This geometrical setting is of particular interest given the situations found in protein pockets and in pores at different materials. To generate the cavities we carved squared holes of different sizes at the center of the monolayer surface by cutting the corresponding number of chains (the bonds of the carbon atoms at the bottom of the hole were saturated so that all chains ended with a methyl group; in other words, they became shorter chain alkanes). Among other cases, we studied holes created when we cut 1, 4, 9, 16, 25, 36, 49 and 64 central chains in order to generate square holes of width 8.5, 12.7, 16.5, 21.3, 25.1, 29.5, 33.8 and 38.0 Å respectively (the width represents the diameters of the hole "mouth", L). Figure 7.6 shows the cases of the holes when we cut 1 chain (width of the hole equal to 8.5 Å) and 25 chains (with a size of 25.1 Å). In all cases, the chains were cut in 5 units, so that the depth of the holes was always of 6.4 Å. We mention

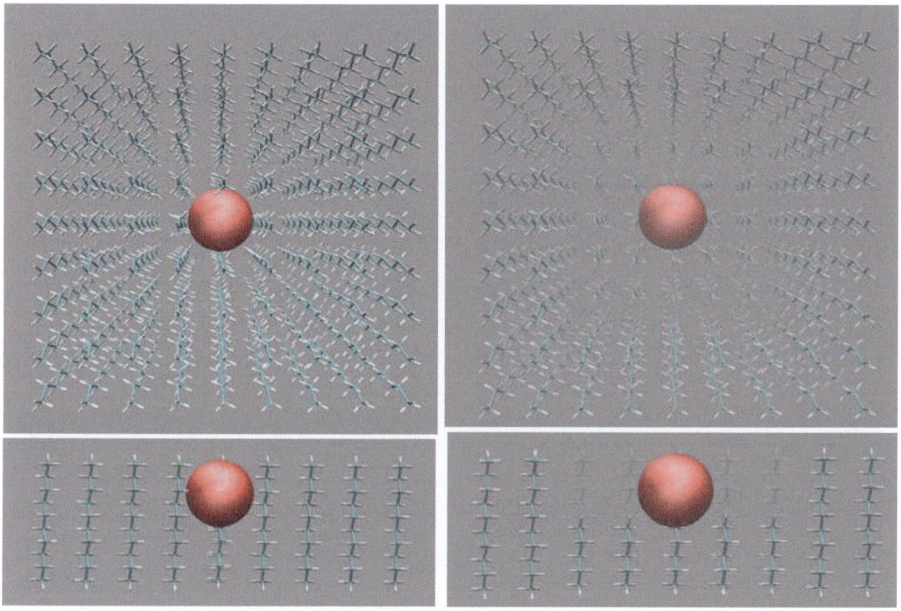

Fig. 7.6 *Left figure*: illustration of the SAM with a hole of width 8.5 Å, where one chain (the central chain) has been shortened in 5 units to create the pore. *Top*: top view of the alkane-like monolayer; *Bottom*: side view of the monolayer, which we have cut at the middle in order to better display the hole. *Right figure*: idem but for the hole created by shortening all the central 25 chains in five units. *Top*: top view. *Bottom*: side (cut) view. The central chains, which were shortened so as to generate the hole, are shown in *light gray*. The *red sphere* indicates the observation volume used

that in order to maintain the shapes of the holes, we restricted positions for all backbone carbon atoms of the chains after equilibration. To study water penetration and hydrophobicity, we calculated the probability distributions for observing N water molecules within spheres located at the "mouth" of the different holes. This means that the observation spheres were located inside the hole, almost tangent to the line defined by the H atoms of the methyls of the alkylic chains framing the holes. Additionally, all spheres were also placed at the center of the corresponding segment that defines the top of each hole. In Fig. 7.6 we indicate the placing of the observation volumes.

We also created tunnels by using a similar procedure and thus we extracted complete chains from the center of the monolayer, so that both the upper and lower faces of the monolayer cube were connected by the tunnel. For comparison, we also simulated single-walled carbon nanotubes with radii 2.0, 3.9, 4.75, 8.15, 12.25, 20.4 and 67.5 Å. Each of the nanotubes was solvated in a box of TIP3P water molecules that extended more than 20 Å away from the nanotube borders in each direction. The axis of the nanotube coincided with the x axis and the length of the nanotubes was 14 Å in all cases.

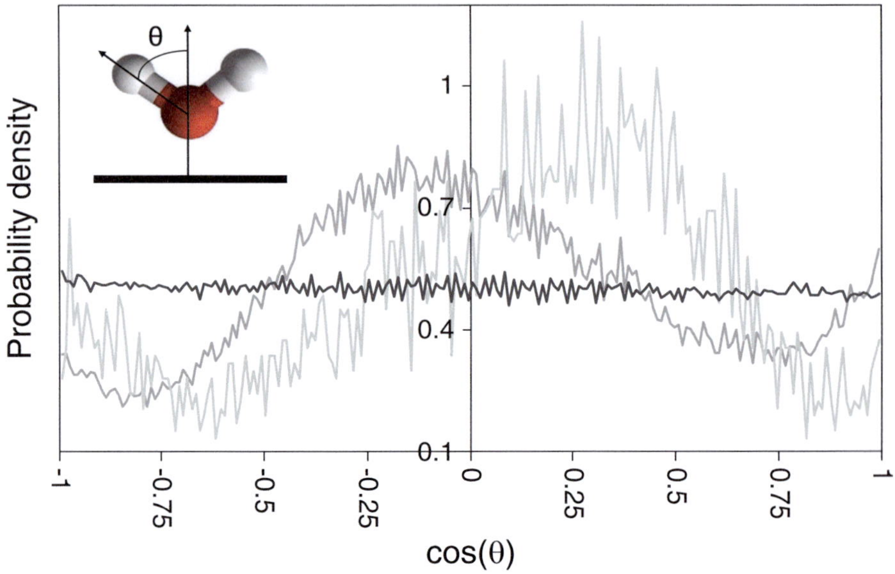

Fig. 7.7 Orientational ordering of water molecules. *Light gray curve* for graphene surface, *dark gray curve* for the alkane-like SAM without hole and *black line* for bulk water

We first studied the orientational ordering of the water molecules at the first peak of the density plot above a perfect SAM (that is, a SAM without a hole). We note that since the SAMs are modeled by using restraints in the heavy atoms, the chains are rigid and water does not penetrate between them. Thus, hydration can be defined at a plane outside the SAM. The results are shown in Fig. 7.7. As in the case of graphene, we can also notice a clear preferential orientation of the water molecules. However, at variance from such case, the water molecules tend to display an angle between 100° and 110°, and thus they tend to orient a face of the tetrahedron parallel to the SAM surface.

In turn, Fig. 7.8 depicts the probability distributions for observing N water molecules within the spherical observation volumes for the different holes. We also include the situation for the perfect monolayer (without hole), where the sphere is located tangent to the surface. From such figure we can learn that the degree of hydrophobicity strongly depends on curvature. The holes are more hydrophobic than the perfect monolayer, but the role of geometry is more conspicuous as the hole size (the diameter of the hole "mouth", L) approaches the subnanometric regime. Such subnanometric-sized holes do not fill with water given the reluctance of the water molecules to lose their hydrogen bond coordination in order to penetrate. Larger holes (up to roughly L = 25 Å), where the water molecules can enter retaining their coordination, are nonetheless more hydrophobic than the perfect monolayer, which means that hydrophobicity is clearly curvature-dependent for concave surfaces. Such result of a geometrically-induced dehydration is interesting, for example, for

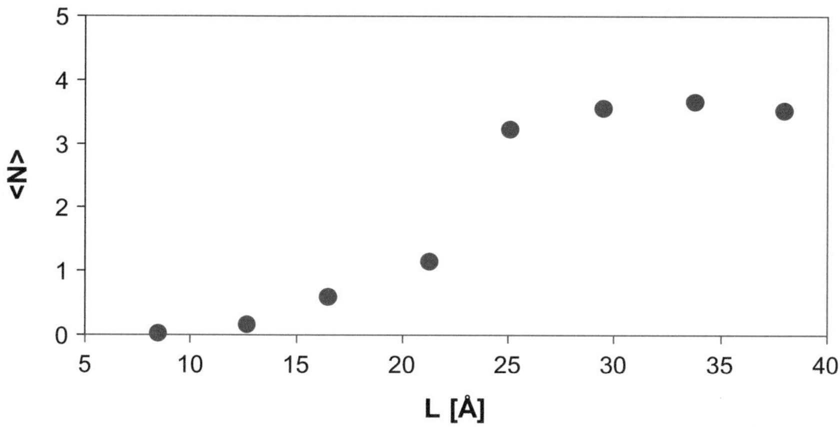

Fig. 7.8 Mean value of the number of water molecules inside the observation volume for the different holes of size L

the context of protein binding, since protein binding pockets are expected to be dry or to contain easily removable water which should be displaced by a ligand upon association (Fernández and Scheraga 2003; Fernández and Scott 2003; Fernández 2010; Qvist et al. 2008; Berne et al. 2009; Young et al. 2007; Wang et al. 2011; Kulp III et al. 2011; Accordino et al. 2012a, b, c, 2013; Sierra et al. 2013; Alarcón et al. 2014; Bogan and Thorn 1998; Li and Liu 2009).

The results of some of these studies also enabled us to demonstrate that the filling of small nanometric cavities (Schulz et al. 2011) is not homogeneous in time but represents a dynamic process with alternation of filled and completely dry states. This is so since the water molecules inside the hydrophobic pore are able to establish some HB interactions but when an interaction is broken it cannot be compensated by other neighboring molecules like in bulk water (where the HB network is continuously rearranging given the many different possible configurations). This fact destabilizes the filling configuration and the pore is desorbed until new water penetration allows a new HB network arrangement inside the cavity. Another interesting result we obtained (Schulz et al. 2011) was that the tunnels carved in the SAMs remain empty until about twice the minimum diameter for a carbon nanotube to get filled (13.5 Å vs. 7.0 Å), thus making evident the more hydrophobic nature of such systems. In fact, the water-wall attractions play a role in the filling of carbon nanotubes, since simulations have shown that the quenching of such interactions induce drying of previously filled nanotubes (Rasaiah et al. 2008). Another feature typical of hydrophobic confinement that we have confirmed in the filling of the tunnels is that the water molecules inside them tend to be engaged in extensive HB networks where the water molecules retain a number of HB partners similar to that at the bulk. Narrower tunnels would demand a restriction in the number of HB partners which prevents them from filling.

7.3 Realistic Contexts. Hydration at Settings with Complex Geometry and Chemistry

Having studied simple systems with different geometries, we now turn to more realistic, complicated systems. A previous step could have been to deal with model systems with simple geometry and controlled chemical topologies, with hydrophilic and hydrophobic regions. This approach has already been taken, as can been seen in the excellent works of the groups of P. G. Debenedetti (Giovambattista et al. 2008, 2012) and S. Garde (Rasaiah et al. 2008) to which we refer the interested reader. Instead, here we shall directly tackle some aspects of the hydration of complex systems of great interest: proteins and membranes.

7.3.1 Protein Hydration Water

In this section we review our work on the behavior of the hydration layers of a protein surface (Accordino et al. 2011a, 2012a, c), a lysozyme molecule, as compared with the behavior at the homogeneous hydrophobic surface of a graphene sheet. In this case the water molecules were modeled by the TIP5P model. We also modeled the water molecules in contact with a graphene sheet by means of the TIP5P model, obtaining very similar results to that of the previous section. Equilibration was carried out in two steps: In a first step we brought the system to the corresponding temperature value with a Langevin thermostat with 0.2 ps timestep and using the SHAKE algorithm; then, we performed a larger equilibration within the NpT ensemble, also using a Langevin thermostat with 2 fs timestep and the SHAKE algorithm. All simulations were done with the AMBER 10 molecular simulation suite and were performed in the NpT ensemble with a Langevin thermostat and by using a long-range-interaction cutoff of 8 Å. We simulated an orthorhombic form of hen egg-white lysozyme (Accordino et al. 2012a, c) placed within a box of 12600 TIP5P water molecules. The force field used in the simulation was ff99SB. We carried simulations at different temperatures (within the normal liquid and supercooled regimes), mean pressure of 1 bar and average density around 1.0 kg/dm^3.

The protein-water radial distribution function (Accordino et al. 2012a, c) presents two well-developed peaks followed by less developed structure. The first small peak corresponds to water molecules very close to the protein which form hydrogen bonds with it. We shall call these molecules as layer 1 molecules, L_1. In order to get a better idea on the arrangement of the molecules of the second peak, we studied the distribution of minimum distances to any atom of the protein of the hydration water molecules of the second peak of the radial distribution function by discriminating between molecules with three and four water-water HBs. The situation is somehow similar to that found for the graphene sheet with two distributions that intersect roughly at a distance of 3 Å. Thus, we again shall distinguish two layers, in this case

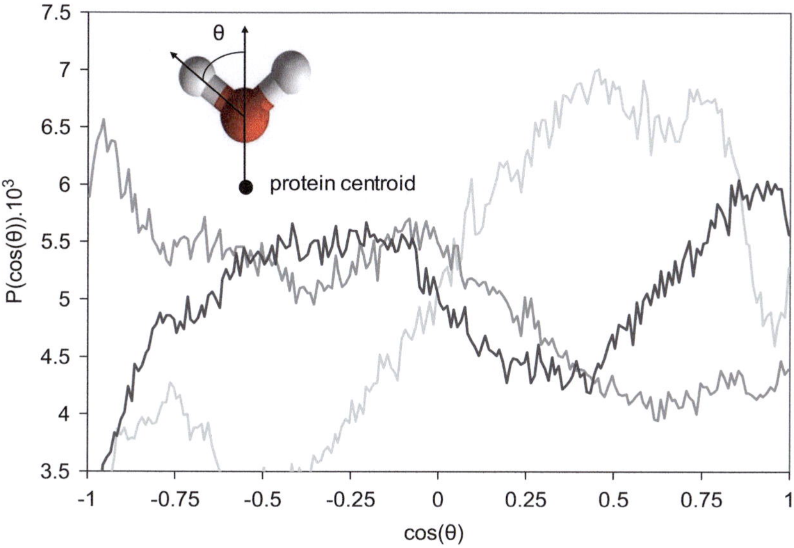

Fig. 7.9 Orientational ordering of the water molecules in the first three layers around the lysozyme molecule. *Light gray*, *dark gray* and *black curves* represent layers 1, 2 and 3 respectively

layer 2, L_2, and layer 3, L_3. In summary, we classify the hydration water molecules in layers: L_1 goes from 0 to 0.2125 Å from the protein, L_2 goes from this value to 3 Å and L_3 includes molecules between 3 and 4.25 Å.

In Fig. 7.9 we show the orientational ordering of the water molecules by calculating the angle θ between the OH and the vector joining the protein centroid and the O atom of the water molecule. While lacking the neat peaks found for the graphene sheet, this figure clearly still exhibits orientational preferences. Even when the situation is more complicated than that at the graphene sheet or the water-air interface, peak positions similar to such both contexts still hold, which again speaks of certain tendency to local ice-like orientation. Hydrophobic portions of the protein are expected to present such local preferences by orienting a vertex of the tetrahedron towards the protein so as to minimize the water-water HB loss. Also when the water molecule hydrogen bonds to the protein it should orient a vertex of the tetrahedron, a situation also consistent with a local orientation similar to an ice surface. Thus, the geometric and chemical complexity does not alter significantly the local orientational preferences found in more simple contexts.

7.3.2 Membrane Hydration

There is mounting evidence (Disalvo et al. 2008, and references therein; see also Arsov 2015; Nickels and Katsaras 2015; Pfeiffer 2015) for the presence of water molecules within phospholipid membranes not only hydrating the polar lipid head groups but also buried within the apolar alkyl chains. This means that (at variance from the simple case of the rigid model SAMs we have previously studied) water penetrates the lipid bilayer towards the region of the carbonyl groups. In turn, the lipid chains are flexible molecules whose dynamics may present different conformational changes. Thus, these systems are expected to evidence a rich behavior for hydration and nanoconfined water.

To learn on the behavior of water at the water-membrane interface we simulated by means of AMBER12 package a Dipalmitoylphosphatidylcholine (DPPC) bilayer. The bilayer was composed of 128 DPPC molecules (64 per monolayer in a 8×8 arrangement). The initial separation between DPPC was 9 Å in a triangular arrangement. The temperature was 323 K (50 °C) and we solvated with a total of 10443 TIP3P water molecules (which means that we had around 81 water molecules per lipid) in order to assure that the system was fully hydrated. We solvated along the Z axis and the system was subject to periodic boundary conditions along the (X, Y) plane. Figure 7.10 shows a typical equilibrated configuration.

All the atoms of the bilayer were fixed during minimization and temperature and pressure stabilization. Then, all restraints were removed and the bilayer was free to establish a lipid-lipid equilibrium distance (we also used a more restricted system where during equilibration we fixed the positions of the last carbon atoms of the lipid tail). After 50 ns equilibration, we obtained an area per lipid of 60.3 for the free bilayers (while this distance was 65.5 Å for the system with anchored tails), a result consistent with experimental values (Nagle and Tristram-Nagle 2000). Figure 7.11 shows the time evolution of the area per lipid.

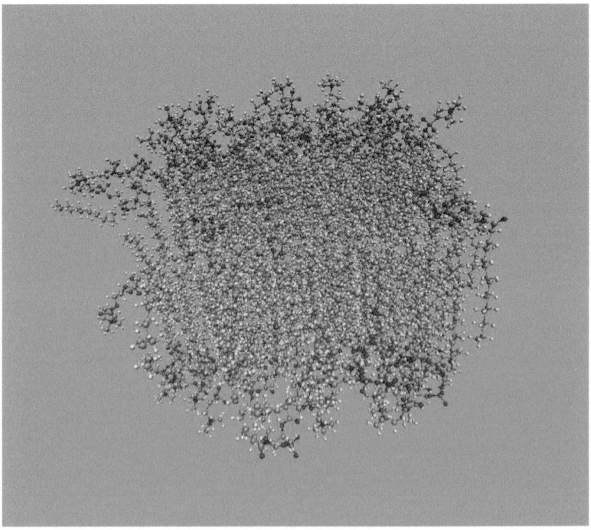

Fig. 7.10 Image of the free bilayer. For the sake of clarity, the water molecules are not shown

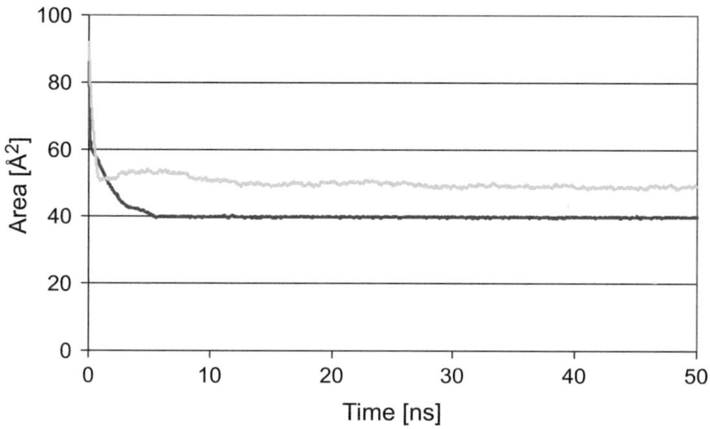

Fig. 7.11 Time evolution of the area per lipid during stabilization

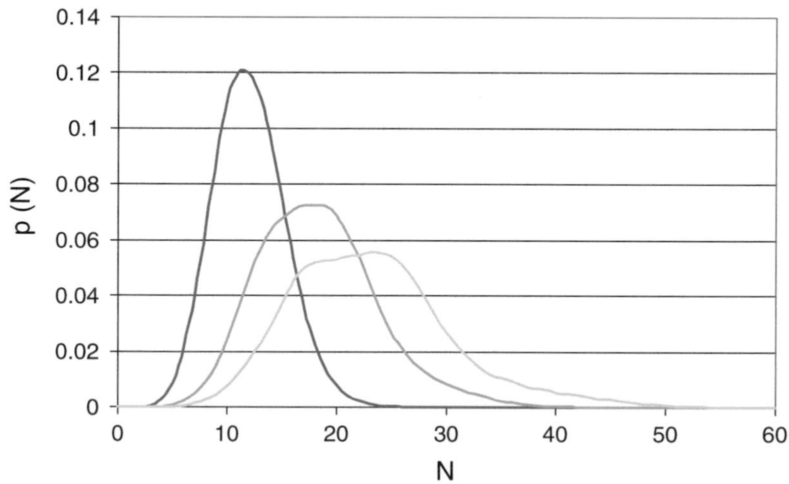

Fig. 7.12 Distribution of the number of water molecules per lipid for threshold values of 3.5, 4.0 and 4.5 Å

Having equilibrated the free bilayers we then proceeded to calculate the number of water molecules per lipid. This was performed by counting all the water molecules whose distance to the lipid is lower than certain threshold value. We choose this value to be 4.0 or 4.5 Å in order to consider only molecules within the first hydration shell. This procedure includes molecules hydrating the exposed polar heads and also water molecules that have penetrated the bilayer. In Fig. 7.12 we present the distribution of the number of water molecules per lipid.

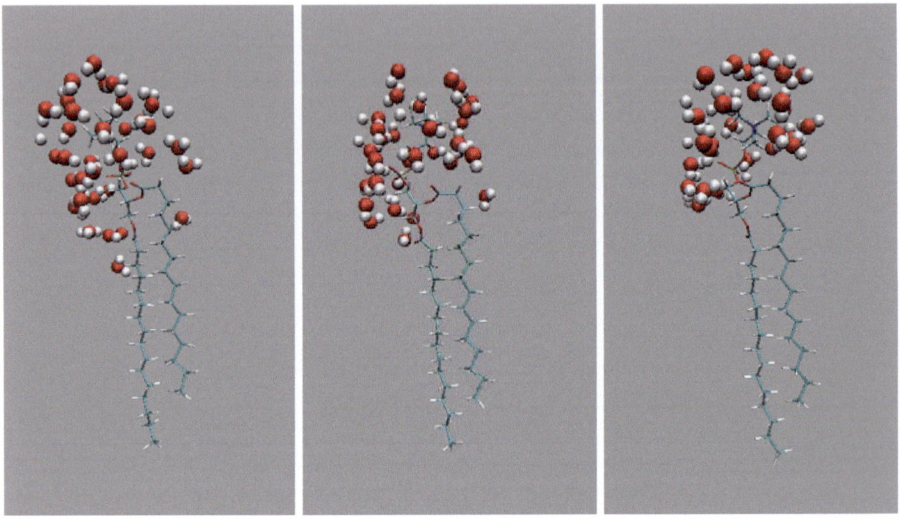

Fig. 7.13 Examples for the location of the water molecules closest than 4.0 Å to a given lipid chain in three different configurations. For simplicity we neither display the rest of the lipid chains nor the rest of the water molecules

For the threshold of 4.0 Å the number of water molecules per lipid is 18.3, while it amounts to 23.2 when such threshold is increased to 4.5 Å. These values are consistent with experimental findings (Cevc 1987).

In turn, in Fig. 7.13 we provide examples of the spatial location of the water molecules around a selected lipid chain. We can see that the water molecules do not only hydrate the polar moieties of the chain head, but also some of them penetrate deeper the bilayer and approach the upper hydrocarbon groups.

Next, we study how these water molecules distribute among the different groups of the lipid chain. To this end, we placed a sphere of radius 4 Å centered at different heavy atoms of the lipid chain where water displays significant population. Figure 7.14 displays the results, which indicate that the phosphate oxygens Op1 and Op2 are the moieties with a higher level of hydration. The amine C atoms are also shown to be highly hydrated. In turn, the C1 and C2 atoms are also well populated by water molecules while the rest of the places display a lower level of hydration. The shapes of the curves are also instructive. Such simple featureless distributions indicate that the conformations adopted at different times are equivalent in which respects to their hydration properties. For example, if a group alternated between two different hydration states at different times, between an exposed and a buried state, the distribution would present clear signs of bimodality. This is not the case for the different groups studied.

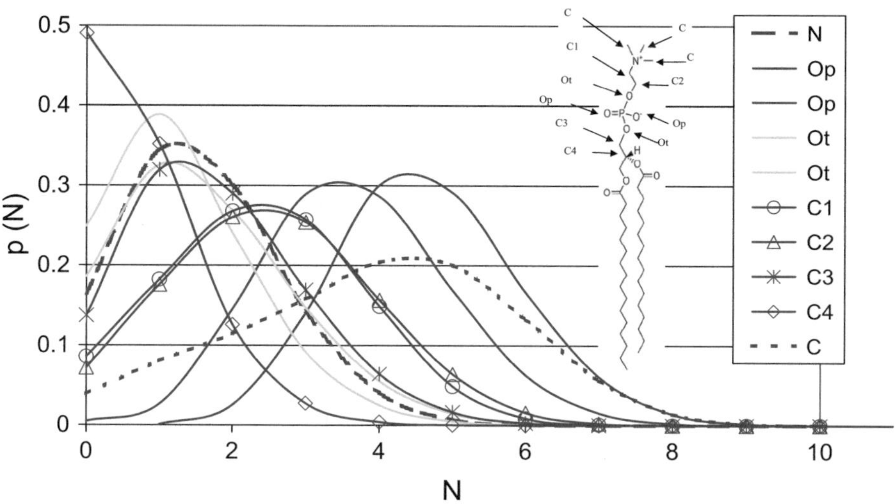

Fig. 7.14 Distribution of water molecules around different groups of the phospholipid

7.4 Quantifying Local Hydrophobicity: Calculations of Water Density Fluctuations

The nanoconfinement that arises upon the interaction of different assembling units, both in biological organization processes and in the supramolecular self-assembly of nanomaterials in a water environment, affects the thermodynamic properties of the hydration water which should be removed for the process to take place (Giovambattista et al. 2012). In realistic contexts, both chemistry and local geometry are expected to define the local hydrophobicity (Giovambattista et al. 2012). Geometrically-induced surface dehydration (by means of water inaccessibly cavities) has been shown to be central for the existence of reactive sites responsible for protein binding (Schulz et al. 2011; Sierra et al. 2013). Such structural packing defects characterized by regions of the protein backbone exposed to the solvent has been shown to promote their local dehydration and to signal binding sites (Fernández and Scheraga 2003; Fernández and Scott 2003; Fernández 2010; Accordino et al. 2012a, b, c, 2013; Sierra et al. 2013). Different approaches have indicated that the hydration properties of protein binding sites play a main role in the binding of ligands or in protein-protein association (Fernández and Scheraga 2003; Fernández and Scott 2003; Fernández 2010; Qvist et al. 2008; Berne et al. 2009; Young et al. 2007; Wang et al. 2011; Kulp III et al. 2011; Accordino et al. 2012a, b, c, 2013; Sierra et al. 2013; Alarcón et al. 2014; Bogan and Thorn 1998; Li and Liu 2009). Ligands are expected to displace hydration water molecules from their binding site and the replacement of so-called “unfavorable” waters by groups of the ligand complementary to the protein surface has been established as a principal driving force for binding (Fernández and Scheraga 2003; Fernández and Scott 2003; Fernández 2010; Qvist et al. 2008; Berne et al. 2009; Young et al. 2007; Wang et al. 2011; Kulp III et al. 2011; Accordino et al. 2012a, b, c, 2013; Sierra et al. 2013; Alarcón et al. 2014; Bogan and Thorn 1998; Li and Liu 2009). In fact, this description has been shown to hold valid for a significant fraction of receptors of pharmaceutical interest (Berne et al. 2009; Young et al. 2007; Wang et al. 2011). Even in some cases, a portion of the receptor active site is so unfavorable for water molecules that it tends to remain practically dry (Berne et al. 2009; Young et al. 2007; Wang et al. 2011). Thus, the estimation of the free energy contribution involved in the displacement of quasilocalized water molecules with unfavorable free energies in the receptor active site constitutes an issue of great interest in computational structure-based drug design (Berne et al. 2009; Young et al. 2007; Wang et al. 2011). Within this same philosophy (Kulp III et al. 2011), a study of fragment clustering of diverse organic probes on hen egg white lysozyme combined with water exclusion (superimposing the different regions targeted by the fragments with the map of water molecules tightly bound to the protein which excluded by blocking possible target regions) has been able of predicting the experimentally known binding site, or hot spot. Thus, a picture of protein binding with regions of easily removable (non-tightly-bound) water molecules at small-molecule binding sites or protein-protein interaction hot spots is emerging. Hence, it becomes relevant to develop a quantitative measure of hydrophobicity useful in determining such regions.

Amongst the different structural, dynamical and thermodynamical measures of hydrophobicity which have been proposed (Giovambattista et al. 2012; Rasaiah et al. 2008), a very appealing one consists in the quantification of water density fluctuations (Rasaiah et al. 2008). It has been demonstrated that superficial water density profiles do not represent a good measure of surface hydrophobicity. This can be expected in terms of the usual knowledge that water abhors vacuum and thus, water molecules tend to hydrate both polar and nonpolar surfaces and, thus, the density profiles normal to the surface plane display similar characteristics, with layering structure in both cases. However, at variance form hydrophilic surfaces where the water molecules are subject to significant attractive interactions, interactions are very weak at hydrophobic surfaces, which makes the hydrating water molecules to display low residence times and to become easily removed. Thus, such hydration layers display enhanced dynamics (Giovambattista et al. 2012; Rasaiah et al. 2008) enhanced compressibility (Giovambattista et al. 2012; Rasaiah et al. 2008) and enhanced density fluctuations (Rasaiah et al. 2008). In particular, the density fluctuations at differently functionalized self-assembled monolayers (SAMs) have been characterized demonstrating that hydrophobic-like surfaces do in fact present much larger density fluctuations than the ones displayed by hydrophilic-like surfaces, thus providing a good quantitative measure of hydrophobicity (Rasaiah et al. 2008). Normalized fluctuations of water number density, $\sigma^2/\langle N\rangle^2$ in small observation volumes (where N is the number of water molecules within such volume) are approximately equal to $2\mu^{ex}/kT$, where μ^{ex} is the free energy of formation of cavity of such radius (Rasaiah et al. 2008). Thus, a high value of the normalized density fluctuations at a given place indicates a favorable work of cavity creation at such place and, thus, a high hydrophobicity. The relevance of this measure is obvious from the above-described scenario: The self-assembling process of nonpolar solutes in water makes use of such dehydration propensity properly accounted for by the density fluctuations measurements, while the binding sites of proteins also exploit this phenomenon since ligands are expected to displace easily removable hydration water upon binding.

We have calculated water density fluctuations at model surfaces, functionalized SAMs, phopholipid membranes and are also studying protein binding sites. In which follows we present some preliminary results. For comparison, we calculated density fluctuation on self-assembled alkane-like monolayers like the ones of Sect. 7.2.2, but functionalized in order to be hydrophobic or hydrophilic. To this end, the chain heads consisted of CH_3 groups or OH groups respectively. At variance from the previous situation where the alkyl chains were rigid, here the chains are flexible. Additionally, differently from previous studies and in order to compare with the situation at the different groups of the phospholipid membranes of Sect. 7.3.2, we calculate water density fluctuations within spheres centered at the heavy atoms of the groups of interest. Thus, such observation volumes move with the atom, being appropriate for studies on flexible molecules. In Fig. 7.15 we display the probability distributions for observing N water molecules, p(N), within a small spherical observation volume of radius 4.0 Å centered at the C of the methyl heads of the hydrophobic SAMS, at the O of the alcohol head groups of the hydrophilic SAMS and at different heavy atoms of the head groups of the phospholipid membranes. A high probability of having zero water molecules in the observation domain, and thus a high value of density fluctuations, implies an enhanced propensity for dehydration or high hydrophobicity. From direct inspection of such figure, we can learn that the phosphate oxygens Op are the most hydrophilic moieties of the lipid bilayer with a hydrophilicity similar (or even higher) to the hydrophilic SAM. The C4 carbons are the most hydrophobic groups presenting density fluctuations similar to the ones of the hydrophobic SAM, while the rest of the groups display an intermediate behavior. As evident from such figure, the different oxygens present behaviors that vary from each other, as also happens for the comparative behavior of different carbonaceous groups. A first lesson emerging from this study is, thus, that the hydrophobicity of a given group can vary significantly depending on its local environment, where both the local chemistry and geometry can play a role. This fact is relevant, for example, for the behavior of binding sites. We are at present performing studies of this kind in protein binding sites. It is known that some binding sites present reactive groups surrounded by (hydrophobic) groups that can locally dehydrate the binding site, as in the O-Ring theories (Bogan and Thorn 1998; Li and Liu 2009). Electrostatic interactions (like hydrogen bonds and ionic interactions) might be strongly enhanced within an environment with a deprived local dielectric given the shielding effect of nonpolar groups (Fernández and Scheraga 2003; Fernández and Scott 2003; Fernández 2010; Accordino et al. 2012a, b, c, 2013; Sierra et al. 2013). Thus, the combination of different non-covalent interactions can produce clearly non-additive effects.

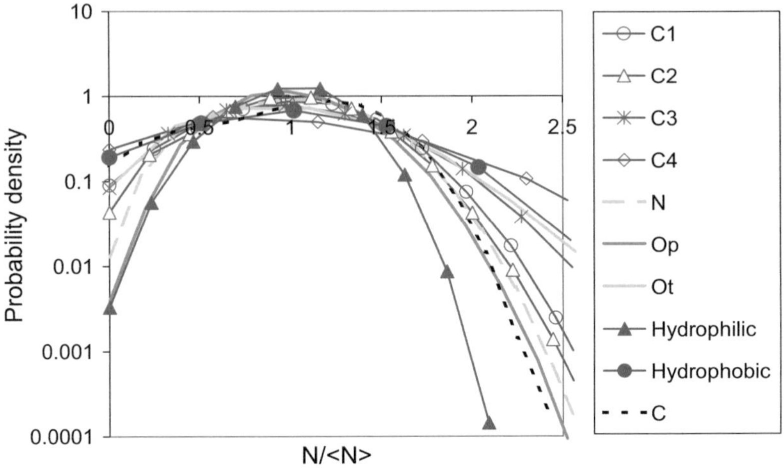

Fig. 7.15 Probability distributions for observing N water molecules, p(N), within a small spherical observation volume of radius 4.0 Å centered at the C atoms of the methyl heads of the hydrophobic SAMS (*black line* and *black triangles*), at the O of the alcohol head groups of the hydrophilic SAMS (*black line* and *black circles*) and at different heavy atoms of the head groups of the phospholipid membranes

7.5 Conclusions and Future Directions

Our study of the hydration layers of different nonpolar surfaces has made evident the existence of an enhancement of water structuring at interfaces. The water molecules present a neat orientational ordering, in some cases resembling an ice-like arrangement. Inclusion of curvature induces a loss in the water structuring. The water molecules tend to exclude the hydrophobe with a minimum loss in HB coordination. Thus, they tend to orient a vertex of the tetrahedron hence conserving the other three first neighbors largely unperturbed. A more complex example, the protein lysozyme still partially displayed this behavior.

We have also shown that subnanometric size cavities carved in hydrophobic SAMs remain "dry", while intermediate size cavities exhibit an alternation of filled and dry states. These facts can be explained in terms of the reluctance of the water molecules to loss HB coordination. The degree of hydrophobicity of the material is also important since small radius carbon nanotubes can be filled while the more hydrophobic alkane-like tunnels of similar size still remain dry. We have also shown that the formation of an extensive HB network is required for filling to occur.

Some preliminary results have also been presented on the hydration and water penetration in phospholipid membranes. We have studied the spatial distribution of water molecules around the different groups of the lipid molecules of the bilayer. Water density fluctuation calculations have also enabled us to test the hydrophobicity of the different moieties.

These results are first steps in a systematic effort to learn how geometry and chemistry affect hydrophobicity and the behavior of nanoconfined water. This knowledge is expected to be relevant for the understanding of biological organization and for supramolecular self-assembly in materials science. In such processes, the nanoconfinement that arises upon the interaction of the different assembling units affects the thermodynamic properties of the hydration water, which usually must be removed for the process to take place. In such contexts, bulk-like knowledge could be not only useless but also misleading and thus a new intuition and new principles are necessary. In this light, it is not surprising that today rational design is practically absent in fields ranging from drug design to the design of soluble self-assembled materials. Most of the non-covalent interactions (mainly the ones which are electrostatic in nature) invoked in such fields would not be operative in bulk water conditions. However, the different non-covalent interactions (like HBs, ionic interactions and hydrophobic interactions) can be clearly non-additive under nanoconfinement, a context-dependent nature which is usually overlooked in design efforts. Thus, a more complete knowledge of the behavior of water under nanoconfinement is expected to open new roads of both academic and technological relevance.

Acknowledgements LMA, JAR-F, MBS, MAM and GAA are research fellows of CONICET. SRA and JMM-O thank CONICET for a fellowship. The authors gratefully thank CONICET and MinCyT for financial support.

References

Accordino SR, Malaspina DC, Rodríguez Fris JA, Appignanesi GA (2011a) Comment on "Glass transition in biomolecules and the liquid-liquid critical point of water". Phys Rev Lett 106:029801

Accordino SR, Rodriguez Fris JA, Sciortino F, Appignanesi GA (2011b) Quantitative investigation of the two-state picture for water in the normal liquid and the supercooled regime. Eur Phys J E 34:48

Accordino SR, Malaspina DC, Rodriguez Fris JA, Alarcón LM, Appignanesi GA (2012a) Temperature dependence of the structure of protein hydration water and the liquid-liquid transition. Phys Rev E 85:031503

Accordino SR, Morini MA, Sierra MB, Rodríguez Fris JA, Appignanesi GA, Fernández A (2012b) Protein packing defects "heat up" interfacial water. Proteins Struct Funct Bioinf 80:1755

Accordino SR, Rodriguez Fris JA, Appignanesi GA, Fernández A (2012c) A unifying motif of intermolecular cooperativity in protein associations. Eur Phys J E 35:59

Accordino SR, Rodriguez Fris JA, Appignanesi GA (2013) Wrapping effects within a proposed function-rescue strategy for the Y220C oncogenic mutation of protein p53. PLoS One 8, e55123

Alarcón LM, Malaspina DC, Schulz EP, Frechero MA, Appignanesi GA (2011) Structure and orientation of water molecules at model hydrophobic surfaces with curvature: from graphene sheets to carbon nanotubes and fullerenes. Chem Phys 388:47

Alarcón LM, Montes de Oca JM, Accordino SR, Rodriguez Fris JA, Appignanesi GA (2014) Hydrophobicity and geometry: water at curved graphitic-like surfaces and within model pores in self-assembled monolayers. Fluid Phase Equilib 362:81

Appignanesi GA, Rodríguez Firs JA, Sciortino F (2009) Evidence of a two-state picture for supercooled water and its connections with glassy dynamics. Eur Phys J E 29:305
Arsov Z (2015) Chapter 6: Long-range lipid-water interaction as observed by ATR-FTIR spectroscopy. In Disalvo EA (ed) Membrane hydration: the role of water in the structure and function of biological membranes. Springer, Cham, pp 127–159
Berne BJ, Weeks JD, Zhou R (2009) Dewetting and hydrophobic interaction in physical and biological systems. Annu Rev Phys Chem 60:85
Bizzarri A, Cannistraro SJ (2002) Molecular dynamics of water at the protein-solvent interface. Phys Chem B 106:6617
Bogan AA, Thorn KS (1998) Anatomy of hot spots in protein interfaces. J Mol Biol 208:1
Cevc G (1987) How membrane chain melting properties are controlled by the polar surface of the lipid bilayers. Biochemistry 26:6305
Choudhury N, Montgomery PB (2005) Dynamics of water trapped between hydrophobic solutes. J Phys Chem B 109:6422
Disalvo A, Lairion F, Martini F, Tymczyszyn E, Frias M (2008) Structural and functional properties of hydration and confined water in membrane interfaces. Biochim et Biophys Acta (BBA)-Biomembranes 1778:2655
Fan Y, Chen X, Yang L, Cremer PS, Gao YQ (2009) On the structure of water at the aqueous/air interface. J Phys Chem B 113:11672
Faul CFJ, Antonietti M (2003) Ionic self-assembly: facile synthesis of supramolecular materials. Adv Mater 15:673
Fernández A (2010) Transformative concepts for drug design: target wrapping. Springer, Heidelberg
Fernández A, Scheraga HA (2003) Insufficiently dehydrated hydrogen bonds as determinants of protein interactions. Proc Natl Acad Sci U S A 100:113
Fernández A, Scott R (2003) Adherence of packing defects in soluble proteins. Phys Rev Lett 91:018102
Gelman Constantin J, Rodriguez Fris JA, Appignanesi GA, Carignano MA, Szleifer I, Corti HR (2011) Structure of supercooled water in clusters and bulk and its relation to the two-state picture of water: results from the TIP4P-ice model. Eur Phys J E 34:126
Giovambattista N, Debenedetti PG, Lopez CF, Rossky PJ (2008) Hydrophobicity of protein surfaces: separating geometry from chemistry. Proc Natl Acad Sci U S A 105:2274
Giovambattista N, Rossky PJ, Debenedetti PG (2012) Computational studies of pressure, temperature and surface effects on the structure and thermodynamics of confined water. Annu Rev Phys Chem 63:179
Huang DM, Chandler D (2000) Temperature and length scale dependence of hydrophobic effects and their possible implications for protein folding. Proc Natl Acad Sci U S A 97:8324
Huang X, Margulis CJ, Berne BJ (2003) Dewetting-induced collapse of hydrophobic particles. Proc Natl Acad Sci U S A 100:11953
Kulp JL III, Kulp JL Jr, Pompliano DL, Guarnieri F (2011) Diverse fragment clustering and water exclusion identify protein hot spots. J Am Chem Soc 133:10740
Li J, Liu Q (2009) 'Double water exclusion': a hypothesis refining the O-ring theory for the hot spots at protein interfaces. Bioinformatics 25:743
Malaspina DC, Rodriguez Fris JA, Appignanesi GA, Sciortino F (2009) Identifying a causal link between structure and dynamics in supercooled water. Europhys Lett 88:16003
Malaspina D, Schulz EP, Alarcón LM, Frechero MA, Appignanesi GA (2010) Structural and dynamical aspects of water in contact with a hydrophobic surface. Eur Phys J E 32:35
Nagle JF, Tristram-Nagle S (2000) Structure of lipid bilayers. Biochim Biophys Acta 1469:159
Nickels JD, Katsaras J (2015) Chapter 3: Water and lipid bilayers. In: Disalvo EA (ed) Membrane hydration: the role of water in the structure and function of biological membranes. Springer, Cham, pp 45–67
Pfeiffer H (2015) Chapter 4: Hydration forces between lipid bilayers: a theoretical overview and a look on methods exploring dehydration. In: Disalvo EA (ed) Membrane hydration: the role of water in the structure and function of biological membranes. Springer, Cham, pp 69–104

Qvist J, Davidovic M, Hamelberg D, Halle B (2008) A dry ligand-binding cavity in a solvated protein. Proc Natl Acad Sci U S A 105:6296

Rasaiah JC, Garde S, Hummer G (2008) Water in nonpolar confinement: from nanotubes to proteins and beyond. Annu Rev Phys Chem 59:713

Rehm TH, Schmuck C (2010) Ionpair induced self-assembly in aqueous solvents. Chem Soc Rev 39:3597

Schulz EP, Alarcón LM, Appignanesi GA (2011) Behavior of water in contact with model hydrophobic cavities and tunnels and carbon nanotubes. Eur Phys J E 34:114

Shen YR, Ostroverkhov V (2006) Sum-frequency vibrational spectroscopy on water interfaces: polar orientation of water molecules at interfaces. Chem Rev 106:1140

Shiratani E, Sasai M (1996) Growth and collapse of structural patterns in the hydrogen bond network in liquid water. J Chem Phys 104:7671

Shiratani E, Sasai M (1998) Molecular scale precursor of the liquid–liquid phase transition of water. J Chem Phys 108:3264

Sierra MB, Accordino SR, Rodriguez Fris JA, Morini MA, Appignanesi GA, Fernández SA (2013) Protein packing defects "heat up" interfacial wáter. Eur Phys J E 36:62

Stanley HE, Kumar P, Xu L, Yan Z, Mazza MG, Buldyrev SV, Chen S-H, Mallamace F (2007) The puzzling unsolved mysteries of liquid water: some recent progress. Phys A 386:729

Vitkup D, Ringe D, Petsko GA, Karplus M (2000) Solvent mobility and the protein glass transition. Nat Struct Biol 7:34

Wang L, Berne BJ, Friesner RA (2011) Ligand binding to protein-binding pockets with wet and dry regions. Proc Natl Acad Sci U S A 108:1326

Young T, Abel R, Kim B, Berne BJ, Friesner RA (2007) Motifs for molecular recognition exploiting hydrophobic enclosure in protein–ligand binding. Proc Natl Acad Sci U S A 104:808

Chapter 8
Aquaphotomics: Near Infrared Spectroscopy and Water States in Biological Systems

Roumiana Tsenkova, Zoltan Kovacs, and Yosuke Kubota

Abstract Aquaphotomics is a new discipline that provides a framework for understanding changes in water molecular system presented as a water spectral pattern, to mirror the rest of the solution and to give a holistic description related to system functionality. One of its main purposes is to identify water bands as main coordinates of future absorbance patterns to be used as a system biomarker. This chapter presents the Aquaphotomics methodology and illustrates a way to identify specific water bands using temperature change and addition of solutions of different ionic strength as perturbations. Rapid and precise measurement of low concentration solutes has been given as a strong evidence of the vast information that "the water spectral pattern as molecular mirror" approach provides. Few applications using near infrared spectroscopy and multivariate analysis as main tools of Aquaphotomics have been presented.

Keywords Water states • Near infrared spectroscopy • Water structure • Effect of salts

8.1 Introduction

Visible light reflected from the surface of the water and the surroundings explains the « water mirror » effect. Next to the visible light in the electromagnetic spectrum between 680 and 2500 nm is the near infrared (NIR) light, which penetrates deep into the water, but does not get fully absorbed. Thus, it makes it possible to measure the spectrum of the light that comes back out of the water. It contains immense

R. Tsenkova (✉) • Y. Kubota
Biomeasurement Technology Laboratory, Kobe University, 1-1 Rokkodai, Nada, 657-8501 Kobe, Japan
e-mail: rtsen@kobe-u.ac.jp

Z. Kovacs
Biomeasurement Technology Laboratory, Kobe University, 1-1 Rokkodai, Nada, 657-8501 Kobe, Japan

Department of Physics and Control, Corvinus University of Budapest, Budapest, Hungary

E.A. Disalvo (ed.), *Membrane Hydration*, Subcellular Biochemistry 71,
DOI 10.1007/978-3-319-19060-0_8

information about the hydrogen bonding in the whole system of water molecules related to the water structure and perturbed not only by the environment, but by the rest of the molecules in the solution (Tsenkova 2009, 2010).

Since the mid '90s, in contemporary biology and life sciences, an explosion of the use of -omes and -omics terms is observed. All the new -omics disciplines consider specific constituents and follow the reduction "analysis" pathway. For further understanding of bio functionality, holistic analysis is required. It would help the next stage of investigation aiming at the "synthesis" pathway. Unfortunately, all of the existing -omics disciplines exclude water despite the fact that it is the matrix of a biological system, its scaffold. In these regards, Aquaphotomics has been proposed (Tsenkova 2009) to fill in this gap and to present a new approach to study complex living and aqueous systems as a whole and as a complimentary to the rest of the -omics disciplines, Fig. 8.1. Aquaphotomics is based on the presumption that all the components of the system shape up the water molecular matrix. Therefore, jumping over the individual constituents and analyzing only the relation between changes of the water matrix (the system "envelope") caused by changes in system functionalities, could explain new phenomena and the "structure-function" relationship. Aquaphotomics means "all about water-light interaction". It is based on the fact that water absorbs energy over the whole electromagnetic spectrum, Fig. 8.2, and water-light interaction can be used as a probe to study changes of water molecular system perturbed by energy of various frequencies. Very informative spectral window to study water-light interaction is the near infrared range. Being located between the visible and infrared range, NIR has proven to be very informative when it comes to studies of water in bio- or aqueous systems as the light is not fully reflected (like in the visible range) or fully absorbed (like in the infrared range). In the near infrared range, water has specific spectral pattern which changes under perturbations. The main concept of Aquaphotomics is to establish a database of all water absorbance bands, called aquaphotome, and to use the spectral patterns based on these bands as biomarkers related to system functionality.

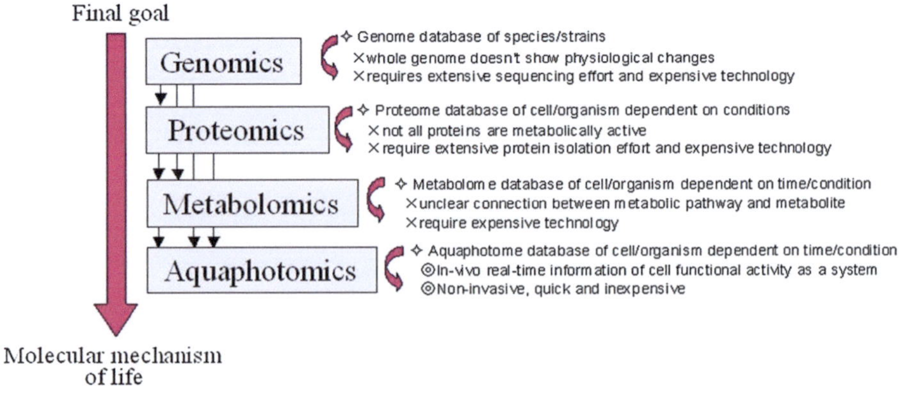

Fig. 8.1 Aquaphotomics in relation with other -omics disciplines

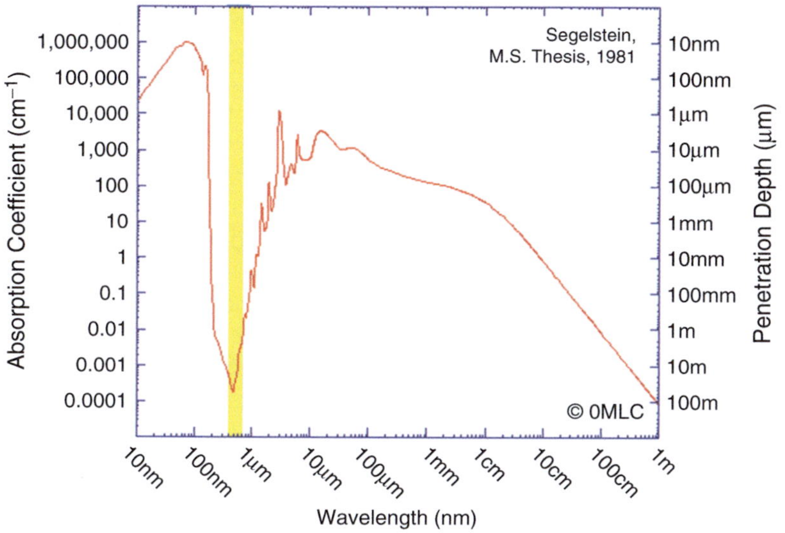

Fig. 8.2 Water absorbance pattern over the whole electromagnetic spectrum (Segelstein 1981)

8.2 Aquaphotomics: Water Spectral Pattern as Holistic Biomarker of System Functionality

8.2.1 Spectroscopy of Aqueous and Biological Systems

Water molecular conformations, for example water dimers, trimers, solvation shells, etc., are known to contribute very specifically to its spectrum (Smith et al. 2005; Robertson et al. 2003; Weber et al. 2000; Franks 1973). They are very sensitive to configuration, hydrogen bonding and charges of the solvated and solvent molecules or clusters. NIR spectrum of the solvent (i.e., water) has been found to contain significant information about its solutes (Tsenkova 2009). In previous research, protein solutions (Murayama et al. 2000), biomolecular water solutions (Murayama et al. 2000; Tsenkova et al. 2004), suspensions of small particles in water and various biological systems such as cells, plants, animal body fluids and tissues have been analyzed with NIR (Tsenkova 2009, 2010; Tsenkova et al. 2001, 2004; Jinendra 2010; Jinendra et al. 2010; Kinoshita et al. 2012; Matija et al. 2012; Nakakimura et al. 2012). In these studies, well known water absorbance bands have been observed under various perturbations (i.e., concentrations of solutes (Nakakimura et al. 2012), nano particles (Matija et al. 2012; Tsenkova et al. 2007a), molecules which don't absorb (Tsenkova et al. 2007a; Gowen et al. 2013) in the NIR range, temperature (Segtnan et al. 2001; Maeda et al. 1995), light illumination, etc.). Dynamic spectra have been acquired and analyzed with multivariate methods. As a result, numerous water absorbance bands have been identified. Many of these bands are in a good agreement with previously reported or calculated overtones of already published water bands in the IR region (Smith et al. 2005; Weber et al. 2000). Some of them have been newly discovered (Tsenkova 2009).

In research carried out at Biomeasurement Technology Laboratory at Kobe University, Japan, water has been perturbed with lactose, human serum albumin (HSA) (Murayama et al. 2000), different isomers of prion protein (PrP) (Tsenkova et al. 2004), NaCl (Gowen et al. 2013), metals (Putra et al. 2010; Tsenkova et al. 2007b) and other solutes at various concentrations. The most prominent wavelengths with highest variations caused by the used perturbations were identified in the spectra through multivariate analysis. It has been found that they are predominantly water absorbance bands. Contrary to the common understanding of overtone spectroscopy (100 to 1000 times lower absorbance than in the mid-IR range), it has been shown that even very small concentrations (at ppb level) (Tsenkova et al. 2004; Gowen et al. 2013) of the solute could be measured with NIR. Measuring low concentrations of polystyrene particles in water is a good example. When the first overtone of water absorbance region at 1300–1600 nm has been used to develop a model for low concentration (1 %–0.0001 %) measurement of polystyrene particles (Tsenkova et al. 2007a), spectral measurements have achieved high accuracy over a wide region of low concentrations. In contrary, with decreasing the concentration, accuracy has decreased substantially when only the polystyrene overtone band at 1680 nm was used. Similar findings have been reported for proteins in solution (Murayama et al. 2000), metals in solution (Putra et al. 2010), etc. Another study has even shown that NIR spectral models for metal concentration measurement (ppb range) in diluted samples are influenced by the method of dilution (serial or direct dilution) (Putra et al. 2010; Tsenkova et al. 2007b). These findings led to the conclusion that the NIR water spectral pattern describes the water matrix in relation to the rest of the molecules in solution in great details. For the first time (Tsenkova 2009), water has been discussed as the common spectral denominator of aqueous and biological systems. In order to systemize the already abundant information concerning the interactions of water in biological and aqueous systems with light at various frequencies, predominantly in the NIR region, a new scientific area called "Aquaphotomics" has been proposed by Prof. Roumiana Tsenkova at the Biomeasurement Technology Laboratory at Kobe University, Japan.

8.2.2 Aquaphotomics: Main Concept and Terms

The idea of Aquaphotomics originates from near infrared studies of raw milk as an output product of the mammary gland. It has been found, using NIR spectroscopy, that milk quality is highly related to the physiological status of the respective mammary gland and highly dependent upon milk constituents and water structure (Putra et al. 2010). The NIR light penetrates deep into the water, but does not get fully absorbed. Thus, it makes it possible to measure the spectrum of the light that comes back out of the water after interacting with all the molecules in the solution.

It contains immense information about the covalent O-H bonds and hydrogen bonding in the whole system of water molecules related to the water structure and perturbed not only by the environment, but by the rest of the molecules in the system. Therefore, all water absorbance bands called **Water Matrix Coordinates (WAMACS)** become « informational hubs ». The combination of the « activated » water bands, at which the light absorbance gets influenced by the perturbations, depicts a characteristic spectral pattern called **Water Spectral Pattern (WASP)**, which reflects the condition of the whole water molecular system. It contains huge amount of physical and chemical information for the solution because the water hydrogen bonding network is easily influenced by any kind of even subtle perturbations (Matija et al. 2012) including the solutes. Graphically, WASP is presented as an **aquagram**, which is a radial graphic of the normalized absorbance at characteristic water bands. Figure 8.3b presents an aquagram depicting the spectral patterns of water and water vapor in the area of the first overtone of water (Fig. 8.3a), 1300–1600 nm when using the main water absorbance bands found in the first overtone of the water. The database of characteristic water bands in the whole electromagnetic spectrum of water is called **aquaphotome**. Each particular system has its own aquaphotome comprised of the activated water absorbance bands under respective perturbation.

8.2.3 Methodology

The main goal of Aquaphotomics is to identify the « activated » water bands, i.e. to build up the aquaphotome of the system of interest and to use further the specific WASP as biomarkers. For that purpose, NIR spectra of the system of interest are acquired under respective perturbations. Subtracted spectra, spectra derivatives, loadings of Principal Component Analysis (PCA), regression vectors of Partial Least Squares (PLS) Regression, discriminating power of Soft Independent Modeling of Class Analogy (SIMCA) extract the bands where spectral variations under controlled and uncontrolled perturbations could be observed, i.e. the variables of interest (Fig. 8.5). The obtained bands define the aquaphotome of the respective system.

In order to build up a database consisting of the « activated » water bands i.e. the WAMACS of various bio- and aqueous systems various experiments have been conducted in the Biomeasurement Technology Laboratory at Kobe University, Japan. Dynamic spectra of various biological systems have been acquired and analyzed using the above described methods to determine the aquaphotome of these systems under different perturbations. Table 8.1 demonstrates the collection of WAMACS of some of the examined systems. In the process of specific water band extraction from the spectral data of various biological and aqueous systems, it has been found that, always, the same twelve wavelengths regions: C1 to C12 were repeatedly activated (in different combination and strength depending on the perturbation and the system) in the first overtone of water, 1300–1600 nm. These

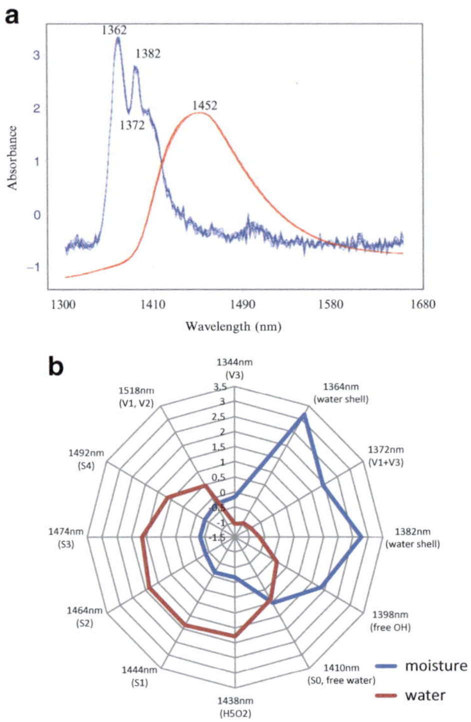

$$A'_\lambda = \frac{A_\lambda - \mu_\lambda}{\sigma_\lambda} \quad (8.1)$$

A'_λ : Value of Aquagram

A_λ : Absorbance after MSC applied on 1st overtone region

μ_λ : Mean of all spectra

σ_λ : SD of all spectra

λ : 12 Wavelengths[(*)]

*<1342, 1364, 1372, 1382, 1398, 1410, 1438, 1444, 1464, 1474, 1492, 1518 nm>

Fig. 8.3 Spectra (**a**) and aquagram (**b**) depicting the spectral patterns of water and water vapor in the area of the first overtone of water, 1300 nm–1600 nm * (Tsenkova 2009)

bands are in a very good agreement with the calculated first overtone of the water bands found by various experiments and molecular dynamic calculations in the IR range.

NIR spectroscopy is a non-destructive, non-invasive, fast method to study biological and aqueous systems as the NIR light is not fully absorbed by water. NIR spectra are characterized by broad and overlapped bands (predominantly water bands). This feature makes the assignment and interpretation of NIR spectra not straight-forward. Therefore, advanced multivariate data analysis techniques are necessary to fully understand the effect of solutes on NIR spectra. Protocol and a number of chemometrics methods have been proposed (Tsenkova 2009, 2010; Seasholtz and Kowalski 1990) for identification of water absorbance bands (Fig. 8.5). Examples include principal component analysis loadings, multivariate curve resolution-alternative least squares (MCR-ALS) (Abdollahi and Tauler 2011), two dimensional correlation spectroscopy, 2DCOS, partial least square regression vectors etc.. PCA is an unsupervised technique in which the major variations in a spectral dataset are compressed into the first few principal components, which are

Table 8.1 WAter MAtrix CoordinateS (WAMACS) of various bio- and aqueous systems (Tsenkova 2009)

C1: 1336~1348	C2: 1360~1366	C3: 1370~1376	C4: 1380~1388	C5: 1398~1418	C6: 1421~1428	C7: 1436~1444	C8: 1448~1454	C9: 1462~1468	C10: 1472~1482	C11: 1482~1495	C12: 1506~1516	
	■		■	■		■						Quarter Milk(Mastitis)
			■	■	■							Quarter milk(Healthy)
■		■	■	■			■				■	Milk-Cell count
		■	■	■							■	Milk-Protein(Mastitis)
			■	■			■				■	Milk-Protein(Healthy)
■	■	■	■	■			■		■			Water
■	■		■	■			■		■			PrP(Mn)
■	■	■		■			■		■			PrP(Cu)
■			■	■			■		■			PrP
■	■			■			■	■		■		Globulin
	■		■									HSA-Low
■				■		■	■			■		HAS-High
■	■	■	■			■		■	■			CHO cells
		■	■	■							■	CHO cells with Aquaporin
				■		■	■	■		■		DNA in water (High cc.)
						■	■	■	■	■		DNA in water (Low cc.)
				■	■	■	■	■	■			DNA irradiated with UV
				■	■	■	■	■	■		■	Lactose(High cc.)
■	■		■	■	■	■		■		■		Lactose(Low cc.)
■	■		■	■	■			■	■		■	KCl(High cc.)
■	■		■	■		■	■		■	■		KCl((Low cc.)
			■	■	■	■			■	■	■	NaCl(High cc.)
■			■	■		■	■		■	■		NaCl(Low cc.)
			■	■	■		■	■		■	■	Acetic acid(High cc.)
	■		■	■	■	■	■	■	■	■	■	Acetic acid (Low cc.)
■	■		■	■	■	■					■	Stored raw milk
	■	■		■	■				■	■	■	Stored defatted milk
	■		■	■			■	■		■		Electrolyzed(acidic)Water
■	■		■	■						■		Amyloid+NaCl solution
				■	■		■	■	■	■	■	Water(HighMineralCont)
	■	■		■	■	■		■				Water(LowMineralCont)
■	■	■		■	■	■	■	■				Bacteria(Probiotic)
	■					■	■	■			■	Bacteria(NonProbiotic)

orthogonal to each other. PCA loadings are useful for interpretation, especially in cases where the perturbation in question is well characterized.

The spectral database of water absorbance bands that have been discovered experimentally and calculated as overtones of already found bands in the IR range has more than 500 bands in the area of the first, second and third overtones of water, i.e. in the near infrared, 700–2500 nm, wavelength region.

Water spectral monitoring and spectral data analysis applied to a time series spectral data of pure water under consecutive illuminations has revealed the existence of a numerous of water absorbance bands underneath the first overtone of water spectrum, 1300 nm–1600 nm, Fig. 8.4. All of the characteristic bands found in this experiment belong to the 12 characteristic water regions (C1–C12) found under the first overtone of water (Tsenkova 2009).

The methodology of Aquaphotomics will be further illustrated by the results of spectral data acquisition and analysis of aqueous system where the only perturbation is temperature (Segtnan et al. 2001). Water near infrared spectra have been acquired at various temperatures in the region from 30 to 65 °C, Fig. 8.6. Broadly speaking, there are two main categories of models for describing water structure: mixture and continuum models. Mixture models consider water to be a multi-component mixture of species with different numbers of hydrogen bonds, which break as temperature is increased (Segtnan et al. 2001; Maeda et al. 1995). Continuum models regard water as a system which is almost completely hydrogen bonded but these bonds weaken with increasing temperature (Smith et al. 2005). A number of studies on the effect of temperature on vibrational spectra of water have been shown to support the mixture model of water structure (Segtnan et al. 2001; Maeda et al. 1995), while the continuum model has also been shown to be consistent with observed data (Smith et al. 2005). Nevertheless, it is agreed that temperature changes result in alterations in hydrogen bonding configurations in water.

The main feature of the NIR spectrum of water is a broad peak around 1450 nm, which is comprised of many overlapped bands, described mainly as overtones of OH stretching and combinations of OH stretching and bending vibrations. The existence of an isosbestic point in the NIR spectra of water measured at different temperatures has been offered as evidence that water can be considered as a two component mixture, comprising of two different types of water: strongly and weakly hydrogen bonded, with the proportion of weakly bonded species increasing as temperature increases (Segtnan et al. 2001). In the middle infrared wavelength range, researchers employing fiber optic evanescent wave spectroscopy detected two isosbestic points: one during the ice-water phase transition at 3280 cm^{-1} (3048.78 nm) and another in the 20–100 °C temperature range located at 3530 cm^{-1} (2832.86 nm) (Raichlin and Katzir 2008). This led them to propose a four component mixture model for liquid water structure: molecules with 4 bonds (ice-like); two 3-coordinate species (either with a broken OH bond or broken lone pair electron bond) and one 2-coordinate species (with two broken bonds). However, studies using Raman spectroscopy and

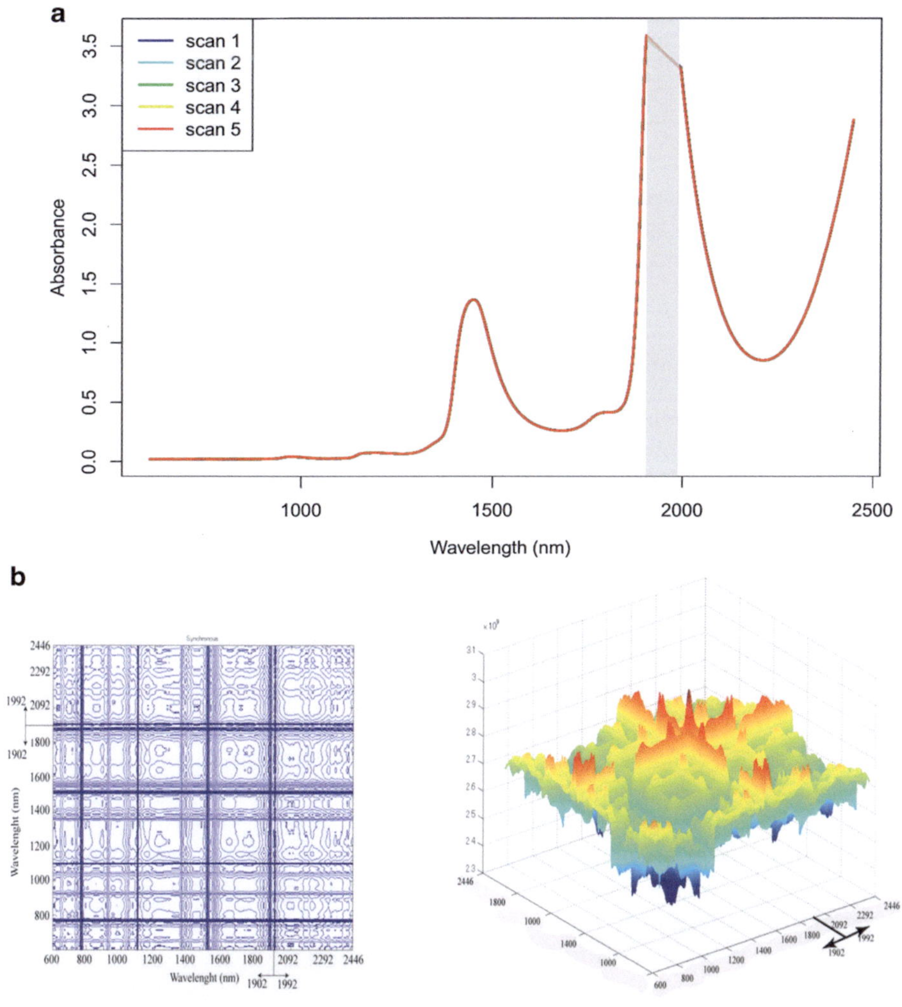

Fig. 8.4 Water spectral monitoring in the spectral range between 600 and 2446 nm (1902–1992 nm was excluded due to undetectable high absorption). (**a**) Raw spectra of water, 1 mm light pathlength. (**b**) Two dimensional correlation spectroscopy plots

Monte Carlo simulations have refuted this hypothesis, by showing that thermal variations of a single species line shape can also generate isosbestic points (Smith et al. 2005). Less hydrogen bonded water is predominant at higher temperature while low temperature spectra are depicted mostly in the area of the hydrogen bonded water where water molecules formed species with one (S1), two (S2), three (S3) and four (S4) hydrogen bonds (Franks 1973).

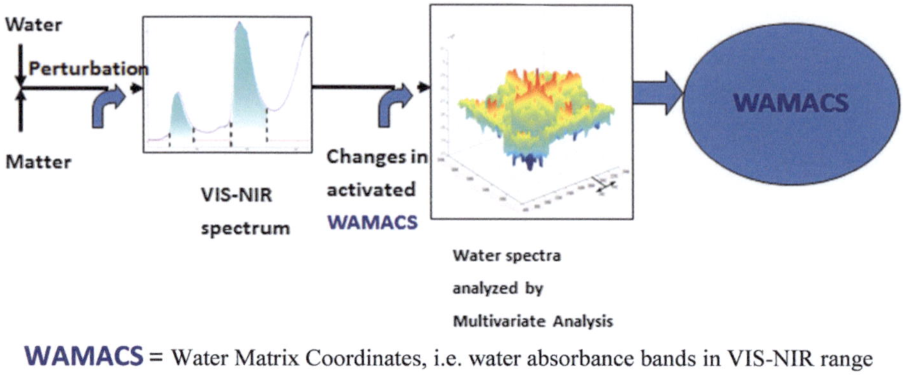

Fig. 8.5 Spectral data analysis flow chart towards finding specific water bands

8.3 Applications

8.3.1 Characterization of WAMACS and WASP for Water Molecular System Using Perturbations, Spectroscopy and Multivariate Analysis as Main Tools in Aquaphotomics

Temperature change and addition of solutions of different ionic strength as water molecular system perturbations were applied to Milli-Q water and compared in terms of activated water molecular structures, i.e. water absorbance bands. Rapid and precise measurement of low concentration solutes were employed to show that it is possible to quantify very low concentrations of contaminants in water and further on, to identify the main absorbance bands of water activated by contaminants in the presence even of another perturbation like temperature.

8.3.1.1 Investigation of the Effects of Temperature on Water Structure

Aquaphotomics aims to relate water spectral pattern as a biomarker to the structure related functions of water in biological systems using spectroscopy. A number of multivariate spectral analysis methods have been proposed for identification of water absorbance bands (i.e. wave bands related to water structure) from near infrared spectra, in order to characterize the so-called "water spectral pattern, WASP". Examples include principal component analysis loadings and partial least square regression vectors. Interpretation of these spectral patterns has to be made very carefully and in an educated manner because of the following reasons:

1. the profile of a PLS regression vector is sensitive to model parameters (e.g. number of latent variables selected, spectral data pretreatment methods etc.);

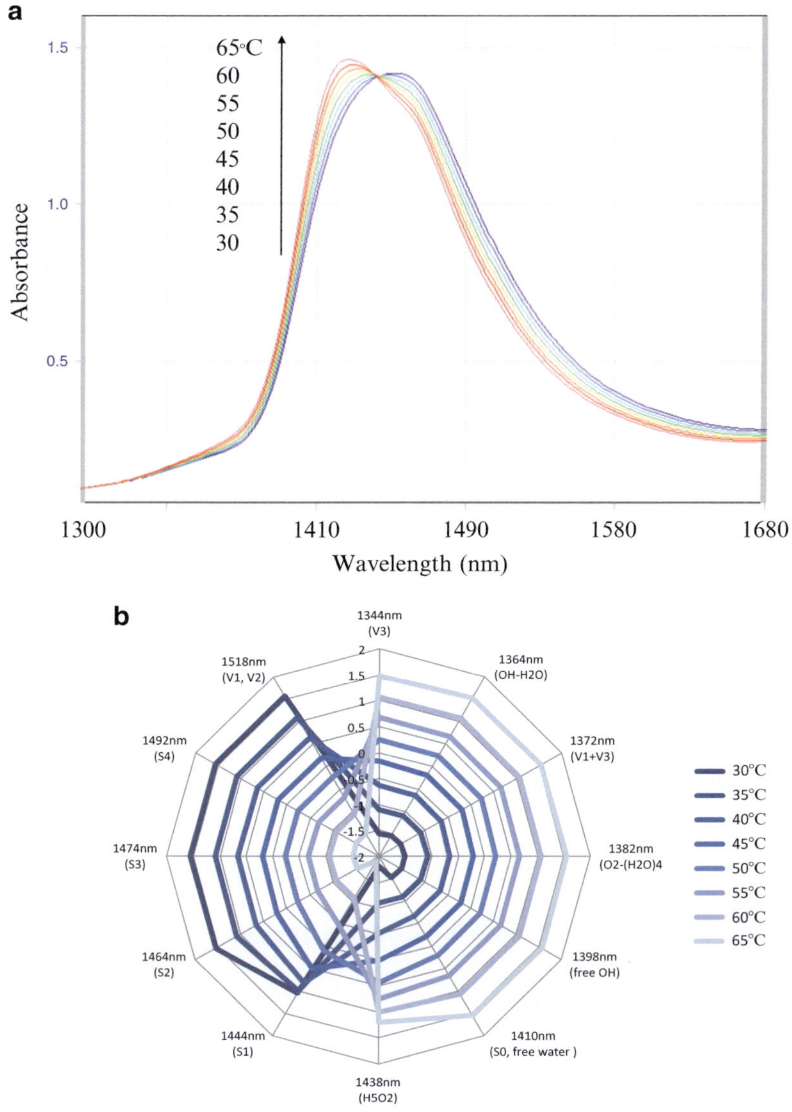

Fig. 8.6 Water spectra at various temperature from 30 to 65 °C, (**a**) in the range of the water first overtone, 1300–1600 nm and (**b**) their respective aquagrams

2. due to the overlapping nature of vibrational overtone and combination bands in the NIR wavelength range, peaks in regression vectors cannot be directly related to absorbance peaks (Seasholtz and Kowalski 1990);
3. interfering factors, such as the strong influence of ambient temperature of NIR spectra on water, may confound the elucidation of spectral responses related to the perturbation of interest.

PCA is more appealing, since it is an unsupervised technique in which the major variations in a spectral dataset are compressed into the first few principal components, which are orthogonal to each other. PCA loadings are useful for interpretation, especially in cases where the perturbation in question is well characterized. One well-known example is aqueous system where the only perturbation is temperature (Segtnan et al. 2001; Maeda et al. 1995). However, in cases where multiple simultaneous (typically unknown) perturbations are present, it may be difficult to interpret the resultant PC loadings. We will concentrate on three different perturbations that might occur simultaneously: temperature, salt type and each salt concentration (Gowen et al. 2013). NIR transmission spectra of deionized water from a Milli-Q water purification system (Millipore, Molsheim, France, resistance = 18 mΩ) at different temperatures, ranging from 30 to 45 °C and its salt solutions at various low concentrations and types of salt were obtained using two types of spectrometers: (1) dispersive grating based spectrophotometer, NIR System 6500 (Foss NIR-System, Laurel, USA), fitted with a quartz cuvette with 1 mm optical path length, wavelength region of 400–2500 nm with 2 nm step acquisition and (2) interferometer based MPA spectrophotometer (Brucker Inc. Billerica, MA) fitted with a quartz cuvette with 1 mm optical path length.

In order to observe water bands activated in the presence of two or more simultaneous perturbations, four different types of salt solutions were analyzed. All salts (NaCl, KCl, $MgCl_2$, $AlCl_3$) solutions had concentrations ranging from 0.2 to 1 mol.L^{-1} in steps of 0.2 mol.L^{-1}. These experiments were carried out at three different temperatures: approximately 28, 38 and 46 °C.

Principle component analysis (PCA) and Multivariate curve resolution (MCR) (Gowen et al. 2013; Abdollahi and Tauler 2011) were applied to the spectral and reference data using the toolbox developed by Jaumot et al. (2005) for MATLAB (The MathWorks, Inc., Natick, MA). The wavelength range of the acquired spectra was trimmed to 1300–1600 nm in order to examine the 1st overtone of the OH stretch vibration in water.

Spectra of pure water at temperatures varying from 30 to 45 °C, measured on the two different spectrometers employed, are shown in Fig. 8.7a, b. Overall, the trends shown by each instrument were very similar, slight variations such as maximum value of log(1/T) and spectral shape, which arise due to the differences in the effective path length and optical configuration of the different spectrometers were observed. As it has been demonstrated by previous research (Segtnan et al. 2001; Maeda et al. 1995), the peak of the first overtone spectra experiences a shift towards lower wavelengths ("blue shift") as temperature is increased, with an apparent isosbestic point around 1440–1442 nm.

Principal component (PC) scores and loadings for the data obtained on each spectrometer are shown in Figs. 8.8 and 8.9, respectively. For each spectrometer, the first two principal components described more than 99 % of the spectral variations in the datasets and the second components varied linearly with temperature. Third component with nonlinear temperature dependence and a structured PC loading was evident. The existence of a third, spectrally distinct component in the evolution of NIR spectra of water with changing temperature has been previously observed

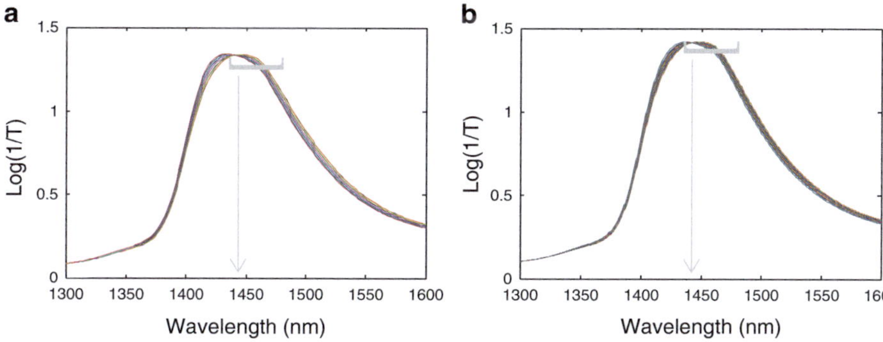

Fig. 8.7 First overtone NIR spectra of water in the temperature range 30–46 °C, obtained using dispersive grating based (**a**) and interferometer based (**b**) spectrometers (first experimental replicate). Region of isosbestic point is marked

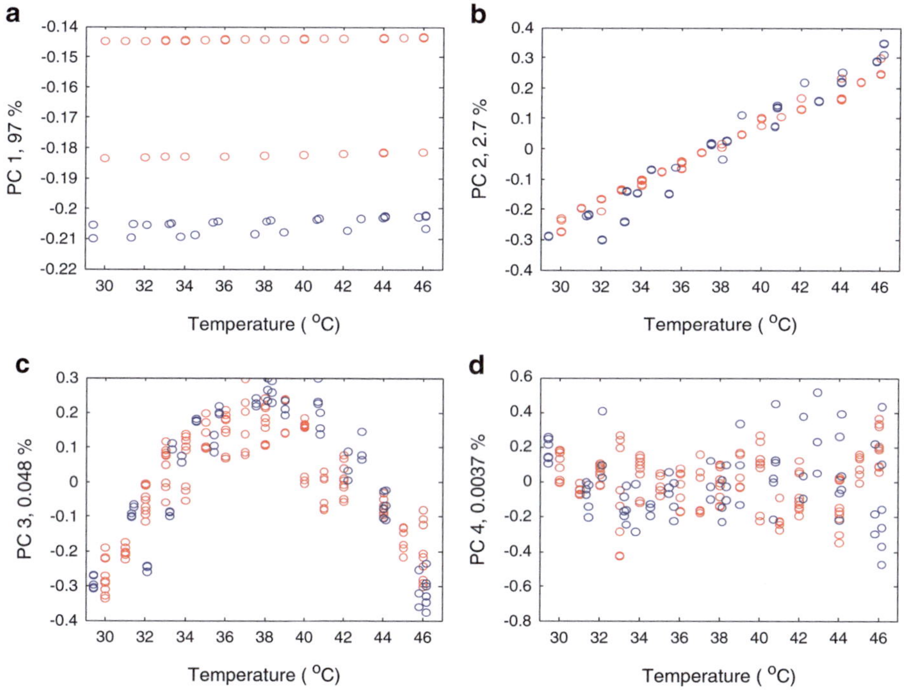

Fig. 8.8 Principal component scores (PC1 (**a**), PC2 (**b**), PC3 (**c**) and PC4 (**d**)) plotted as a function of temperature for first overtone spectra of water in the range 30–46 °C obtained using dispersive grating based spectrometer (*shown in blue*) and interferometer based spectrometer (*shown in red*)

(Segtnan et al. 2001). These experiments indicated an increase in the third PC score with temperature, to a peak at around 37 °C, followed by a decrease to 45 °C, indicating that this represents an intermediate state of water.

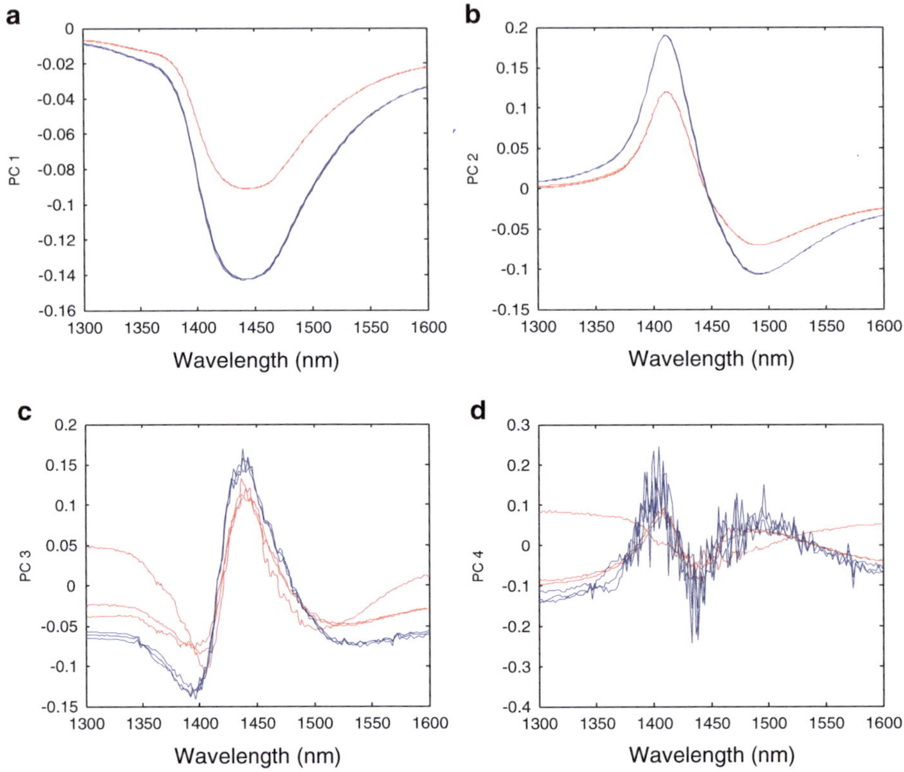

Fig. 8.9 Principal component loadings (PC1 (**a**), PC2 (**b**), PC3 (**c**) and PC4 (**d**)) plotted as a function of wavelength for first overtone spectra of water in the range 30–46 °C obtained using dispersive grating based spectrometer (*shown in blue*) and interferometer based spectrometer (*shown in red*)

The locations of the main turning points in each of the first three PC loadings are summarised in Table 8.2. These can be regarded as the characteristic water matrix coordinates, WAMACS, for the perturbation of changing temperature applied to the water matrix (Gowen et al. 2013). The table contains the main WAMACS for salt solutions, too, which will be discussed further.

The first PC exhibited a single broad peak at 1440–1443 nm, representing a combination of the first overtone of the OH bending and fundamental OH asymmetric stretching vibration ($2v_2 + v_3$) (Lin and Brown 1993). The second PC loading exhibited a positive peak at 1412 nm and a negative one at 1492 nm. These peaks have been previously assigned by various authors. The feature at 1412 nm has been attributed to a form of water with relatively weak hydrogen bonding (Segtnan et al. 2001), a combination mode of the fundamental symmetric and asymmetric OH stretching vibrations ($v_1 + v_3$) for water molecules with no hydrogen bonds (S_0) (Franks 1973; Maeda et al. 1995) or the first overtone of the

Table 8.2 Locations of peaks in principal component (PC) loadings for temperature and salts applied as perturbations to water

Samples	PC1 loading bands	PC2 loading bands	PC3 loading bands
Water	1440-3 (−)	1412 (+), 1492 (−)	1393-1400(−), 1438 (+)
NaCl 28 °C	1448-50 (+)	1396 (+), 1430(−), 1526(+)	1406(+), 1466(−)
NaCl 36 °C	1442 (+)	1396 (+), 1432(−), 1524(+)	1406(+), 1468(−)
NaCl 46 °C	1436-8 (+)	1396 (+), 1434(−), 1528(+)	1406(+), 1470(−)
KCl 28 °C	1448-50 (+)	1396 (+), 1430(−), 1524–6(+)	1404(+), 1468(−)
KCl 36 °C	1440-2 (+)	1398 (+), 1430(−), 1524(+)	1406(+), 1464-72(−)
KCl 46 °C	1436-8 (+)	1398 (+), 1434(−), 1530(+)	1406(+), 1478(−)
MgCl2 28 °C	1450-2 (+)	1404(−), 1492(+)	1364(+), 1400(−), 1428(+)
MgCl2 36 °C	1444-6 (+)	1404(−), 1496(+)	1368(+), 1400(−), 1428(+)
MgCl2 46 °C	1440 (+)	1406(−), 1494(+)	1368(+), 1400(−), 1428(+)
AlCl3 28 °C	1452 (+)	1408(−)	1404-22 (+), 1470-1494 (−)
AlCl3 36 °C	1446-8 (+)	1408(−)	1404-22 (+), 1470-1494 (−)
AlCl3 46 °C	1442-4 (+)	1408(−)	1404-22 (+), 1470-1494 (−)

antisymmetric stretch for free OH in the free water molecule (Czarnik-Matusewicz and Pilorz 2006). The feature at 1492 nm has been attributed to a water species with stronger hydrogen bonds (Franks 1973), a combination mode of the fundamental symmetric and asymmetric OH stretching vibrations ($v_1 + v_3$) for water molecules with four hydrogen bonds (S_4) (Franks 1973; Maeda et al. 1995). The feature at 1412 nm relates to weaker hydrogen bonding and that at 1492 nm relates to stronger hydrogen bonding. The opposing signs of these features in the second loading indicate that as one species increases in concentration, the other decreases. In other words, as the temperature increases, the proportion of weakly bonded water increases and that of strongly bonded water decreases. As for the third PC loading, a small negative feature was observed at 1393–1400 nm and a relatively larger, positive feature at 1438 nm. The feature at 1393–1400 nm is probably due to the changing concentration of weakly hydrogen bonded water in the system (Franks 1973; Segtnan et al. 2001; Maeda et al. 1995), while that at 1438 nm has been previously assigned as a third species of water (Segtnan et al. 2001) or a protonated water dimer in gas phase (Tsenkova 2009). The evidence for a nonlinear system dynamics regarding temperature as water perturbation and the existance of more than two water matrix coordinates suggests the existance of water molecules in transitional state that maintain the balance of the two populations of strongly bonded and weakly bonded water molecules. With this investigation, it was possible to establish the aquaphotome of the Milli-Q water in the 30–45 °C temperature range. Three characteristic water bands were found to be related to temperature perturbation, in this temperature range: C5, C7 and C11. One additional bands was found, too: 1393 nm. In the C7 range, there 2 bands related to temperature: 1438 nm and 1440 nm. This is how the study of this aqueous system brought a contribution to Aquaphotomics in terms of a new specific water absorbance band and finding two specific bands in one region. All together, five WAMACS were identified and all the loadings depict the specific WASP, i.e. the weights of each of those bands with the change of the temperature in the 30–45 °C range. The simultaneous activation of the found WAMACS would be always identified with temperature changes in water.

8.3.1.2 Investigation of the Effects of Multiple Perturbations on Water Structure; Low Concentrations of Salts and Temperature

Another perturbation that illustrates changes in the OH bonding of the water molecular system is the effect of low concentrations of salts diluted in water (Gowen et al. 2013; Inoue et al. 1984; Molt et al. 1997; Frost and Molt 1997). However, we are aware that unavoidable, though minor, fluctuations in ambient laboratory conditions (mainly temperature) also have an influence on water structure, manifest as a shift in the peak position of the OH first overtone peak (located at around 1450 nm). It can be seen, Fig. 8.10, that upon the addition of salt, the so-called isosbestic point of water spectrum becomes even more blurred than was the case for pure water. It can also be seen that changing temperature by 20 °C had a far greater effect on the water matrix than did adding salts up to 1 Mol concentration. In addition, the spread among spectra at each temperature is informative: it can be seen that $AlCl_3$ had the greatest effect on the spectrum, as compared to the other salts. This is due to its relatively higher Lewis acidity. The trend of the shift in the peak position due to temperature was similar to that observed in the temperature perturbed pure water (i.e. the peak moves towards lower wavelengths as temperature is increased), but the actual peak locations were different to those for the pure water. In order to diminish the temperature effects, spectra of control samples of pure water were measured during the course of each experiment. The principal variations in these spectra were obtained by applying PCA and extracting the 1st PC loading as an "interferent spectrum". Then extended multiplicative scatter correction (EMSC) (Afseth and Kohler 2012) was applied to the spectra of salt solutions, to correct the data for the interferent spectrum. This effectively diminished the influence of ambient conditions on the spectra of salt solutions. Consequently, when applying PCA to the EMSC-corrected data, the 1st PCA loading could be related to the perturbation of interest (i.e. addition of salt), and would thus present us with a reasonable estimate of the WAMACS and WASP related to salts, i.e. the aquaphotome of salts could be established.

8.3.1.3 Evaluate the Performance of NIRS Combined with Chemometrics Modelling for Prediction of Low Concentrations of Salt in Water

In order to investigate the influence of the environmental when concentrating on the extraction of WAMACS related to salts, spectral data acquired at four different test locations (Gowen et al. 2015) were used to construct the respective WASP of different salts. As salts do not absorb NIR light, accurate measurement of even

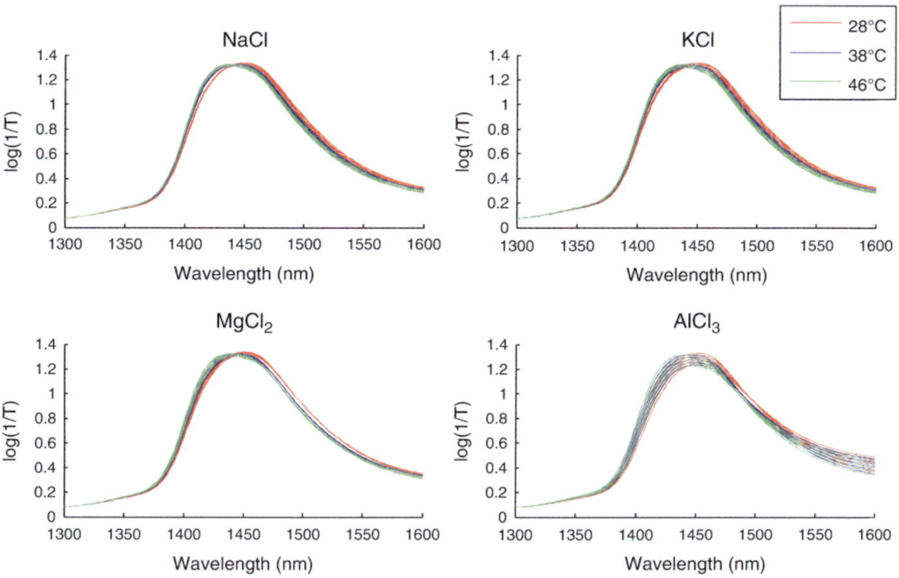

Fig. 8.10 Absorbance (Log(1/T)) spectra of salt solutions of different concentrations, grouped by temperature

low concentration of salts means that salts change the surrounding water molecular structure according to the number of the solvent molecules in the solution. High prediction of salt concentration when using the respective solution spectra means that specific WASP has been found for each salt. In this process, firstly, after spectra for each location studied and each experimental day were pooled, spectral data in the 1300–1600 nm wavelength region (the so-called "first overtone region of water") were extracted. EMSC (Afseth and Kohler 2012) was applied to this data, using the 1st principal component loading of the control (Milli-Q water) spectra as an interferent spectrum. Finally, principal component analysis was applied to the SNV + EMSC pre-treated data, and the 1st principal component loading was extracted. In order to avoid sign and scale ambiguities, this loading was squared and scaled to the range [0,1]. The resultant scaled loading was employed as a water spectral pattern (WASP) for each salt/location, Fig. 8.12. The combination of investigated spectral pre-treatment, wavelength range, salt and test day resulted in a rather large number of competing models (432) for each test location. In order to compare the effects of spectral pre-treatments and wavelength ranges on model performance, the Root Mean Squared Error of Prediction (RMSEP) values corresponding to models for each test set and location were averaged. The average RMSEPs obtained are plotted for each salt as a function of wavelength range and spectral pre-treatment in Fig. 8.11. It is evident from this figure that in most cases the first overtone of water, 1300–1600 nm wavelength range, resulted in the lowest prediction error, while the most effective spectral pre-treatment was EMSC.

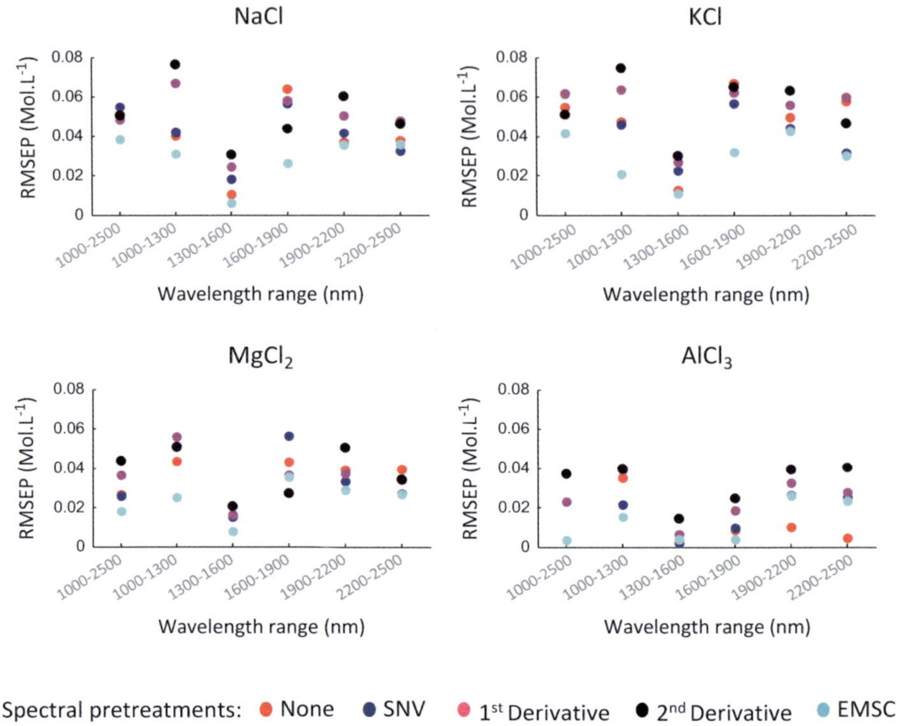

Fig. 8.11 Root mean squared error of prediction (RMSEP) averaged over permuted experimental test days and test locations for each salt, pre-treatment and wavelength range studied (*SNV* standard normal variate, *EMSC* extended multiplicative signal correction)

When RMSEP values from the individual test locations were evaluated separately, similar findings were obtained. Calibration models built in this wavelength range and employing the EMSC pre-treatment resulted in highly accurate prediction with RMSEP values ranging from 0.004 (for $AlCl_3$) to 0.01 (for KCl) $Mol.L^{-1}$. Water absorbance spectra first PC loadings for each salt/experimental location are plotted in Fig. 8.12. Each of the salts' WASP exhibited peaks around 1400, 1430–1450 and 1500–1520 nm, which can be related to different hydrogen bonding strengths in the water molecules (bond strength decreasing with decreasing wavelength) perturbed by salts. Similar to the previous results for water perturbed by temperature, mainly three ranges underneath the first overtone of water were activated, but their location was different. Although the peak locations, the WAMACS, in the WASP were similar for each salt studied, the shapes of the profiles varied markedly with the number of Chloride ions present.

The presence of ions in solution affects hydrogen bonding in water, i.e. different ion concentration works as a perturbation that can be observed directly by NIR spectroscopy (NIRS) (Weber et al. 2000). Consequently, NIRS has been used to estimate the concentration of OH^- and H^+ ions in solution, i.e. pH (Molt et al.

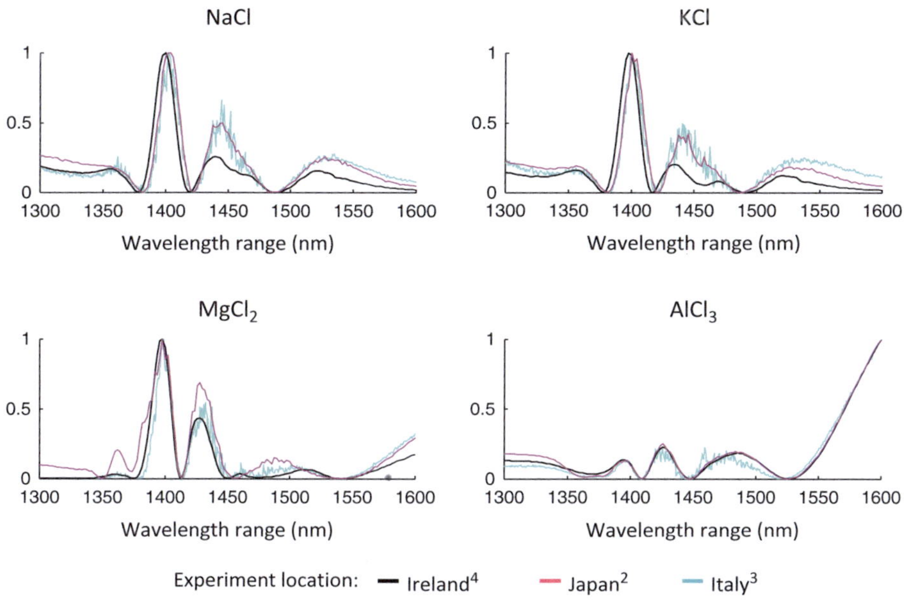

Fig. 8.12 Water absorbance pattern (1st principal component loading of SNV followed by EMSC pre-treated data) for each salt and test location

1997). It was demonstrated that binary mixtures of inorganic ionic salts (e.g. NaCl, KCl, $MgCl_2$ and $AlCl_3$) can be quantitatively analyzed with NIR due to the specific effects of cation/anion OH interactions (Frost and Molt 1997), also known as the Lewis acidity effect (McNaught and Wilkinson 1997). Therefore, identification of water absorbance bands related to respective perturbations would lead to further use of these bands as coordinates of a specific spectral patterns that can be exploit as biomarkers. The action of salts during dissociation in water is a complex phenomenon. Structure induced shifts are highly dependent on several parameters, such as the valence and charge distribution of ions in solution, the electron affinity and the size of the ions. This dependence can be observed by the changes in the pH observed in different solutions of the same ionic salt at different concentration levels.

Shifts according to the salt employed have been observed: for NaCl or KCl the peak locations ranged from 1436 to 1450 nm, for $MgCl_2$, they ranged from 1440 to 1452 nm and for $AlCl_3$ they ranged from 1442 to 1452 nm. The appearance of the principal components PC2 and PC3 loadings differed according to the salt used and also differed substantially in profile from those obtained for the temperature perturbation, indicating that the addition of salts to water perturbed the hydrogen bond network in a manner distinct from that of changing only temperature. Considering the second PC, both NaCl and KCl exhibited positive features at 1396 nm and 1526 nm and negative features at 1430–1434 nm; $MgCl_2$ and $AlCl_3$ exhibited negative features at around 1404–1408 nm, and $MgCl_2$ exhibited a positive feature at 1492 nm. As for the third PC, both NaCl and KCl exhibited positive features at 1404–1406 nm and negative ones at around 1468–1478 nm, while $MgCl_2$ exhibited positive features at 1364–1368 nm and 1428 nm and a negative one at 1400 nm. $AlCl_3$ exhibited a positive feature at 1404–10,422 nm and negative ones around 1470–1494 nm.

It has been shown that multivariate analysis can be applied to characterize the effects of different perturbations separately and simultaneously (i.e. temperature and addition of salts) to the water matrix in terms of principal spectral components and concentration profiles. Based on the temperature perturbation experiments, it has been found that three distinct components with varying temperature dependence are present in water in the temperature range 30–45 °C, a range which is significant for biological systems. The salt solutions studied exhibited similar trends with respect to the temperature perturbation, while the locations of WAMACS varied according to the salt used. This gives insight into the differing characteristics of these salts and their respective effects on the hydrogen bonded network in water. Each salt and concentration level affected water in a different way, according to its acidity. The results indicate that water structure can be reasonably interpreted as a multi-state system, and are in line with the traditional definition of salts as structure makers/breakers.

8.3.2 *Other Applications*

The first publications where water bands have been used to understand and present new phenomena have been on diagnosis of mammary gland inflammation (Tsenkova et al. 2001), prion protein isoforms identification (Tsenkova et al. 2004) and water functional analysis based on real time monitoring of the water bands in the spectra of prion protein solution (Tsenkova 2010). Recent publications have shown at various levels of system organization that WAMACS are real « informational hubs ». Research on soybean species identification related to their cold resistance abilities have proved that genetically modified soybean plants with different resistance can be characterized by different respective water absorbance pattern. It has been found that there are specific WASP for different level of cold resistance (Jinendra 2010; Jinendra et al. 2010). Further on, soybean mosaic virus diagnosed with near infrared spectroscopy using the Aquaphotomics concept showed that for the sick plants, there was an inhibition of water absorbance at the high energy bands, i.e. less hydrogen bonded water (Jinendra 2010). Another interesting application is the use of Aquagrams to illustrate the specific WASP of urine when Giant Panda is in estrus (Kinoshita et al. 2012). Aquaphotomics has been proposed in the research of nanomaterials where Aquagrams showed water spectral pattern as characterization of hydrogenated nanomaterial (Matija et al. 2012). With Aquaphotomics, we have found that successful bacteria identification with NIRS is highly accurate not because of the difference in bacteria cells, but because of the metabolites that change the conformation of the water molecular system (Nakakimura et al. 2012) surrounding them.

8.4 New Frontiers

Water in biology is still underestimated, underdeveloped subject. On the contrary, for the last few years, water has been « noticed » and like an avalanche, the number of papers on water are increasing very fast. Near Infrared Spectroscopy is a powerful tool to study water: rapid and noninvasive. Acquisition of immense number of spectra of various systems in real time as time series and under various possible perturbations became an important research subjects of other laboratories, too. We, together, are building up the Aquaphotome of water and other aqueous systems in order to « see » the « hubs » where we can harvest information from the « water computers ». Without chemometrics, multivariate analysis and other new methods for spectral data analysis (to be developed), it would be impossible to utilize all the spectral data that we can acquire. New, super computers will help meeting the experimental data with the results of computer simulation and modeling, which will show expected consistency and will open new horizons in understanding unknown phenomena in various fields. There are so many areas related to water, but especially life science will gain most. Non-invasive diagnosis, identification and understanding of DNA, cells, tissue, whole organisms is not a dream anymore. Abnormality in other molecular pathways is "reflected" by the water molecular system as a mirror. We can even foresee that having information from the water molecular system in the body, in real time, gives the chance to think about how can we influence and change it in order to maintain its balance and get it back to the normal condition. It would open a completely new field of water molecular network manipulation. The definition of « normal », « healthy » from the point of view of Aquaphotomics is the ever standing challenge.

References

Abdollahi H, Tauler R (2011) Uniqueness and rotation ambiguities in Multivariate Curve Resolution methods. Chemom Intell Lab Syst 108:100–111. doi:10.1016/j.chemolab.2011.05.009

Afseth NK, Kohler A (2012) Extended multiplicative signal correction in vibrational spectroscopy, a tutorial. Chemom Intell Lab Syst 117:92–99. doi:10.1016/j.chemolab.2012.03.004

Czarnik-Matusewicz B, Pilorz S (2006) Study of the temperature-dependent near-infrared spectra of water by two-dimensional correlation spectroscopy and principal components analysis. Vib Spectrosc 40:235–245. doi:10.1016/j.vibspec.2005.10.002

Franks F (1973) Water: a comprehensive treatise. Plenum Press, New York, pp 276–279

Frost VJ, Molt K (1997) Analysis of aqueous solutions by near-infrared spectrometry (NIRS) III. Binary mixtures of inorganic salts in water. J Mol Struct 410-411:573–579. doi:10.1016/S0022-2860(96)09707-4

Gowen AA, Amigo JM, Tsenkova R (2013) Characterisation of hydrogen bond perturbations in aqueous systems using aquaphotomics and multivariate curve resolution-alternating least squares. Anal Chim Acta 759:8–20. doi:10.1016/j.aca.2012.10.007

Gowen AA, Marini F, Tsuchisaka Y, De Luca S, Bevilacqua M, O'Donnell C, Downey G, Tsenkova R (2015) On the feasibility of near infrared spectroscopy to detect contaminants in water using single salt solutions as model systems. Talanta 131:609–618. doi:10.1016/j.talanta.2014.08.049

Inoue A, Kojima K, Taniguchi Y, Suzuki K (1984) Near-infrared spectra of water and aqueous electrolyte solutions at high pressures. J Solut Chem 13:811–823

Jaumot J, Gargallo R, de Juan A, Tauler R (2005) A graphical user-friendly interface for MCR-ALS: a new tool for multivariate curve resolution in MATLAB. Chemom Intell Lab Syst 76:101–110. doi:10.1016/j.chemolab.2004.12.007

Jinendra BM (2010) Near infrared spectroscopy and aquaphotomics: novel approach for rapid in vivo diagnosis of soybean. Ph.D thesis, Kobe University

Jinendra B, Tamaki K, Kuroki S, Vassileva M, Yoshida S, Tsenkova R (2010) Near infrared spectroscopy and aquaphotomics: novel approach for rapid in vivo diagnosis of virus infected soybean. Biochem Biophys Res Commun 397:685–690. doi:10.1016/j.bbrc.2010.06.007

Kinoshita K, Miyazaki M, Morita H, Vassileva M, Tang C, Li D, Ishikawa O, Kusunoki H, Tsenkova R (2012) Spectral pattern of urinary water as a biomarker of estrus in the giant panda. Sci Rep 2:856. doi:10.1038/srep00856

Lin J, Brown CW (1993) Near-IR spectroscopic measurement of seawater salinity. Environ Sci Technol 27:1611–1615. doi:10.1021/es00045a017

Maeda H, Ozaki Y, Tanaka M, Hayashi N, Kojima T (1995) Near infrared spectroscopy and chemometrics studies of temperature-dependent spectral variations of water: relationship between spectral changes and hydrogen bonds. J Near Infrared Spectrosc 3:191–201. doi:10.1255/jnirs.69

Matija LR, Tsenkova RN, Miyazaki M, Banba K, Muncan JS (2012) Aquagrams: water spectral pattern as characterization of hydrogenated nanomaterial. FME Trans 40:51–56

McNaught AD, Wilkinson A (1997) Terminology, compendium of chemical (the "Gold Book"), 2nd edn. Blackwell Scientific Publications, Oxford

Molt K, Niemöller A, Cho YJ (1997) Analysis of aqueous solutions by near-infrared spectrometry (NIRS) II. Titrations of weak and very weak acids with strong bases. J Mol Struct 410–411:565–572. doi:10.1016/S0022-2860(96)09706-2

Murayama K, Czarnik-Matusewicz B, Wu Y, Tsenkova R, Ozaki Y (2000) Comparison between conventional spectral analysis methods, chemometrics, and two-dimensional correlation spectroscopy in the analysis of near-infrared spectra of protein. Appl Spectrosc 54:978–985. doi:10.1366/0003702001950715

Nakakimura Y, Vassileva M, Stoyanchev T, Nakai K, Osawa R, Kawano J, Tsenkova R (2012) Extracellular metabolites play a dominant role in near-infrared spectroscopic quantification of bacteria at food-safety level concentrations. Anal Methods 4:1389. doi:10.1039/c2ay05771a

Putra A, Santo R, Kuroki S, Tsenkova R (2010) Robust spectral model for low metal concentration measurement in aqueous solution reveals the importance of water absorbance bands. In: Proceedings of the 14th international conference on near infrared spectroscopy, Bangkok, Thailand, pp 831–835

Raichlin Y, Katzir A (2008) Fiber-optic evanescent wave spectroscopy in the middle infrared. Appl Spectrosc 62:55A–72A

Robertson WWH, Diken EEG, Price EAE, Shin J-W, Johnson MA (2003) Spectroscopic determination of the OH- solvation shell in the OH-. (H2O)n clusters. Science 299:1367–1372. doi:10.1126/science.1080695

Seasholtz M, Kowalski B (1990) Qualitative information from multivariate calibration models. Appl Spectrosc 44:1337–1348

Segelstein D (1981) The complex refractive index of water. M.S. thesis, p 167

Segtnan VH, Šašić Š, Isaksson T, Ozaki Y (2001) Studies on the structure of water using two-dimensional near-infrared correlation spectroscopy and principal component analysis. Anal Chem 73:3153–3161. doi:10.1021/ac010102n

Smith JD, Cappa CD, Wilson KR, Geissler PL, Cohen RC, Saykally RJ (2005) Unified description of temperature-dependent hydrogen-bond rearrangements in liquid water. Proc Natl Acad Sci 102:14171

Tsenkova R (2009) Introduction aquaphotomics: dynamic spectroscopy of aqueous and biological systems describes peculiarities of water. J Near Infrared Spectrosc 17:303–314. doi:10.1255/jnirs.869

Tsenkova R (2010) Aquaphotomics: water in the biological and aqueous world scrutinised with invisible light. Spectrosc Eur 22:6–10

Tsenkova R, Atanassova S, Toyoda K (2001) Near infrared spectroscopy for diagnosis: influence of mammary gland inflammation on cow's milk composition measurement. Near Infrared Anal 2:59–66

Tsenkova RN, Iordanova IK, Toyoda K, Brown DR (2004) Prion protein fate governed by metal binding. Biochem Biophys Res Commun 325:1005–1012. doi:10.1016/j.bbrc.2004.10.135

Tsenkova R, Iso E, Parker M, Fockenberg C, Okubo M (2007a) Aqua-photomics: a NIRS investigation into the perturbation of water spectrum in an aqueous suspension of mesoscopic scale polystyrene spheres. In: 13th international conference on near infrared spectroscopy. Umea-Vasa, Sweden & Finland, pp A–04: 72

Tsenkova R, Fockenberg C, Koseva N, Sakudo A, Parker M (2007b) Aqua-photomics: water absorbance patterns in NIR range used for detection of metal ions reveal the importance of sample preparation. In: 13th international conference on near infrared spectroscopy. Umea-Vasa, Sweden & Finland, 03–02: 73

Weber JMJ, Kelley J, Nielsen S, Ayotte P, Johnson M (2000) Isolating the spectroscopic signature of a hydration shell with the use of clusters: superoxide tetrahydrate. Science 287:2461–2463. doi:10.1126/science.287.5462.2461

Chapter 9
Hydration in Lipid Monolayers: Correlation of Water Activity and Surface Pressure

E. Anibal Disalvo, Axel Hollmann, and M. Florencia Martini

Abstract In order to give a physical meaning to each region of the membrane we define the interphase as the region in a lipid membrane corresponding to the polar head groups imbibed in water with net different properties than the hydrocarbon region and the water phase. The interphase region is analyzed under the scope of thermodynamics of surface and solutions based on the definition of Defay-Prigogine of an interphase and the derivation that it has in the understanding of membrane processeses in the context of biological response. In the view of this approach, the complete monolayer is considered as the lipid layer one molecule thick plus the bidimensional solution of the polar head groups inherent to it (the interphase region). Surface water activity appears as a common factor for the interaction of several aqueous soluble and surface active proteins with lipid membranes of different composition. Protein perturbation can be measured by changes in the surface pressure of lipid monolayers at different initial water surface activities. As predicted by solution chemistry, the increase of surface pressure is independent of the particle nature that dissolves. Therefore, membranes give a similar response in terms of the determined surface states given by water activity independent of the protein or peptide.

Keywords Lipid monolayers • Surface pressure • Water activity • Defay-Prigogine model • Peptide interaction

E.A. Disalvo (✉) • A. Hollmann
Laboratorio de Biointerfases y Sistemas Biomimeticos, Centro de Investigacion y Transferencia de Santiago del Estero, Universidad Nacional de Santiago del Estero-Consejo Nacional de Investigaciones Científicas y Técnicas, 4200 Santiago del Estero, Argentina
e-mail: disalvoanibal@yahoo.com.ar

M.F. Martini
Instituto de Química y Metabolismo del Fármaco, IQUIMEFA UBA-CONICET, Facultad de Farmacia y Bioquímica, Universidad de Buenos Aires, Junín 956 PP (1113), Buenos Aires, Argentina
e-mail: flormartini@yahoo.com.ar

E.A. Disalvo (ed.), *Membrane Hydration*, Subcellular Biochemistry 71,
DOI 10.1007/978-3-319-19060-0_9

9.1 Introduction

The acceptance that water is part of the membrane structure has updated the view of a lipid bilayer as a composite element consisting of the lipid bimolecular leaflet with adjacent hydrated polar head groups (Jendrasiak 1996; Milhaud 2004; Disalvo et al. 2008).

In order to give the proper physical meaning to each region of the membrane we define the interphase as the region in a lipid membrane corresponding to the polar head groups imbibed in water with net different properties than the hydrocarbon region and the water phase.

In a first approximation, this region is confined between ideal planes, called interfaces. One is the Gibbs plane defined at the carbonyl level that determines the limit between the hydrocarbon region and the polar interphase (McIntosh et al. 1989). The other is at the external plane along of the hydrated groups of the phospholipids (Gibbs 1986; Tatulian 1987).

Temptatively, the inner interface is the plane at which dipoles of the carbonyl groups and water polarized by them are aligned to define the dipole potential (Smaby and Brockman 1990; Franklin and Cafiso 1993; Brockman 1994; Diaz et al. 1999). The outer interface is the ideal plane at which the zeta potential is measured to determine the surface charge potential (Seelig et al. 1987; McLaughlin 1989; Lairion and Disalvo 2009).

In this chapter, the physical chemical properties of the interphase region will be analyzed under the scope of thermodynamics of surface and solutions. The proposal is mainly based on the definition of Defay-Prigogine of an interphase and the derivation that it has in the understanding of membrane processes in the context of biological response (Davies and Rideal 1961; Defay and Prigogine 1966; Damodaran 1998; Adamson 1967; Harkins 1952).

The description of the interphase region in relation to water states is schematically described in Fig. 9.1.

It is observed that, according to a wide range of methodologies, structural analyses indicate an uneven distribution of water molecules facing different types of chemical groups (see Chap. 7). This situation makes the description more complicated because in the interphase region sub states of water, according to the environment in which it is, could be found due to the number and type of hydrogen bondings. In other words, not all the water molecules located at the interphase region bear the same free energy excess.

However, the description of Fig. 9.1 is still limited. It denotes the membrane as a sequence of compartments that are sharply separated from each other by the ideal planes defined above. This is a rough approximation because water can penetrate as fingers in the hydrocarbon phase (trace at the top of the Fig. 9.1) and the external plane may have different topological arrangements due to the protrusion and conformational changes of the polar groups of the different types of lipids as shown in the right side of the Figure (Simon and McIntosh 1986; Disalvo et al. 2008; Marrink et al. 2009).

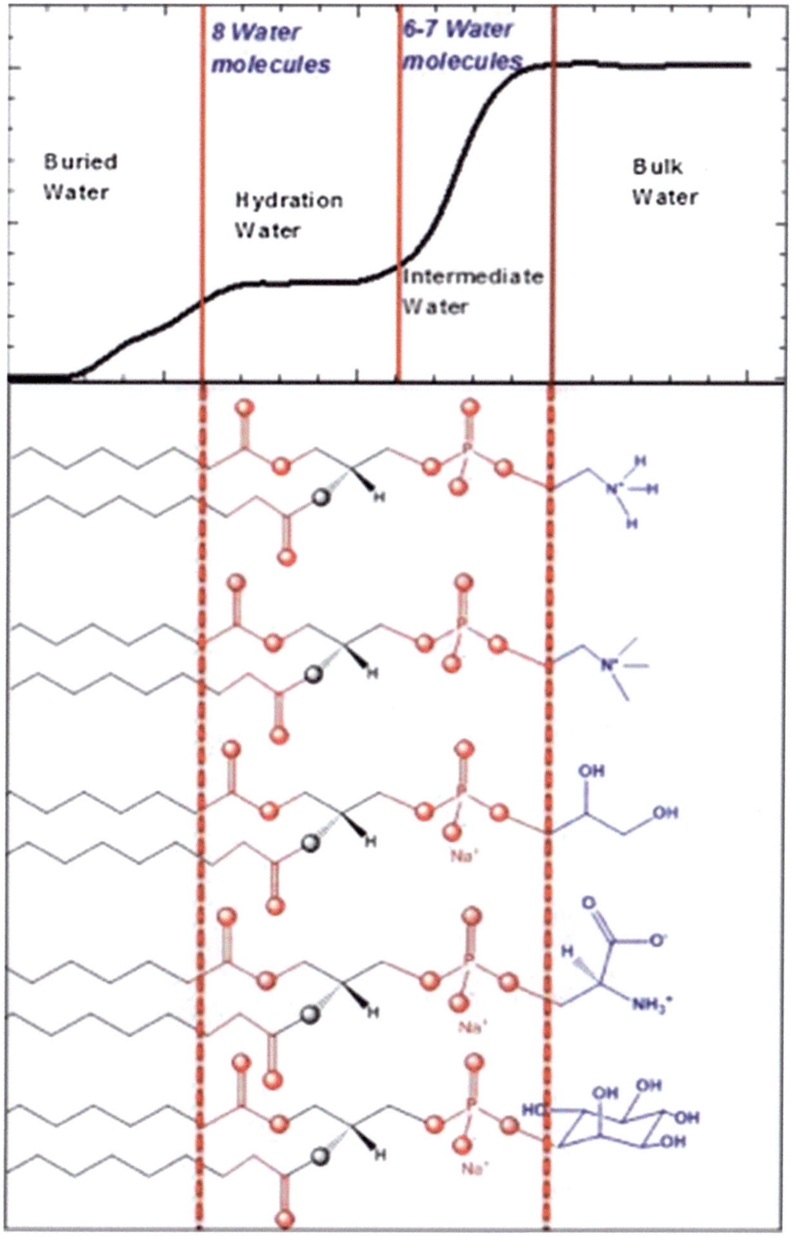

Fig. 9.1 Schematic representation of the interphase region in a lipid membrane. *Vertical dotted lines* denotes the ideal planes separating the interphase from the hydrocarbon region and the bulk water. *Upper traces* denotes the level of hydration in the different region (Adapted from Disalvo et al. 2014)

This may provide a variety of water sub states and a broad distribution of domains of water with different excess free energy, i. e. surface tension (Lam et al. 2002; Goldsmith and Martens 2009; Berkowitz and Vácha 2012).

With these limitations in mind we will explore the response of the lipid membrane considering that the interphase region is a bidimensional solution composed by the polar head groups dissolved in water. In this system, the polar head groups behave similarly to ions or polar molecules in a bulk aqueous solution. They have a first hydration layer, usually tightly bound to the group, with a structure different than bulk water due to the strong interaction of the water dipoles with the groups. The magnitude of this interaction and the changes imposed to water structure will depend on the nature of the polar or charged group (Cevc 1991; Marrink et al. 1996; Jendrasiak and Smith 2004).

In this regard, according to several laboratories, the number of water molecules bound to the lipids is in the range between 9 and 12 (McIntosh 1996; Goñi and Arrondo 1986; Fitter et al. 1999; Nagle and Tristram-Nagle 2000). This would be the tightly bound core of the hydration shell. Beyond that, other water molecules to complete the 22–24 water/lipid usually reported by calorimetry and found to contribute to the permeability barrier can be present (Disalvo and de Gier 1983).

Part of this water may be unevenly distributed in defects as suggested by Nagle and Tristram-Nagle (2000), (see Fig. 2.1 in Chap. 2), because of imperfect bilayer alignment or by topological features, such as curvature, protrusion, etc. (Marrink et al. 2009; Disalvo et al. 1996; Frias et al. 2007).

Another point to consider is that the different water subpopulations could be exchanged in different magnitudes and rates according to the compound that interacts with the membrane. At this point, at least two processes can be distinguished: one in which a solute (such as a non electrolyte, free aminoacid, aminoacid motifs of a protein or small peptides) can enter the interphase region causing an imbalance in the chemical potential of water in the interphase solution. In this case, water in the bulk will be driven towards the interphase producing its swelling. In this case swelling corresponds to the expansion of the bidimensional solution in the monolayer at constant surface pressure.

The other possibility is that a given solute may compete with water in the hydration shell of some of the polar groups. The final result would be a reaccommodation of the lipid head groups due to the intercalation of the solute in different structural arrays (see changes in the dipole potential produced by trehalose, for example, Luzardo et al. 2000 and Chap. 11).

In addition to these interactions at the interphase region, water can also be reorganized due to hydrophobic interactions in the non polar region (Heerklotz and Epand 2001; Sharp and Madan 1997).

Between these possibilities several intermediate and combined cases can be guessed according to the water-water, water-solute, lipid-lipid, lipid-water and solute-lipid interaction forces. In addition, it must be clear that these interactions are not between pairs but arises from the cooperativity and synergism between components in a surface.

The point here is that in each of these possibilities water is a part of the chemical balance of the mechanisms of interaction, contributing to the total free energy exchange with enthalpic and entropic changes, which is not always considered.

Analogous to what happens with the relation between volume and pressure in a tridimensional gas phase a relation between surface pressure and area can be also analyzed in the bidimensional solution (Ruckenstein and Li 1995; Pallas and Pethica 2009). As a consequence of the entrance/exit of water at constant pressure, any of those water exchanges between lipids in the membrane and the bulk solution can occur with changes in the area per lipid. Equivalently, if changes are done at constant area, an increase in pressure should be expected (Marsh 1996).

The process in lipid monolayers has been treated considering the lipid as a bidimensional gas to which the law of real gases can be applied (Pallas and Pethica 2009). However, this view does not explicit consider the contact with water and neglects the contribution of the aqueous phase to the expansion/compression phenomena (Jyoti et al. 1997). Moreover, in the view of Defay-Prigogine, the complete monolayer should be considered as the lipid layer of one molecule thick plus the bidimensional solution of the polar head groups inherent to it (the interphase region). Thus, a different thermodynamic frame should be considered.

Therefore, it is of interest to find a way to follow these changes in the monolayer during the different processes of permeation, insertion or penetration of solutes into the membrane. For this, we found it adequate to use the Defay-Prigogine model.

9.2 The Defay Prigogine Model of an Interphase

The proposal made by Defay-Prigogine for an interphase (Defay and Prigogine 1966; Damodaran 1998; Rao and Damodaran 2004; Disalvo et al. 2013a, b) allows to ascribe measurable thermodynamic properties to the lipid surface. It is based on the premise that the surface pressure is related to the decrease in the surface tension of water, and this in turn reflects changes in the surface free energy. In contrast to the visualization of the lipid monolayer as a bidimensional gas composed of lipid molecules spread on a surface, the proposal considers the effects of the lipids on the surface tension of water (Klopfer and Vanderlick 1996). Hence, the response of the monolayer is measured by changes in the water properties. In this regard, it must be stressed that liquid water should be present in biological systems in order to attain a measurable thermodynamic response, i.e. free energy changes.

Considering the results in regard to the organization of water in lipid membranes, the surface free energy is a consequence of the stability of the different arrays of water around the different membrane groups (see Chap. 7 by Alarcon et al.).

The bidimensional solution region, confined between the carbonyl group plane and external plane tangent to the phosphates depicted in Fig. 9.1, is composed by the carbonyl groups, the hydrated phosphates, water around ether oxygens and the choline groups (Bernik et al. 2001).

From the thermodynamic point of view, the surface tension of pure water can be defined as

$$\gamma^0 A = RT \ln\left(\frac{a_w^i}{a_w^b}\right) \quad (9.1)$$

where γ^0 is the surface tension of pure water, A is the average area per mole of water in the interphase region, a_w^i is the activity of water in the interphase of pure water and a_w^b is the water activity in the bulk phase. When a lipid monolayer is spread on the water surface, the surface tension changes to

$$\gamma A = RT \ln\left(\frac{a_w^L}{a_w^b}\right) \quad (9.2)$$

where a_w^L is the surface water activity in the presence of lipids, i.e. in the interphase region.

Thus, the difference between the surface tension of pure water (γ^0) and surface tension of water with lipids spread on it forming a monolayer (γ), i.e. the surface pressure of the monolayer (π) is expressed as a function of the surface water activities as:

$$\pi A = (\gamma^o - \gamma) A = RT \ln \frac{a_w^i}{a_w^L} \quad (9.3)$$

This equation clearly denotes that the surface pressure (π) increases when a_w^L decreases below 1, i.e. when the excess of lipids increases, and becomes zero when $a_w^i = a_w^L$, i.e. the activity of pure water when lipids coverage is zero. In that condition: $\gamma = \gamma^0$. In consequence, the surface pressure of an insoluble monolayer is a direct measure of the surface water activity (Damodaran 1998; Disalvo et al. 2013a, b).

An important consequence of Eq. 9.3 is that the surface pressure increases with the amount of lipids at the interface at constant area. This provides a method to regulate surface pressure by adding lipids to the air-water surface (Lairion and Disalvo 2007). In the present work, the different initial surface pressures before the addition of perturbant solutes to the subphase are adjusted by adding known amounts of lipids to an air-water surface in a Langmuir trough. This method allows to fix the initial water activity at the interphase and has an extra benefit in relation to the thermodynamic state of the monolayer as compared to that in which the surface excess and hence, the surface pressure is varied by decreasing the area at constant lipid amounts. In this last case, lipids are forced to pack by a lateral external force that may cause distortions in the head group region (Koenig et al. 1997; Marrink et al. 2009; Disalvo and Frias 2013).

It is clear that with the Defay-Prigogine definition, the thermodynamic parameter of surface pressure can be related to the water organization, which is implicit in the water activity term. At equilibrium, the chemical potential of water at the interphase (μ_{wi}) will be equal to the chemical potential of water in the bulk water phase (μ_{wb})

$$\mu_{wi} = \mu_{wb} \tag{9.4}$$

When a solute from the bulk water dissolves in the interphase region, the water activity (a_w^L) decreases. This decrease in water activity with respect to bulk promotes a flow of water into the interphase region. The film pressure can be described as a difference in osmotic pressure, over a thickness of the bidimensional solution, between the interphase at the monolayer and the bulk phase (Defay-Prigogine 1966; Kinnunen 2000). In consequence, the surface pressure increases when the solute solution enters the interphase.

In terms of solution chemistry, in principle, this should be independent of the particle nature that dissolves in it, at least in dilute systems according to the definition of colligative properties. Therefore, independent of the protein or peptide used, lipid membranes should give a similar response if the target water activity conditions are achieved. Most probably, deviations from the ideal behavior should be included in the activity coefficient different from 1.

As pointed out above, another way in which the dynamics of water exchanged can be manifested is when the penetrant molecule or residue (free aminoacid, aminoacid residue, aminoacid motifs of a peptide or protein) displaces water from the hydration shell due to a strong interaction with the lipid chemical groups.

Following Pfeiffer (see Chap. 4), a number of exchanged water molecules n_W for a schematic reaction as

$$\mathrm{M}\,(\mathrm{n_w}) + \mathrm{S}\,(\mathrm{n_{w'}}) \rightleftarrows +\mathrm{MS} + \mathrm{n_w} + \mathrm{n_{w'}}$$

can be defined, where M stands for the membrane phase, S for any solute (aminoacid residue or aminoacid motifs of a protein or peptide in the bulk aqueous phase) and n_w and $n_{w'}$ the respective number of water molecules hydrating each of them, respectively. In principle, there is no restriction to admit that n_w can be water attached to the lipids and solutes either to hydrophilic or hydrophobic regions. These are other deviations of the colligative behavior and introduces specific interactions forming new association in the surface. All of them will be reflected in the surface pressure changes via changes in water activity.

9.3 Experimental Determination of Surface Pressure Changes Induced by Peptides and Proteins

The changes in surface pressure at different initial surface pressures, induced by defined concentrations of peptides in the subphase, has been usually interpreted by the plots of $\Delta\pi$ *vs* π as shown in Fig. 9.3 (Martini and Disalvo 2007; Hollmann

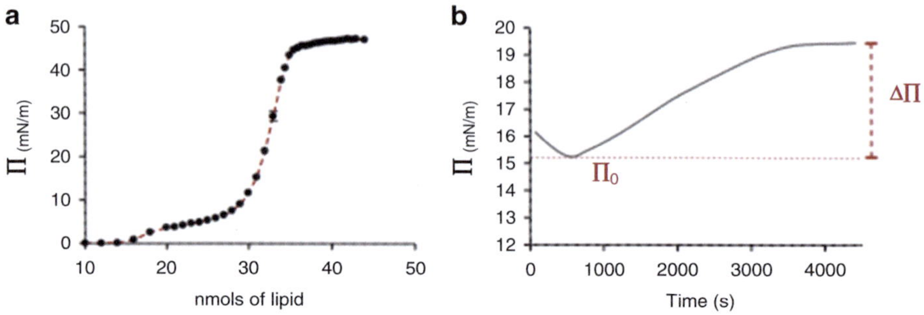

Fig. 9.2 (**a**) Formation of a monolayer on the surface of an aqueous solution by titration with known amount of lipids. (**b**) Injection of a known amount of perturbant compounds to the subphase of a monolayer of a selected surface pressure adjusted by titration as in part **a**. The final asymptotic value stabilized at long times is π_p

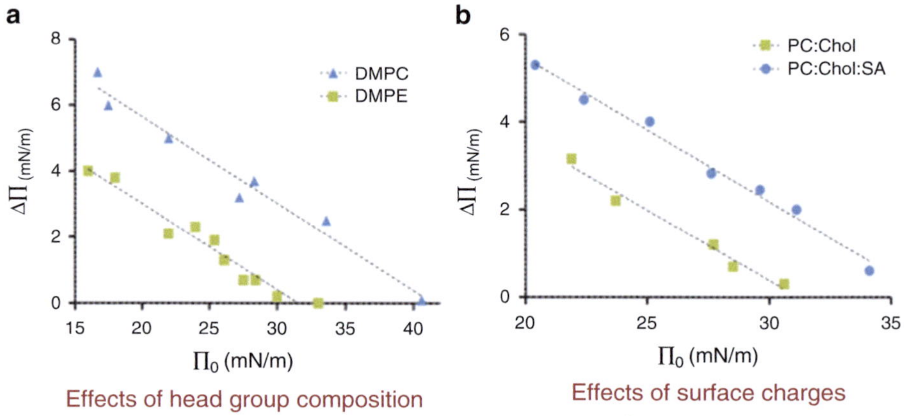

Fig. 9.3 Perturbation caused by solutes in the subphase at different initial surface pressures. (**a**) Effect of proteases on DMPE (Δ) and DMPC (▲) monolayers at 25 °C. (**b**) Effects of S-layer from Lactobacillus on PC: cholesterol and PC: Cholesterol: stearylamine monolayers

et al. 2007). In this methodology, a known amount of lipids is spread on the water surface of known area until a desired surface pressure is obtained (Fig. 9.2a). After stabilization at a given surface pressure, a known amount of the challenging solute is added to the subphase (Fig. 9.2b).

The injection produces a change in the surface pressure which stabilizes in a constant value as shown in Fig. 9.2b. The difference between the initial surface pressure obtained with pure lipids and the final surface pressure obtained after the stabilization after the addition of the perturbant compounds ($\Delta\pi$) is plotted vs the initial surface pressure (π_o). The procedure is repeated for different initial surface pressures as in Fig. 9.3.

In this figure, we summarize published results obtained with different kinds of lipids and two different aqueous soluble proteins (protease of *Mucor miehei* and S-layer extracted from *lactobacilli*) (Martini and Disalvo 2007; Hollmann et al. 2007).

9.3.1 Effect of the Interphase Region

It is observed that the cut-off (or critical pressure defined as the pressure at which no perturbation is observed) depends on the head group region composition. The straight lines corresponding to phosphatidylcholine (PC) and phosphatidylethanolamine (PE) in part A differ in the cut off value, being lower for PE than for PC (Fig. 9.3a).

A similar behavior is found in another completely different system when monolayers containing a charged component such stearylamine (SA) are compared with PC/cholesterol monolayers (Fig. 9.3b). In this case, the cut-off of the monolayer with SA is larger than those with neutral lipids only.

From these two examples, it is clear that the cut-off depends on the lateral interactions at the head group region. In PE, the strong intermolecular H bonding between PO group of the phosphate and amine groups reduces the cut off area. Similarly, the charge repulsion between the net charge of SA group increases the available area for the lipids and hence the cut-off.

However, in all the cases, a visible perturbation can only be found when the monolayers are relaxed from this cut off. In other words, at the critical pressure lipids have reached the maximum packing in relation to the excluded volume corresponding to its hydration layer. In order to produce a perturbation, i.e. the monolayer be sensitive to the presence of the perturbing compound, a lower surface pressure than that corresponding to a saturated monolayer is required. This means that the surface tension of the interphase should be higher, i.e. towards the value of pure water.

This is attained if an area larger than that corresponding to the first hydration layer is created. Presumably, water beyond the tightly bound hydration molecules is now entering the interphase. However, the response is not due to the magnitude of the area but to the quality of the interphase that is created, probably with a different water tension. This conclusion is driven from the fact that the area increase is equivalent to at most two to five water molecules (Almaleck et al. 2013; Bagatolli et al. 1998).

The lower limit of tightly bound water is headgroup structural water, shown in Fig. 2.7. of Chap. 2. The larger lipid areas require more water, according to Tristram-Nagle (Chap. 2).

It is interesting to observe that the change of PC by PE displaces the curve but the slope remains unaltered. This means that water entering the interphase is of the same quality (in thermodynamic terms) in spite of being incorporated around different head groups and hence difference water arrays.

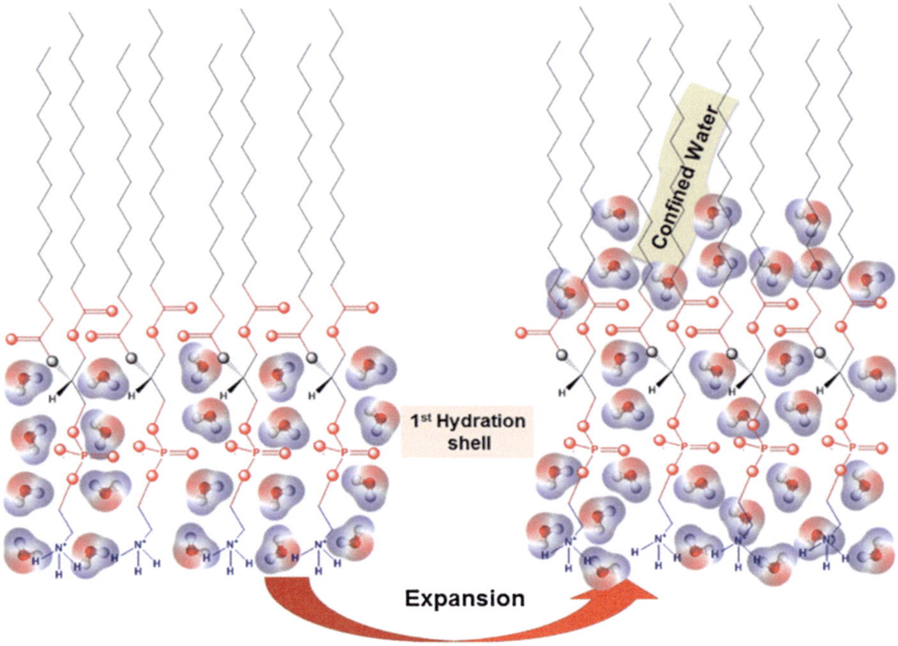

Fig. 9.4 Schematic description of the expansion of a monolayer at the cut off pressure in Fig. 9.3, in which only hydration water is found, to a larger area in which water beyond the hydration shell can penetrate (confined water)

This finding can be explained by considering that the additional water is seeing a similar environment in both situations. As the acyl chains of the PC and the PE are the same in these assays, a reasonable conclusion is that water beyond the hydration shell corresponds to water facing the non polar regions of the monolayer.

This has been denoted as confined water, to differentiate it from the hydration water (Disalvo et al. 2008; Alarcon et al. Chap. 7). It has been identified with water species in terms of H-bonding facing different residues of the lipid molecules (Disalvo and Frias 2013). A schematic picture of water incorporation upon a slight expansion is shown in Fig. 9.4.

A similar reasoning can be done with the results found when charges are introduced. Moreover, the slope of the PE curves nearly overlap with that for PC/cholesterol, and the curve for pure PC is displaced to that for PC/cholesterol/SA.

9.3.2 Effects of Acyl Chain

In another series of experiments the changes in the non polar region was tested. In Fig. 9.5, PCs with different acyl chains showed that the cut off were similar within

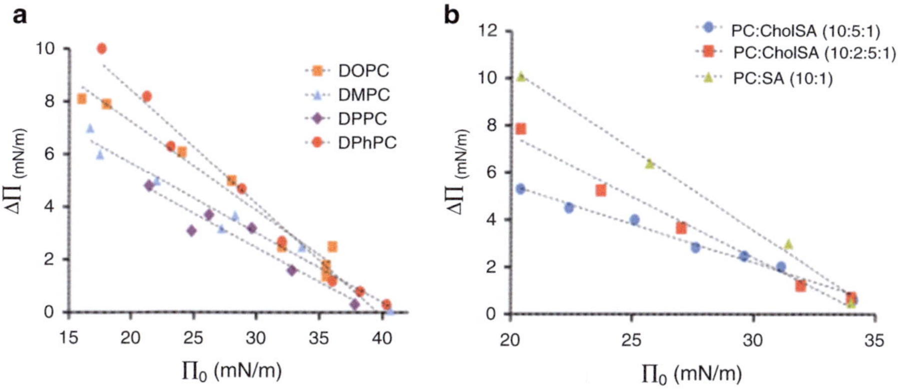

Fig. 9.5 (**a**) Effect of protease on monolayers composed by dimyristoylPC (DMPC) (▲), dipalmitoylPC (DPPC) (♦), dioleoylPC (DOPC) (■), diphytanoylPC (DPhPC) (•). (**b**) Effect of S layer on monolayers of PC: SA containing 0 % cholesterol, 2. 5 % cholesterol, and 5 % cholesterol

the experimental error. It is observed that the slope is steeper in function of any of these variables: higher unsaturation; longer acyl chain and presence of ramification. This sustains the previous proposal that the perturbations observed below the cut-off reflects the entrance of water beyond the hydration water is within the acyl chain region.

The results of both A and B parts of the recently described figure (Fig. 9.5) take us to the same conclusion. When cholesterol is added, the slope decreases meaning that less water is entering the monolayer in the presence of cholesterol, when a similar expansion of the monolayer with respect to the cut off is made.

9.4 Monolayer Perturbation and Water Activity

The difference between the critical surface pressure (π_c) and a chosen initial surface pressure achieved by the lipid surface excess is related to the surface pressure decrease necessary to make the lipid interphase sensitive to solutes in the subphase. The meaning of this expansion can be explained in the following way: suppose the monolayer stabilized at a surface pressure π quite below the critical cut off. If the surface pressure is increased by a $\Delta\pi = \pi_c - \pi$, adding more lipids, the monolayer reaches a state at which it does not respond to any perturbant solute, i.e. no excess surface energy is available. If at the same π_o, a perturbant solute is added instead of lipids, the surface pressure evolves to a value approaching π_c, i.e. the solute uses the excess free energy to reach the saturation point. It may be though that the final pressure obtained when perturbant is added should be the same as the critical pressure. However, there is a little differences which is beyond the purpose of this chapter. For details see (Disalvo et al. 2008).

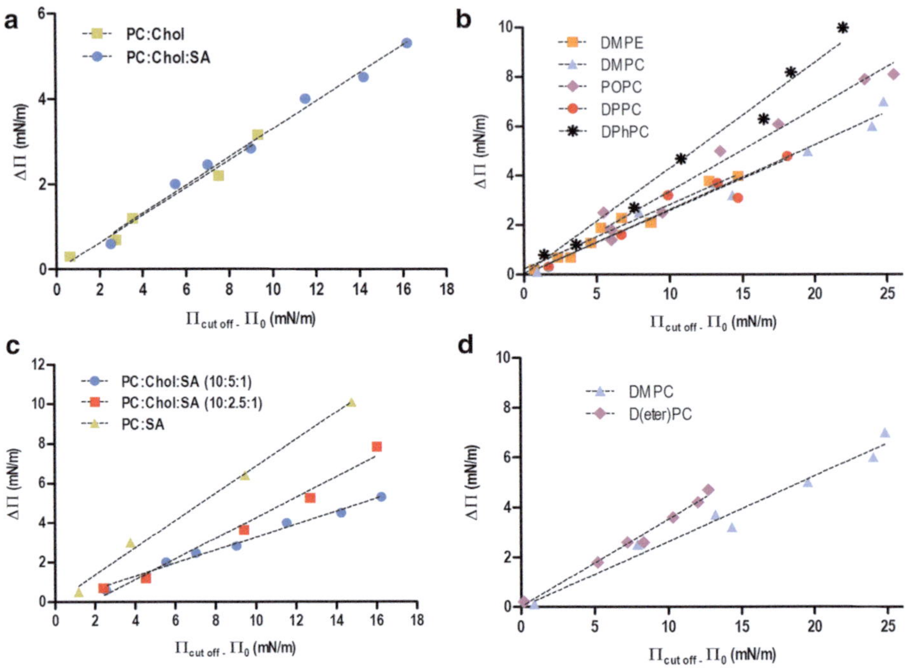

Fig. 9.6 Perturbation of compounds in the subphase as a function of the excess surface pressure. (**a**) S layer injected to the subphase of PC: Cholesterol and PC:Cholesterol:SA. (**b**) Protease injected to DMPC, DOPC, DPPC, DPhPC monolayers. (**c**) S layer injected to PC:SA monolayers with 0 % cholesterol, 2.5 % cholesterol, and 5 % cholesterol. (**d**) Protease injected to DMPC and ether PC monolayers

The plots of Fig. 9.5 can be phenomenologically described by

$$\Delta\pi = k\,(\pi_c - \pi) \tag{9.5}$$

where $\Delta\pi$ defines the perturbation of the initial surface pressure of the monolayer induced by the proteins added to the subphase (Fig. 9.6), and the difference $\pi_c - \pi$ represents the departure from the critical pressure.

If the disturbance and excess surface pressure were the same, the slope obtained should have been one. However, it is observed that the slope depends on the nature of the hydrocarbon region and perturbing compound concentration.

The initial surface pressure is related to the water activity at the interphase, according to Eq. 9.1, and is modulated by the amount of lipids added to the water surface at values below π_c.

The value of the slope k is clearly a function of the acyl chain composition including the presence of carbonyl groups, according to data in Fig. 9.6b, d. Specifically, the increase in branching or unsaturation and the depletion of cholesterol and the existence of carbonyl groups increases the slope. The absence of carbonyl groups in the lipid affects the slope but not the cut off, which allow to conclude that the carbonyl groups are contributing to the so called second hydration layer (or confined water in terms defined above). The slope of the ether lipid is higher than that corresponding to the ester ones, i.e. a similar effect as adding unsaturation to the acyl chain. A direct conclusion could be that the magnitude of the perturbation is related to the kinks formation due to the rotational isomers of the acyl chains and the cooperativity (Träuble 1971). However, since those membrane conformers imply water penetration (Disalvo et al. 2013a, b), it is plausible to analyze these results in terms of the effects that those lipid components may cause on the water activity of the surface, following the hypothesis of [Damodaran 1998] and the formalism of Defay Prigogine (1966) described in the introduction.

The physical meaning of k in Eq. 9.5 is clearly related to phase state of the monolayer. It is interesting to observe the effect of cholesterol. In natural systems, the liquid condensed phase is physiologically relevant. Cholesterol, mainly found in the plasma membranes of eukaryotic cells is considered a passive modulator of membrane physical properties. Biophysical studies in phospholipid:cholesterol model systems have been carried out in the range of 20 % molar ratio (Marsh 2009).

In the liquid condensed phase, molecules have high diffusivity parallel to the plane of the membrane and undergo rapid rotational diffusion about the axis perpendicular to the plane of the membrane. In the absence of cholesterol, this enhanced diffusion is always accompanied by the onset of conformational freedom of the acyl chains, i.e., low orientational order, so that the normal fluid phase of pure lipid systems is appropriately described as the 'liquid expanded' (l_d) phase.

The departure of the surface pressure values with respect to the critical surface pressure denoted in the abscissa can, in principle, be related to area changes concomitant to the onset of conformational freedom of acyl chains (Fig. 9.5). The area per lipid molecule corresponds to the area excluded by the lipid head group and the immobilized hydration shell. This area, calculated from monolayers studies, and from X-ray diffraction is around 64 $Å^2$ for DOPC (Nagle and Tristram-Nagle 2000; Disalvo et al. 2008). Note that this area is larger than that reported for collapsed monolayers, and corresponds to lipids at the interphase at saturation without compression (i. e. constant area).

The excess area beyond that occupied by a lipid molecule with its hydration shell at saturation is difficult to justify by the ideal formalism of an increase in geometrical space.

Therefore, the increase in surface pressure promoted by the proteins cannot be interpreted with the simple geometrical criterion in which the protein intercalates with the lipids and therefore increases the surface pressure. The water molecules beyond the hydration shell of the phospholipids and the phospholipid themselves defines a thermodynamic state of the interphase. The monolayer expansion gives place to surface sites, in terms of surface excess free energy, reactive for the amino acid residues of the proteins. The perturbation can thus be expressed by the difference of free energy between the final state of the monolayer with protein and the free energy of the initial state of the monolayer (prior to protein addition).

The resolution of k in terms of Eq. 9.1 can now be done considering that the surface pressure produced by the protein in the monolayer is

$$\pi A = RT \ \ln \frac{a_w^i}{a_w^p} \tag{9.6}$$

where a_w^p is water activity of interphase after protein addition.

Thus, considering that π_p is a surface pressure of monolayer in the final state after protein addition (see Fig. 9.2b), the pressure perturbation ($\Delta\pi$) can be re-written as

$$\Delta \pi A = (\pi_p - \pi_o) A = RT \ \ln \frac{a_w^L}{a_w^p} \tag{9.7}$$

This equation denotes that there is no perturbation when $a_w^L = a_w^p$. This condition is achieved when the whole surface is occupied by the lipids, i. e. π_c is reached. In addition, $\Delta\pi > 0$ when $a_w^p < a_w^L$. That is, the protein insertion reduces the water activity at the interface.

On the other hand, the difference of the surface pressures with respect to the critical one (π_c) can be expressed as

$$A \ \pi_c = RT \ \ln \frac{a_w^i}{a_w^{Lc}} \tag{9.8}$$

From which

$$(\pi_c - \pi_o) \ \mathrm{A} = \mathrm{RT} \ln \frac{\mathrm{a_w^L}}{\mathrm{a_w^{LC}}} \tag{9.9}$$

It is clear that for $\pi_c - \pi > 0$, the water activity at the interphase for any lipid concentration should be higher than at the critical a_w^{Lc}, given by the limit of packing of the lipids with its hydration shells.

Thus, dividing member by member (9.8) by (9.9) we have:

$$k = \frac{\left(\ln a_w^L - \ln a_w^p\right)}{\left(\ln a_w^L - \ln a_w^{Lc}\right)} \tag{9.10}$$

The increase of the slopes due to the increase of unsaturation or the depletion of cholesterol, as shown in Fig. 9.6, means that protein insertion depends on the difference in a_w^L with respect to a_w^P, for a given departure from a_w^{Lc}.

Multiplying and dividing by RT and knowing that chemical potential (μ) can be defined as: $\mu = \mu^0 + RT \ln a$

$$\mathrm{k} = \frac{(\mu_{\mathrm{w}} - \mu_{\mathrm{wp}})}{(\mu_{\mathrm{w}} - \mu_{\mathrm{wc}})}. \tag{9.11}$$

Assuming that there are no significant differences between the standard chemical potentials in the different conditions, Eq. 9.11 clearly denotes that the process is driven by the difference in the chemical potential of water in the different states of the interphase.

For a given value of $\mu_w - \mu_{wc}$, the perturbation increases with the unsaturations and cholesterol depletion, which is reflected in a greater difference between the chemical potential of water at the pure lipid interphase and that with proteins that grows with the increase in water spaces.

The change in surface pressure, i. e. chemical potential of water of 6–8 mN/m is equivalent to an energy shift for protein adsorption of 6 kJ/mol, which matches with the energy of one H-bonding and is 6 times higher than a dispersion force. This is near to the free energy reported for Cytochrome interaction with PC bilayers around 10 kJ/mol (Heerklotz and Epand 2001). These numbers are indicative that the changes in surface pressure (surface tension) are energetically comparable with reported values for protein-membrane interactions. These interactions would take place within the water- accessible region that is about three methylene groups. The energy of the interaction calculated for 100 $Å^2$ amounts an equivalent of two CH_2 groups. The value agrees with literature data on hydrocarbons and amphiphiles where the group contributions per methylene were two chains, burying 20 % of the surface (Heerklotz and Epand 2001). Thus, the energy changes measured by surface tension are related to the lipid membrane groups exposed to water that determines the water activity at the interphase.

Thus, the k values are related to chain conformation and packing on the interaction enthalpy and serves to explain a variety of effects reported on membrane binding. The structural counterpart of these responses is given by changes in the saturation/unsaturation ratio, presence or absence of cholesterol and carbonyl depletion (ether vs esther phospholipids). In other words, an excess of free energy can be obtained in relation to water organization around the lipids being these modulated by metabolic factors affecting membrane composition.

9.5 Comparison of Monolayers and Bilayers

With this background it is possible to analyse the kinetics of the interactions of compounds with membranes under another approach. The hypothesis is based on the fact that the surface pressure of a lipid monolayer is a direct measure of the thermodynamic activity of water at the lipid-water interface.

Recalling the results and interpretation of van Zoelen et al. (1976) in regard to water and solute permeation in bilayers of liposomes, several common pints can be highlighted.

When water penetrates lipid vesicles, they swell until the chemical potentials of the solutions are equal inside and outside the structure (Van Zoelen et al. 1976). Meanwhile, the chemical potential in the outer media is given by the concentration of the components; those inside are determined by the concentration and the pressure exerted by the network of the bilayer on the solution. The viscoelastic properties of the bilayer may then contribute to the permeation processes.

In the case of liposomes, the pressure can be exerted by the bilayer surrounding the internal solution. This is given by the cohesive forces maintaining the phospholipids in the structure.

The variation of the membrane elasticity may be related to surface tension effects: when permeant solutions are able to perturb the cohesive forces, more swelling would occur (Evans and Skalak 1980; Jyoti et al. 1997; Kubo et al. 2001).

At a 4 % volume increase, the swelling process results in a release of the liposomal content due to an increase in surface area. Thus, bilayer selectivity will vary according to the value of the reflection coefficient in the swelled structure a point also discussed by Nickels and Katsaras in Chap. 3.

The cut off pressure is given by the contact of the head group hydration water defined as the critical surface pressure, at which no effect of aminoacids, peptides or proteins on monolayer surface pressure is found. The perturbation of surface pressure is produced when the area per lipid is above just 4 % larger than that corresponding to the hydration shell of the phospholipid head groups found in the cut –off, and therefore surface pressure is decreased with respect to the cut off. The possibility that this area increase is related with the increase in water beyond the hydration shell, as anticipated by Bagatolli (Chap. 6) and studied by Arsov (Chap. 7) is discussed. The change in interfacial water activity is related with the surface pressure according to the Defay-Prigogine interphase model, which considers the interphase region as a bidimensional solution of head groups in water. As predicted by solution chemistry, the increase of surface pressure is independent of the protein nature but depends on the water surface state determined by the lipid composition.

The osmotic imbalance has been also pointed as one of the factors that make the membrane a responding structure (Sparr and Wennerstrom 2001). In our context this imbalance is connected through surface tension changes produced by water exchanged induced by solute interactions.

9.6 Conclusions

A link between the thermodynamics of lipid interfaces with the organization of water at the different regions of the membrane to explain the interactions of aqueous soluble protein with lipid membranes according to its hydration state has been demonstrated.

Surface water activity appears as a common factor for the interaction of several aqueous soluble and surface active proteins with lipid membranes of different composition. Under the thermodynamic approach of the Defay-Prigogine protein perturbation can be measured by changes in the surface pressure of lipid monolayers at different initial water surface activities. As predicted by solution chemistry, the increase of surface pressure is independent of the particle nature that dissolves.

Therefore, membranes give a similar response in terms of the determined surface states given by water activity independent of the protein or peptide.

In real systems, in which area and lipids are maintained constant, the excess of free energy necessary for peptide or protein insertion can be produced by fluctuations in curvature and packing according to the viscoelasticity of the membrane system.

References

Adamson AW (1967) The physical chemistry of surfaces, 2nd edn. Interscience, New York

Almaleck H, Gordillo GJ, Disalvo A (2013) Water defects induced by expansion and electrical fields in DMPC and DMPE monolayers: contribution of hydration and confined water. Colloids Surf B Biointerfaces 102:871–878

Bagatolli LA, Gratton E, Fidelio GD (1998) Water dynamics in glycosphingolipid aggregates studied by LAURDAN fluorescence. Biophys J 75:331–341

Berkowitz ML, Vácha R (2012) Aqueous solutions at the interface with phospholipid bilayers. Acc Chem Res 45:74–82

Bernik DL, Zubiri D, Tymczyszyn E, Disalvo EA (2001) Polarity and packing at the carbonyl region of lipid bilayers. Langmuir 17:6438–6442

Brockman H (1994) Dipole potentials of lipid membranes. Chem Phys Lipids 73:57

Cevc G (1991) Hydration force and the interfacial structure on the polar surface. J Chem Soc Faraday Trans 87:2733–2739

Damodaran S (1998) Water activity at interfaces and its role in regulation of interfacial enzymes: a hypothesis. Colloids Surf B Biointerfaces 11:231–237

Davies JT, Rideal EK (1961) Interfacial phenomena. Academic, New York

De Disalvo EA, Gier J (1983) Contribution of aqueous interphases to the permeability barrier of lipid bilayer for non-electrolytes. Chem Phys Lipids 32:39–47

Defay R, Prigogine I (1966) Surface tension and adsorption. Wiley, New York

Diaz S, Amalfa F, Biondi de Lopez AA, Disalvo EA (1999) Effect of water polarized at the carbonyl groups of phosphatidyl- cholines on the dipole potential of lipid bilayers. Langmuir 15(15):5179–5182

Disalvo EA, Frias MA (2013) Water state and carbonyl distribution populations in confined regions of lipid bilayers observed by FTIR spectroscopy. Langmuir 29(23):6969–6974. dx.doi.org/10.1021/la304390r

Disalvo EA, Viera LI, Bakás LS, Senisterra GA (1996) Lysophospholipids as natural molecular harpoons sensoring defects at lipid membranes. J Colloid Interface Sci 178:417–425

Disalvo EA, Lairion F, Martini F, Tymczyszyn E, Frías M, Almaleck H, Gordillo GJ (2008) Structural and functional properties of hydration and confined water in membrane interfaces. Biochim Biophys Acta 1778:2655–2670

Disalvo EA, Hollmann A, Semorile L, Martini MF (2013a) Evaluation of the Defay-Prigogine model for the membrane interphase in relation to biological response in membrane-protein interactions. Biochim Biophys Acta 1828:1834–1839

Disalvo EA, Bouchet AM, Frias MA (2013b) Connected and isolated CH_2 populations in acyl chains and its relation to pockets of confined water in lipid membranes as observed by FTIR spectrometry. Biochim Biophys Acta 1828:1683–1689. http://dx.doi.org/10.1016/j.bbamem.2013.02.007

Disalvo EA, Martini MF, Bouchet AM, Hollmann A, Frías MA (2014) Structural and thermodynamic properties of water–membrane interphases: significance for peptide/membrane interactions. Adv Colloid Interface Sci 211:17–33

Evans E, Skalak R (1980) Mechanics and thermodynamics of biomembranes. CRC Press, Boca Raton
Fitter J, Lechner RE, Dencher NA (1999) Interactions of hydration water and biological membranes studied by neutron scattering. J Phys Chem B 103:8036–8050. doi:10.1021/jp9912410
Franklin JC, Cafiso DS (1993) Internal electrostatic potentials in bilayers: measuring and controlling dipole potentials in lipid vesicles. Biophys J 65:289–299
Frías MA, Nicastro A, Casado N, Gennaro AM, Díaz S, Disalvo EA (2007) Arbutin blocks defects in the ripple phase of DMPC bilayers by changing carbonyl organization. Chem Phys Lipids 147:22–29
Gibbs JW (1986) The collected works of J.W. Gibbs, vol I. Longmans, Green and Co, New York (1931) 219
Goldsmith J, Martens CC (2009) Effect of boundary conditions on the structure and dynamics of nanoscale confined water. J Phys Chem 113:2046–2052
Goñi FM, Arrondo JL (1986) A study of phospholipid phosphate groups in model membranes by Fourier-transform infrared-spectroscopy. Faraday Discuss Chem Soc 81:117–126
Harkins WD (1952) Physical chemistry of surface films. Reinhold, New York
Heerklotz H, Epand R (2001) The enthalpy of acyl chain packing and the apparent water-accessible apolar surface area of phospholipids. Biophys J 80:271–279
Hollmann A, Delfederico L, Glikmann G, De Antoni G, Semorile L, Disalvo EA (2007) Characterization of liposomes coated with S-layer proteins from Lactobacilli. Biochim Biophys Acta 1768:393–400
Jendrasiak GL (1996) The hydration of phospholipids and its biological significance. J Nutr Biochem 7(11):599–609
Jendrasiak GL, Smith RL (2004) The effect of the choline head group on phospholipid hydration. Chem Phys Lipids 131(2):183–195
Jyoti A, Prokop RM, Neumann AW (1997) Manifestation of the liquid- expanded/liquid-condensed phase transition of a dipalmitoylphosphatidylcholine monolayer at the air-water interface. Colloids Surf B Biointerfaces 8:115–124
Kinnunen PKJ (2000) Lipid bilayers as osmotic response elements. Cell Physiol Biochem 10:243–250
Klopfer KJ, Vanderlick TK (1996) Isotherms of dipalmitoylphosphatidylcholine (DPPC) monolayers: features revealed and features obscured. J Colloid Interface Sci 182:220–229
Koenig BW, Strey HH, Klaus Gawrisch K (1997) Membrane lateral compressibility determined by NMR and X-ray diffraction: effect of acyl chain polyunsaturation. Biophys J 73:1954–1966
Kubo S, Adachi H, Maeda AS (2001) Phosphatidylcholine monolayers observed with Brewster angle microscopy and P-A isotherms. Thin Solid Films 393:80–85
Lairion F, Disalvo EA (2007) Effect of arbutin on the dipole potential and area per lipid of ester and ether phosphatidylcholine and phosphatidylethanolamine monolayers. Biochim Biophys Acta 1768:450–456
Lairion F, Disalvo A (2009) Effect of dipole potential variations on the surface charge potential of lipid membranes 1607. J Phys Chem B 113:1607–1614
Lam P, Wynne KJ, Wek GE (2002) Surface-tension-confined microfluidics. Langmuir 18:948–951
Luzardo MC, Amalfa F, Núñez A, Díaz SB, de López AC, Disalvo EA (2000) Effect of trehalose and sucrose on the hydration and dipole potential of lipid bilayers. Biophys J 78:2452–2458
Marrink SJ, Tieleman DP, van Buuren AR, Berendsen HJC (1996) Membranes and water- an interesting relationship. Faraday Discuss 103:191–201
Marrink SJ, de Vries AH, Tielemann DP (2009) Lipids on the move: simulations of membrane pores, domains, stalks and curves. Biochim Biophys Acta 1788:149–168
Marsh D (1996) Lateral pressure in membranes. Biochim Biophys Acta 1286:183–223
Marsh D (2009) Cholesterol-induced fluid membrane domains: a compendium of lipid-raft ternary phase diagrams. Biochim Biophys Acta Biomembr 1788(10):2114
Martini MF, Disalvo EA (2007) Superficially active water in lipid membranes and its influence on the interaction of an aqueous soluble protease. Biochim Biophys Acta 1768:2541–2548

McIntosh TJ (1996) Hydration properties of lamellar and non-lamellar phases of phosphatidylcholine and phosphatidyl ethanolamine. Chem Phys Lipids 81(2):117–131
McIntosh TJ, Simon SA, Dilger JP (1989) Location of the water-hydrocarbon interface in lipid bilayers. In: Benga G (ed) Water transport in biological membranes, vol I. CRC Press Inc, Boca Raton
McLaughlin S (1989) The electrostatic properties of membranes. Ann Rev Biophys Biophys Chem 18:113–136
Milhaud J (2004) New insights into water-phospholipid model membrane interactions. Biochim Biophys Acta 1663:19–51
Nagle JF, Tristram-Nagle S (2000) Structure of lipid bilayers. Biochim Biophys Acta Rev Biomembr 1469(3):159–195
Pallas NR, Pethica BA (2009) Intermolecular forces in lipid monolayers. Two-dimensional virial coefficients for pentadecanoic acid from micromanometry on spread monolayers at the air/water interface. Phys Chem Chem Phys 11:5028–5034
Rao CS, Damodaran S (2004) Activation of sphingomyelinase in lipid monolayer is related to interfacial water activity: evidence from two disparate systems. Colloids Surf B Biointerfaces 45:49–55
Ruckenstein E, Li JB (1995) A surface equation of state based on clustering of surfactant molecules of insoluble monolayers. Langmuir 11:3510–3515
Seelig J, Mac Donald PM, Scherer PG (1987) Phospholipid head groups as sensors of electric charge in membranes. Biochemistry 26:7535–7541
Sharp K, Madan B (1997) Hydrophobic effect, water structure and heat capacity changes. J Phys Chem B 101:4343–4348
Simon SA, McIntosh TJ (1986) Depth of water penetration into lipid bilayers. Methods Enzymol 127:511–521
Smaby JM, Brockman HL (1990) Surface dipole moments of lipids at the argon–water interface. Similarities among glycerol-ester- based lipids. Biophys J 58(1):195–204
Sparr E, Wennerström H (2001) Responding phospholipid membranes-interplay between hydration and permeability. Biophys J 81(2):1014–1028
Tatulian SA (1987) Binding of alkaline-earth metal cations and some anions to phosphatidylcholine liposomes. Eur J Biochem 170:413
Träuble H (1971) The movement of molecules across lipid membranes: a molecular theory. J Membr Biol 4(1):193–208
Van Zoelen EJJ, Blok MC, De Gier J (1976) An improved method for the description of non-electrolyte permeation through liposomes, based on irreversible thermodynamics. Biochim Biophys Acta (BBA) Biomembr 436(2):301–306

Chapter 10
Water at Biological Phase Boundaries: Its Role in Interfacial Activation of Enzymes and Metabolic Pathways

Srinivasan Damodaran

Abstract Many life-sustaining activities in living cells occur at the membrane-water interface. The pertinent questions that we need to ask are, what are the evolutionary reasons in biology for choosing the membrane-water interface as the site for performing and/or controlling crucial biological reactions, and what is the key physical principle that is very singular to the membrane-water interface that biology exploits for regulating metabolic processes in cells? In this chapter, a hypothesis is developed, which espouses that cells control activities of membrane-bound enzymes through manipulation of the thermodynamic activity of water in the lipid-water interfacial region. The hypothesis is based on the fact that the surface pressure of a lipid monolayer is a direct measure of the thermodynamic activity of water at the lipid-water interface. Accordingly, the surface pressure-dependent activation or inactivation of interfacial enzymes is directly related to changes in the thermodynamic activity of interfacial water. Extension of this argument suggests that cells may manipulate conformations (and activities) of membrane-bound enzymes by manipulating the (re)activity of interfacial water at various locations in the membrane by localized compression or expansion of the interface. In this respect, cells may use the membrane-bound hormone receptors, lipid phase transition, and local variations in membrane lipid composition as effectors of local compression and/or expansion of membrane, and thereby local water activity. Several experimental data in the literature will be reexamined in the light of this hypothesis.

Keywords Enzyme activation • Interfacial water • Oscillatory reactions • Membrane bound receptors

---

S. Damodaran (✉)
Department of Food Science, University of Wisconsin-Madison, 1605 Linden Drive, Madison, WI 53706, USA
e-mail: sdamodar@facstaff.wisc.edu

E.A. Disalvo (ed.), *Membrane Hydration*, Subcellular Biochemistry 71,
DOI 10.1007/978-3-319-19060-0_10

10.1 Introduction

A majority of biological reactions in a cell occur at the membrane-water interface. Many control mechanisms that regulate energy production (mitochondrion), protein biosynthesis (endoplasmic membrane), and protein secretion are located in membranes. Membranes also control communication between inside and outside of cells. This is usually carried out through pumping of ions through ion channels or through interaction of hormones with receptors on the outer leaflet of plasma membrane and transducing this signal through conformational changes in transmembrane enzymes/proteins. For instance, binding of hormones and neurotransmitters (e.g., insulin and epinephrine) to receptor proteins on the outer leaflet of plasma membrane results in initiation of a cascade of cell-specific metabolic reactions. Enzymes that shuttle between cytosol and the membrane also facilitate communication between inside and outside of a cell. One of the interesting characteristics of these enzymes is that membrane binding modulates their activities. The enzymes that fall into this *interfacial activation* category include lipases (Schmid and Verger 1998; Thuren et al. 1991; Carriere et al. 1997; Roussel et al. 1998), phospholipases (James et al. 1997; Verger et al. 1976; Lio and Dennis 1998; Moreau et al. 1988; Hirche and Ulbrich-Hofman 1999), sphingomyelinase (Jungner et al. 1997), neuraminidase (Perillo et al. 1994), protein kinase C (Souvignet et al. 1991), and cholesterol oxidase (Gronberg and Slotte 1990). These enzymes act on specific membrane components and the products produced in turn act as messengers for activation or termination of biological functions. For instance, hydrolysis of sphingomyelin, which is present predominantly in the outer leaflet of plasma membrane (Verkleij et al. 1973), by sphingomyelinase produces ceramide, which initiates a cascade known as "ceramide signalling". This 'signal' has been shown to control several cell regulation processes, such as cell proliferation and differentiation, growth, growth arrest, inflammation, apoptosis, and cytokine biosynthesis (Obeid and Hannun 1995; Auge et al. 1996; Sasaki et al. 1995; Chen et al. 1995; Chmura et al. 1996).

10.1.1 Interfacial Activation of Enzymes

The realization that a majority of biological reactions in cells are essentially heterogeneous enzyme catalysis at the lipid-water phase boundary has stimulated a growing interest in the elucidation of the mechanism of interfacial activation of enzymes. Since a lipid monolayer at the air-water interface represents a useful model of the exoplasmic or cytoplasmic leaflets of biomembranes, several investigators have used lipid monolayers as substrates to elucidate the mechanism of 'interfacial activation' of enzymes (Thuren et al. 1991; Carriere et al. 1997; Roussel et al. 1998; James et al. 1997; Verger et al. 1976; Lio and Dennis 1998; Moreau et al. 1988; Hirche and Ulbrich-Hofman 1999; Jungner et al. 1997; Perillo et al. 1994;

Souvignet et al. 1991; Gronberg and Slotte 1990; Tsujita and Brockman 1987; Cernia et al. 1996; Muderhwa and Brockman 1992; Peters et al. 2000; Wilcox et al. 1993; Rogalska et al. 1995; Marguet et al. 1999; Ransac et al. 1997; Tanaka et al. 2000). The spread lipid monolayers at the air-water interface as substrates offer several advantages over other forms of lipid dispersions. For instance, once formed, the thermodynamic state of lipids in a micellar aggregate or lipid vesicle cannot be altered at will. This is not the case with spread monolayers at the air-water interface. In this system, the thermodynamic properties of the interface, i.e., the surface concentrations of the substrate and other constituents and interfacial tension can be varied at will by compressing or expanding the monolayer. Thus, the spread monolayer system enables one to study the influence of these variables on *interfacial activation* of enzymes. One of the most intriguing findings of these studies is that the activities of lipolytic enzymes are dependent on surface pressure, Π. For example, the activities of lipases have been reported to have a strong dependence on surface pressure of the monolayer (Thuren et al. 1991; Carriere et al. 1997; Roussel et al. 1998). For phospholipases, sphingomyelinase and neuraminidase the optimum surface pressure for maximum activity is about 15–25 $mN{\bullet}m^{-1}$ (James et al. 1997; Verger et al. 1976; Lio and Dennis 1998; Moreau et al. 1988; Hirche and Ulbrich-Hofman 1999; Jungner et al. 1997; Perillo et al. 1994). In contrast, the critical surface pressure for optimal activity of protein kinase C is between 30 and 35 $mN{\bullet}m^{-1}$ (Souvignet et al. 1991).

Several theories have been proposed to explain surface pressure dependency of interfacial activation of enzymes. These theories can be grouped into two categories, namely the '*substrate theory*' and the '*enzyme theory*'. The proponents of the '*substrate theory*' argue that an increase in the activities of enzymes at the lipid-water interface is related to the concentration, the hydration state, and conformation (i.e., ideal orientation of head groups in the two-dimensional plane) of substrate molecules in monolayers (Brockman et al. 1973; Tsujita et al. 1989; Muderhwa and Brockman 1990, 1992; Smaby et al. 1994). It has been pointed out, however, that an increase in the rate of enzyme catalysis at the interface by a factor of 10^3–10^4 compared to that in the bulk phase cannot be explained only by concentration of the substrate (Verger et al. 1976; Verger and de Haas 1976). Furthermore, if the high concentration of the substrate in monolayers is the main factor for the observed increase in enzyme catalysis, then why should the rate of the reaction decrease at surface pressures >20 $mN{\bullet}m^{-1}$ for phospholipases where the substrate concentration is greater than that at $\Pi < 20$ $mN{\bullet}m^{-1}$? To account for this anomaly, it has been suggested that it is not only the concentration, but also the physical properties of the substrate in monolayers, particularly phase transition of lipids at the interface, which is surface pressure-dependent, also plays a regulatory role in interfacial activation of enzymes (Honger et al. 1997). Dahmen-Levison et al. (1998) reported that phospholipase A_2 has maximum activity on phospholipids of different chain lengths (e.g., dipalmitoylphosphatidyl choline, disteroylphosphatidyl choline, and dimyristylphosphatidyl choline) when the monolayers had coexisting regions between liquid-expanded and liquid-condensed states. In addition, lateral lipid distribution (Muderhwa and Brockman 1992) and lateral phase separation (Gomez-Fernandez et al. 1990; Koynova et al. 1987) in mixed lipid monolayers also have been suggested to play a regulatory role in interfacial activation of enzymes.

If we critically examine each of these explanations in the context of biological systems, they do not stand up to the task. For instance, although the suggestion that phase transition and lateral lipid distribution play a regulatory role in the control of enzyme activity at the membrane-water interface is appealing, it is questionable if these indeed are the control mechanisms that prevail in biological systems. It should be recognized that the time scales of phase transitions and lateral lipid diffusion in membranes are typically far slower than the time scales of external stimuli-mediated cellular processes, which are typically on the order of microseconds to milliseconds. In a "fight or flight" situation, the control mechanisms at the cell level must respond to the situation within microseconds to milliseconds. This cannot happen if cells depend on a slow phase transition or lateral lipid diffusion process to activate metabolic processes. Furthermore, the surface pressure of membrane bilayers in cells is close to 72 mN m^{-1} (i.e., the interfacial tension of a typical biological membrane is close to zero) (de Haas et al. 1997), which is far higher than the surface pressures where phase transition in lipid monolayers are observed in *in vitro* model systems. Thus, in normal cell, the phase state of various lipids at various regions in a membrane might be globally invariant.

The '*enzyme theory*' postulates that the mechanism of interfacial activation of enzymes at the lipid-water interface is related to a conformational change in the enzyme upon binding to the interface (Entressangies and Desnuelle 1974). Recent crystallographic studies provide ample evidences for this theory (Scott et al. 1990; Brady et al. 1990; Schrag et al. 1991; Derewenda et al. 1992). The crystallographic structures of most lipases show a 'flap' or a 'lid' consisting of a short amphiphathic α-helix covering the substrate-binding site. It has been suggested that this lid may open upon binding of the enzyme to the lipid-water interface, allowing access of the substrate to the catalytic site (van Tilbeurgh et al. 1993). The '*enzyme theory*' is more attractive than the '*substrate theory*' because it provides a logical justification for *interfacial activation*, i.e., an increase in the specific activity of the enzyme, which is a conformation-related phenomenon. However, although the crystallographic structure provides a strong evidence for conformational-change-mediated interfacial activation of lipolytic enzymes, two unresolved questions remain (Damodaran 1998): *What is the mechanism by which the opening of the lid is effected at the lipid-water interface*, and *why is this process surface pressure-dependent?*

It should be pointed out that since the underlying assumptions in the 'substrate theory' and 'enzyme theory' are very different, they are mutually exclusive. That is, if activation is due to surface pressure-dependent conformation change in the enzyme, then the underlying mechanism of this process cannot be simultaneously invoked to argue in support of conformation or orientation of the head groups of substrate molecules or substrate concentration at the interface. Although there is a general agreement that interfacial activation of several enzymes is dependent on surface pressure, the molecular basis for this dependence has not been deciphered yet (James et al. 1997).

One of the explanations put forth for the surface pressure-dependent enzyme activation is based on the notion that surface pressure is a two-dimensional lateral pressure of the lipid layer, and this acts as a mechanical force causing structural changes in the adsorbed enzyme. For instance, to explain the mechanism of action of anesthetics, Cantor (1997a, b) suggested that incorporation of amphiphilic solutes (e.g., anesthetics) into the membrane bilayer might increase the lateral pressure selectively near the aqueous interface. If the opening of the ion channel in a postsynaptic ligand-gated protein involves a conformational change, then the anesthetic-induced increase in the lateral pressure near the aqueous interface would prevent such a conformational change and opening of the gate, since channel opening would require greater work against this higher pressure (Cantor 1997a). Although most lipolytic enzymes that become activated upon adsorption to the lipid-water interface are not integral membrane proteins, an extension of Cantor's suggestion would imply that the surface pressure of the monolayer or bilayer can act as a physical/mechanical force to cause conformational changes in enzymes at the lipid-water interface. However, there is a major obstacle to accepting this hypothesis to explain interfacial activation of enzymes: A careful examination of the data in the literature indicates that despite conformational differences, hydrolytic enzymes such as phospholipases (James et al. 1997; Verger et al. 1976; Lio and Dennis 1998; Moreau et al. 1988; Hirche and Ulbrich-Hofman 1999), neuraminidase (Perillo et al. 1994), and sphingomyelinase (Jungner et al. 1997), which act on various lipid substrates, show maximal activity at about a surface pressure of 20 $mN\ m^{-1}$. Given that the physicochemical and structural properties of these enzymes are very different, it is inconceivable that all these enzymes would attain the catalytically active optimal conformation at a lateral mechanical force corresponding to a surface pressure of about 20 $mN{\bullet}m^{-1}$. This raises a very fundamental question: Is surface pressure really related to a *mechanical* lateral pressure exerted by the lipid layer or does it represent some other thermodynamic state or property of the interface that is essential for enzymes to adopt the catalytically most optimal structure? Or, as Verger and De Haas (1973) had suggested, is it related to the "quality of the interface" as affected by the organization of lipids? If it is the latter, then one of the most important components that affect the "quality of the interface" is the activity of water in the interfacial region and therefore it is imperative to include the thermodynamic state of water in any discussion on the mechanism of membrane function.

A much broader question that is raised in this paper is '*what are the evolutionary reasons for biology to choose the lipid-water boundary as the site for performing and/or controlling crucial biological reactions? What is the universal physical principle that is very singular to the lipid-water interface that biology exploits as a 'key' for regulating the activities of membrane-bound enzymes?*

10.2 Physical Nature of Surface Pressure

When an insoluble spread monolayer of a surfactant is formed on an aqueous surface, the decrease in the tension at the surface is given by,

$$\gamma = \gamma_o - \Pi \tag{10.1}$$

where γ is the surface tension in the presence of the spread monolayer, γ_o is the surface tension of pure water, and Π is the experimental surface pressure,. According to this conventional view, the net reduction in the surface/interfacial tension of the system is the result of two opposing forces, viz., the lateral surface pressure exerted by the kinetic energy of surfactant molecules in the monolayer acting against the surface tension of the aqueous substrate (Myers 1991; Hiemenz 1986). Implicit in this argument is that the free energy of the water surface γ_o is invariant in the presence of an adsorbed surfactant monolayer, and the reduction in tension at the surface is simply the algebraic sum of these two diametrically opposing forces.

Van der Waals Equation of State According to this conventional view, at low monolayer densities a monolayer is considered to behave as a two-dimensional ideal gas and obeys the van der Waals equation of state

$$\Pi A = kT \text{ or } \Pi = kT\Gamma \tag{10.2}$$

where A is the area occupied per molecule, Γ (=1/A) is the number of molecules per unit area, k is the Boltzmann constant, and T is the temperature. For ideal monolayers, Π in Eq. 10.1 corresponds to Π_{kin}, i.e., the pressure arising from kinetic energy only. For monolayers obeying the two-dimensional ideal gas law, the ratio $\Pi/kT\Gamma$ is expected to be unity. Deviations from ideality, which is normal rather than exception, have been attributed to other molecular forces in the monolayer, such as cohesive interactions between the alkyl chains due to van der Waals forces, and electrostatic repulsive interactions between ionic head groups of the adsorbed molecules. Thus, the modified equation of state for a spread monolayer at an interface is expressed as (Davies 1956)

$$(\Pi_{kin} - \Pi_{coh} + \Pi_e)\ (A - A_o) = kT \tag{10.3}$$

where Π_{kin} is the surface pressure from kinetic energy, Π_{coh} is the decrease in lateral pressure due to van der Waals cohesive interactions between the apolar chains, Π_e is the contribution from repulsive electrostatic forces between the head groups to lateral pressure, and A_o is the excluded area related to the cross-sectional area of the surfactant molecule. For uncharged monolayers at the air-water interface (e.g., myristic acid in the unionized state at pH <2, where $\Pi_e = 0$), Π_{coh} at a given Γ is estimated by subtracting Π_{kin} (Eq. 10.2) from experimental Π (Davies 1956). In these types of analyses, Π_{coh} is assumed to be a function of A and T only. The Π_{coh}

thus determined at a range of A empirically followed the relation (Davies 1956),

$$\Pi_{\mathrm{coh}} = b/\mathrm{A}^{3/2} \tag{10.4}$$

where b is a coefficient related to the lateral intermolecular cohesive forces in the monolayer. On the other hand, Gershfeld (1970) reported that Π_{coh} showed a better fit with A^{-2} for aliphatic alcohols, acids, nitriles and ethyl esters. In the case of charged monolayers, the electrostatic contribution Π_{e} to the surface pressure is determined from the Davies equation based on the Gouy's ionic double layer model (Davies 1963),

$$\Pi_e = 6.1c^{1/2}\left\{\cosh\sinh^{-1}\left(134/A_{ele}c^{1/2}\right) - 1\right\} \tag{10.5}$$

where c is the concentration of a univalent electrolyte in the substrate, and $\mathrm{A}_{\mathrm{ele}}$ is the area per charge in the monolayer. It should be noted that according to Eq. 10.3 the contributions of Π_{kin}, Π_{coh}, and Π_{e} to the total pressure of monolayer is taken as additive, and that Π_{e} arising from repulsion between the head groups does not influence the magnitude of Π_{coh} arising from cohesive interactions between the nonpolar chains.

Experimental studies on spread monolayers of various surfactants and lipids have shown that they do not obey the van der Waals equation of state. For most surfactants, the ratio ΠA/kT is usually not equal to unity even in very dilute monolayers (Davies 1956; Gershfeld 1970; Pallas and Pethica 2009; Middleton et al. 2011; Mingins, et al. 1992), where the system is expected to obey the two-dimensional ideal gas law. For instance, the ratio ΠA/kT of spread monolayer of sodium octadecyl sulfate on aqueous NaCl solution (0.001–0.01 M) in the surface concentration range up to 3×10^{-2} molecules/nm^2 and in the surface pressure range of 0–300 μN m^{-1} was greater than unity (Fig. 10.1) (Middleton et al. 2011). This deviation is generally attributed to contribution of Π_{e} to the experimental surface pressure. However, it was shown that the Gouy model (Eq. 10.5) completely failed to account for the electrostatic contribution arising from head group repulsion (Middleton et al. 2011). On the other hand, the ΠA/kT ratio was found to be lesser than unity for a very dilute monolayer of n-pentadecanoic acid on a 0.01 M HCl substrate (where $\Pi_{\mathrm{e}} = 0$) (Fig. 10.2) (Pallas and Pethica 2009). According to conventional wisdom, ΠA/kT < 1 might be due to the negative contribution of Π_{coh} to the surface pressure (Eq. 10.3). However, when the data were analyzed using the van der Waals equation of state (Eqs. 10.3 and 10.4), the values of A_{o} and b determined from a plot of the second virial coefficient B_2 versus 1/T gave an unrealistically high value (by two orders of magnitude) for A_{o} and an uncharacteristic sixfold monotonic decrease in the b coefficient as the temperature was increased from 15 to 30 °C (Pallas and Pethica 2009). If Π_{coh} were responsible for ΠA/kT < 1, then one would expect an increase in b coefficient with increase of temperature. These abnormal predictions clearly indicate that the van der Waals equation of state is unable to adequately describe the experimental Π-A isotherms

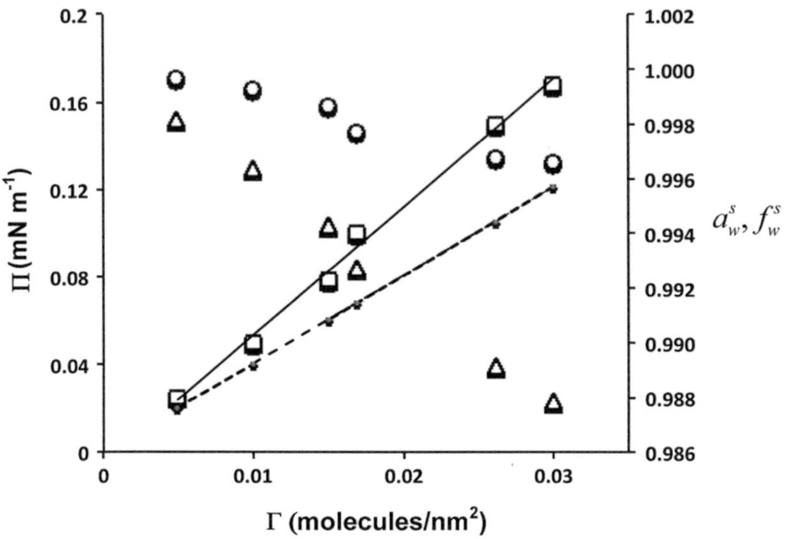

Fig. 10.1 Π versus Γ relationship (□) of sodium octadecyl sulfate spread monolayer at the 0.001 M NaCl-air interface at 20 °C. The data points were extracted from Middleton et al. (2011). The *dotted line* represents theoretical curve predicted by Eq. 10.2. Δ, determined from Eq. 10.21; ○, determined from Eq. 10.22

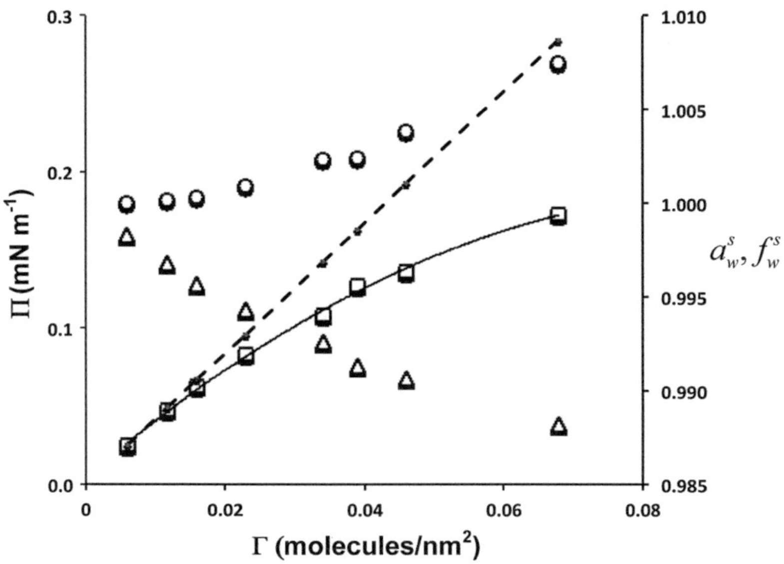

Fig. 10.2 Π versus Γ relationship (□) of n-pentadecanoic acid spread monolayer at the 0.01 M HCl-air interface at 30 °C. The data points were extracted from Pallas and Pethica (2009). The *dotted line* represents theoretical curve predicted by Eq. 10.2. Δ, determined from Eq. 10.21; ○, determined from Eq. 10.22

even for the very dilute monolayer (Pallas and Pethica 2009). Empirical corrections based on Π_{coh} and Π_e and/or application of the two-dimensional virial expansion to fit the experimental Π-A data will undoubtedly provide fitting coefficients; however, it does not necessarily mean that the system truly obeys the van der Waals equation of state as there is no thermodynamic or conceptual underpinnings in such empirical formulations. This is not to say that the van der Waals equation of state is wrong, but it simply implies that the origin of the net reduction in surface/interfacial tension is not linked to the lateral pressure arising from kinetic motions of the adsorbed molecules. In other words, the definition of Π as lateral surface pressure is misconstrued.

As the lipid bilayer-water interface plays an important role in biological systems, a fundamental understanding of the nature of the molecular forces operating at this interface is vital to understanding how some of the membrane-associated metabolic pathways in cells are potentially controlled by these interfacial forces. One of shortcomings of the two-dimensional gas law approach is that it “a priori” assumes or imposes that the reduction in surface tension in the presence of a spread lipid monolayer is related to the lateral pressure exerted by the kinetic motions of the molecules in the spread monolayer. The genesis of this argument can be traced back to one of the earliest observations: Consider a trough filled with water and a freely movable barrier or a float placed at the center of the trough, dividing the water surface (Fig. 10.3). When a monolayer of an insoluble surfactant is spread on one side of the barrier, the barrier moves away from the side to which the surfactant has been added. This is generally interpreted as due to a two-dimensional lateral pressure from the surfactant side, pushing the barrier to move, and it is termed as lateral surface pressure of the film. The logic of this argument is debatable. Because, it also can be argued that the barrier moves not because of a lateral pressure from the solute, but due to the free energy difference between water surfaces on both sides of the barrier. The free energy difference between the two sides of the water surface occurs as a result of interaction of surface water molecules on the surfactant side with the polar head group of the surfactant molecules. Since the water surface on both sides of the barrier is connected through the bulk, the system ‘sees’ a free energy gradient across the surface and therefore the side with higher surface free

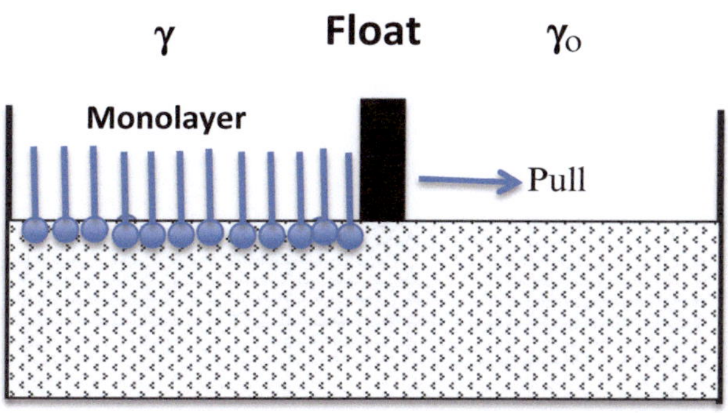

Fig. 10.3 Schematic representation of surface tension differential between water surfaces with and without a spread surfactant monolayer separated by a float. γ_o is the surface tension of pure water, and γ is the surface tension of water with a spread monolayer. $\gamma_o-\gamma$ is the force pulling the float

energy 'pulls' the barrier as it tries to contract the surface area. Since the spread monolayer is insoluble in the aqueous phase, it cannot 'see' a chemical potential gradient on the other side of the barrier. Thus, the movement of the barrier away from the surfactant side should be attributed to a 'pull' or contractile force from the high-energy water surface on the other side of the barrier. The barrier will move until the chemical potential of surface water molecules on both sides of the barrier is the same. An extension of this argument is that when the barrier is made immovable, the net reduction in surface tension (i.e., Π) on the surfactant side of the barrier relative to the pure water surface on the other side should be regarded as due to a net change in the thermodynamic activity of surface water molecules, and not due a lateral pressure arising from kinetic motions of the surfactant molecules. If such a paradigm shift in our interpretation of surface pressure is accepted, then all surface pressure-dependent biological phenomena at biological phase boundaries can be coherently explained in terms of the thermodynamic activity of interfacial water.

Gibb's Formulation The thermodynamics of fluid-fluid interfaces is described by the Gibbs' adsorption equation

$$d\gamma = -\sum_i \Gamma_i d\mu_i \tag{10.6}$$

where Γ_i is the surface excess amount of component i and μ_i is its chemical potential in the bulk phase. In Gibbs' formulation of the interface, the dividing line between the two fluid phases is chosen such that the surface excess of the solvent component $\Gamma_1 = 0$ (Prosser and Frances 2001). For this condition, Eq. 10.6 becomes,

$$d\gamma = -\sum_{i=2}^{n} \Gamma_i d\mu_i \tag{10.7}$$

In the case of an insoluble spread monolayer, the chemical potential in Eq. 10.7 defines the thermal energy (kT) of the surfactant molecules and the equation reduces to the two-dimensional gas law equation. By setting the dividing line at $\Gamma_1 = 0$, the Gibbs' equation stipulates that the reduction in surface tension is related to the surface excess of the solute component only. No consideration is given to the potential contribution of the change in the chemical potential of surface water to surface tension reduction. However, conceptually, as surface tension of pure water arises from the excess free energy of surface water molecules compared to those in the bulk phase, the reduction in surface tension in the presence of an adsorbed solute must also be related to a reduction in the excess free energy of surface water molecules, potentially as a result of its interaction with the adsorbed solute molecules.

Molecular dynamics simulations have shown that the density of water increases from zero at the surface to 1.0 g cm^{-3} at a depth of 6–8 Å (Broadskaya et al. 1996). What this means is that the surface tension arises solely from the excess free energy

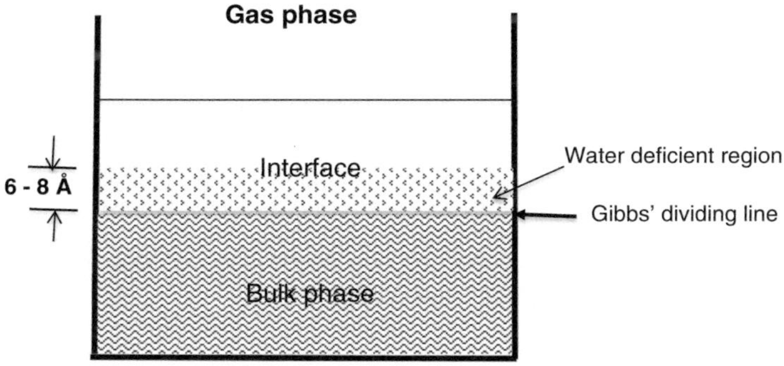

Fig. 10.4 Modified representation of Gibbs' dividing line at the air-water interface. $\Gamma_w = 0$ at the Gibbs' dividing line in relation to the bulk water, whereas Γ_w is low in relation to the bulk water above the dividing line

of water molecules in the low density region corresponding to 6–8 Å surface layer. Each water molecule in this region is hydrogen bonded to ≤ 3 other water molecules compared to 4 in the bulk phase. If the Gibbs' dividing line is set at $\Gamma_1 = 0$, then this dividing line is located at 6–8 Å below the water surface (as shown in Fig. 10.4), where the density of water is same as in the bulk. Above this dividing line there is no surface excess, but there is surface deficiency of water. This water-deficient surface region is an integral part of the interface and therefore the change in the chemical potential of water in this region in the presence of solute molecules must be taken into consideration to account for the reduction in surface tension.

10.2.1 Interfacial Water Activity Concept

Based on the above considerations, a solution that is in contact with its vapor phase can be divided into three phases, namely, bulk liquid (b), interface (s), and the gas (g) phase. The interfacial region has a finite thickness, which includes the 6–8 Å region where the density of water is less than 1 g cm^{-3}. The thermodynamics of this system can be examined as follows (Damodaran 1998). At equilibrium, the chemical potential of a component i in each phase of the system at constant temperature, pressure, and n_j can be written as

$$\mu_i^b = \frac{\partial G^b}{\partial n_i^b} = \frac{\partial F^b}{\partial n_i^b} + P\frac{\partial V_i^b}{\partial n_i^b} \tag{10.8}$$

$$\mu_i^s = \frac{\partial F^s}{\partial n_i^s} = P\frac{\partial V_i^s}{\partial n_i^s} + \gamma\frac{\partial A_i^s}{\partial n_i^s} \tag{10.9}$$

The third term on the right hand side of Eq. 10.9 refers to contribution from the surface force field, where γ is the surface tension, n_i is the number of moles of component i, and A is the surface area. Since the chemical potential of an ideal solution at constant temperature and pressure is

$$\mu_i^b = \mu_i^{o,b} + RT \ln a_i^b \tag{10.10}$$

where μ_i^b is the standard chemical potential component i in the bulk phase, and a_i^b is its activity. Upon substitution of Eq. 10.10 in 10.8,

$$\frac{\partial F^b}{\partial n_i^b} = \mu_i^{o,b} + RT \ln a_i^b - P\frac{\partial V_i^b}{\partial n_i^b} \tag{10.11}$$

Assuming that the interface behaves as an ideal two-dimensional solution, an equation similar to Eq. 10.11 can be written for the interface as

$$\frac{\partial F^s}{\partial n_i^s} = \mu_i^{o,s} + RT \ln a_i^s - P\frac{\partial V_i^s}{\partial n_i^s} \tag{10.12}$$

Substitution of Eq. 10.12 in 10.9 gives

$$\mu_i^s = \mu_i^{o,s} + RT \ln a_i^s - \gamma\frac{\partial A_i^s}{\partial n_i^s} \tag{10.13}$$

The partial differential term in Eq. 10.13 can be defined as the partial molar surface area, ω_i. That is

$$\mu_i^s = \mu_i^{o,s} + RT \ln a_i^s - \gamma\omega_i \tag{10.14}$$

At equilibrium, $\mu_i^s = \mu_i^b$. Thus, from Eqs. 10.10 and 10.14,

$$\gamma\omega_i = \mu_i^{o,s} - \mu_i^{o,b} + RT \ln \frac{a_i^s}{a_i^b} \tag{10.15}$$

If component i refers to water (component 1), the standard chemical potential of water in the surface phase ($\mu_i^{o,s}$) cannot be the same as the standard chemical potential of water in the bulk phase ($\mu_i^{o,b}$) because the density of water in the surface region is lower than that in the bulk phase. The surface tension of water principally arises because of this difference between the standard chemical potentials, and therefore the term $\mu_i^{o,s} - \mu_i^{o,b}$ in Eq. 10.15 can be defined as $\gamma_i\omega_i$, where γ_i is the surface tension of pure water. Accordingly, Eq. 10.15 can be rewritten as,

$$\gamma_1 - \gamma = -\frac{RT}{\omega_1} \ln \frac{a_1^s}{a_1^b} \tag{10.16}$$

Where γ_1 ($= \gamma_o$) is surface tension of pure water and γ is the surface tension in the presence of a solute monolayer. In the case of a spread monolayer of an insoluble solute, the activity of water in the bulk phase is unity, and by defining $\gamma_o - \gamma = \Pi$, Eq. 10.16 takes the form

$$\Pi = -RT\Gamma_w \ln a_w^s \tag{10.17}$$

where Γ_w ($=1/\omega_1$) is the moles of water per unit area of the interfacial phase. Equation 10.17 stipulates that the reduction in surface tension in the presence of a lipid monolayer is fundamentally related to a reduction in the thermodynamic activity of water in the interfacial region. This reduction in surface tension is not due to two-dimensional lateral pressure exerted by the lipid monolayer, but due to a reduction in the free energy of surface water molecules as a result of interaction with various chemical groups in the monolayer.

The Γ_w can be determined fairly accurately from the known physical constants of water. The enthalpy of formation of a unit area of water surface at constant volume is given by

$$H^s = \gamma_1 - T\left(\frac{\partial \gamma}{\partial T}\right)_V \tag{10.18}$$

The value of the partial differential is about –0.1529 mN m^{-1} K^{-1} in the temperature range 5–40 °C. The H^s is the energy required to bring Γ_w moles of water from the bulk phase to the surface against cohesive forces in the bulk liquid phase. The cohesive energy, E_{coh}, of water in the bulk liquid phase is given by

$$E_{coh} = \Delta H_v - RT \tag{10.19}$$

where ΔH_v is the heat of evaporation of one mole of water at temperature T. Assuming that the magnitude of attractive forces from the bulk phase acting on surface water molecules is one-half of that in the bulk phase, the energy required to bring one mole of water from the bulk phase to the surface would be equal to $E_{coh}/2$. Then, the energy spent to bring Γ_w moles of water from the bulk phase to the surface is

$$H^s = \frac{\Gamma_w E_{coh}}{2}$$

$$\text{or } \Gamma_w = \frac{2H^s}{E_{coh}} \tag{10.20}$$

Equation provides a means to estimate moles of water per unit area at the surface. For instance, at 25 °C, the heat of formation (H^s) of water surface and the cohesive energy of water is 118.42 erg cm^{-2} and 9.93 Kcal/mol, respectively. Substitution of these values in Eq. 10.20 gives a value of $\Gamma_w = 5.708 \times 10^{-10}$ mol cm^{-2}. If we assume that the thickness of the interfacial phase where this water resides is

6 Å, then the average density, ρ_w^s, of water across the interfacial phase is about 0.17 g cm^{-3}. The average surface area occupied by a water molecule is 29.1 Å^2, which is much larger than the area 9.65 Å^2 determined from the molecular volume of water (Fowkes 1962). This indicates that there is a large spatial separation between water molecules at the surface and the vacant area (≈20 Å^2) between water molecules at the surface might act as binding sites for surfactant molecules.

Several inferences can be drawn from Eq. 10.17: It stipulates that surface pressure of a monolayer film is a measure of the thermodynamic activity of interfacial water. The net reduction in the surface tension of water in the presence of a monolayer solely arises from a reduction in the free energy of surface water molecules and it is not related to a lateral pressure exerted by the monolayer. This is somewhat analogous to the hydrophobic effect, in which the free energy change for the transfer of a hydrocarbon from an aqueous phase to any organic solvent emanates solely from the free energy change in the aqueous phase and it is independent of the properties of the nonpolar solute as well as the organic solvent to which it is transferred (Damodaran and Song 1986). Figure 10.5 shows a plot of surface pressure versus interfacial water activity, a_w^s, at 25 °C, predicted by Eq. 10.17. The fact that the a_w^s approaches zero as the surface pressure approaches 72 dynes cm^{-1} (the value of the surface tension of pure water) clearly indicates that the Γ_w value (=5.708 × 10^{-10} mol cm^{-2} at 25 °C) determined from Eq. 10.20 is fairly accurate. Equation 10.17 can be rewritten as

$$\Pi = -RT\Gamma_w \ln x_w^s f_w^s \tag{10.21}$$

$$\text{or } \Pi = -RT\Gamma_w \ln \left\lfloor \left(1 - x_2^s\right) f_w^s \right\rfloor \tag{10.22}$$

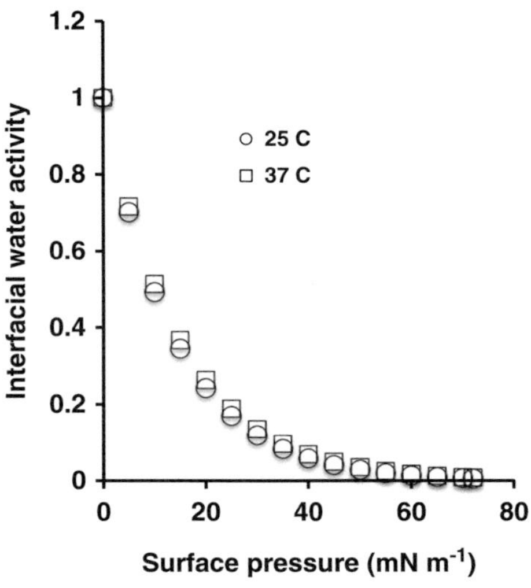

Fig. 10.5 The relationship between surface pressure and interfacial water activity predicted by Eq. 10.17 at 25 and 37 °C

where x^s_w is the mole fraction of water, x^s_2 is the mole fraction of the solute, and f^s_w is the activity coefficient of water in the surface region. In the case of a very dilute spread monolayer, if $x^s_2 \ll 1$, and if the activity coefficient of water can be assumed to be unity, then Eq. 10.22 becomes

$$\Pi = RT\Gamma_w x^s_2 \tag{10.23}$$

If Γ_2 is the number of moles of the solute per unit area of the surface, then $x^s_2 = \Gamma_2 / (\Gamma_w + \Gamma_2)$ and

$$\Pi = RT\Gamma_w \frac{\Gamma_2}{\Gamma_w + \Gamma_2}$$

When $\Gamma_2 \ll \Gamma_w$, Eq. 10.23 is reduced to

$$\Pi = RT\Gamma_2 \tag{10.24}$$

which is same as the ideal two-dimensional gas law. However, it should be categorically emphasized that in the above formulation Π is not the lateral pressure, but explicitly defined as the net reduction in the free energy of surface water molecules caused by its interaction with Γ_2 mol of the solute at the surface.

Experimental studies have shown that the Π - Γ_2 isotherms of insoluble fatty acid monolayers do not follow Eq. 10.24 (and Eq. 10.2) even at very low surface density: This is shown in Figs. 10.1 and 10.2 for octadecyl sulfate and n-pentadecanoic acid monolayers, respectively. This implies that we cannot assume that $f^s_w = 1$ (Eq. 10.22) even for a very dilute monolayer and the change in f^s_w seems to be non-trivial. The changes in a^s_w and f^s_w as a function of Γ_2, calculated using Eqs. 10.21 and 10.22, are shown Figs. 10.1 and 10.2 for octadecyl sulfate and n-pentadecanoic acid, respectively.

For the situations where $\gamma > \gamma_o$ (i.e., for $-\Pi$), which is the case for salt solutions (Pogram and Record 2007), Eq. 10.21 predicts that the activity of surface water molecules in those cases will be greater than the water activity at the pure water surface. This increase in surface water activity ($a^s_w > 1$) might be either due to an increase in Γ_w or due to an increase in the activity coefficient f^s_w, or both. In the case of the former, it can be inferred that $\Gamma'_w > \Gamma_w$, where Γ'_w is moles of water per cm^{-2} at the surface in salt solutions, might occur as a consequence of injection of more water molecules into the surface phase by the salt and depletion of the salt ions from the interfacial region due to repulsive dispersion interaction with the gas phase.

10.2.2 The Potential Role of Interfacial Water in Biology

Although it is well recognized that structural evolution of proteins and formation of lipid vesicles and cell membranes are simple manifestations of the hydrophobic effect, i.e., a consequence of energetics of interaction of water with the apolar moieties of these molecules, the possibility of water playing a vital role in the very functioning of these biological systems is often overlooked. Biochemists often tend to explain biological activities at the membrane-water interfaces in terms of specific effects of substrates or other small molecule activators/modulators on conformation of membrane bound proteins/enzymes. The ubiquitous role of interfacial water in membrane-associated biological activities, such as interfacial activation of enzymes or as an integral part of the signal transduction systems in biology, is often overlooked and/or ignored.

In biological systems, enzymes generally catalyze non-aqueous reactions in an aqueous milieu. Some of these enzyme-catalyzed reactions may be inefficient and their precise control may be impossible in a bulk aqueous environment. In this context, if one examines the internal and external environments of a cell, the membrane-water interface is the only domain where control of the (re)activity of water is possible. Equation 10.17 instructs that the changes in the interfacial tension (i.e., surface pressure) at the membrane-water interface are a measure of the changes in water activity at the interface. The interfacial tension, and thereby the thermodynamic activity of interfacial water, can be manipulated by compression or expansion of the membrane. Thus, all surface pressure-dependent biological processes at the membrane-water interface must be regarded as processes innately controlled by interfacial water activity.

As mentioned earlier, several lipolytic enzymes become activated when they bind to the lipid-water interface, and their activity is strongly dependent on the surface pressure. For instance, the specific activities of several lipases (Thuren et al. 1991; Carriere et al. 1997; Roussel et al. 1998; Salah et al. 2001; Fendri et al. 2004; Tanaka et al. 2000), phospholipases (Pattus et al. 1979; Bianco et al. 1992; Rao and Damodaran 2004; James et al. 1997; Hirche and Ulbrich-Hofman 1999), neuraminidase (Perillo et al. 1994), and sphingomyelinase (Jungner et al. 1997; Rao and Damodaran 2005) exhibit a maximum in the range of 10–25 mN m^{-1} surface pressure in spread lipid monolayers. In contrast, protein kinase C exhibits a maximum specific activity in the surface pressure range 30–35 mN m^{-1} (Souvignet et al. 1991). If we accept the paradigm that surface pressure is a direct measure of interfacial water activity, then according to Eq. 10.17, the surface pressure range 15–25 mN m^{-1}, where most the lipolytic enzymes exhibit maximum specific activity, corresponds to an interfacial water activity of about 0.17–0.5. That is, these hydrolytic enzymes seem to acquire a conformation with maximum catalytic activity when the interfacial water activity is in the range of 0.2–0.5. The variation in water activity requirement by these various hydrolytic enzymes might reflect their structural differences. Above and below this interfacial water activity range the conformations of these enzymes change to either an inactive or a less active form.

The effect of interfacial water activity (determined from Eq. 10.17 at various experimental surface pressures) on the specific activity of phospholipase A_2 and sphingomyelinase is shown in Figs. 10.6 and 10.7, respectively. In lipid monolayers at the air-water interface, both these lypolytic enzymes show maximum specific

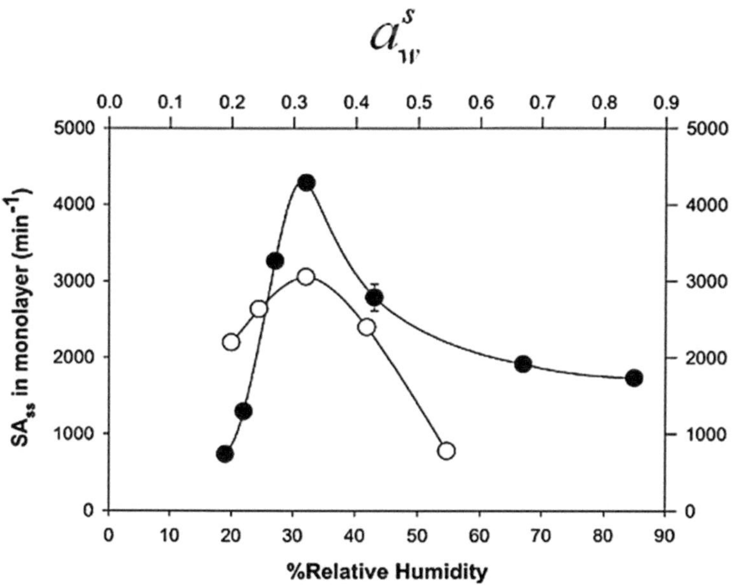

Fig. 10.6 (○) Relationship between specific activity of phospholipase A_2 (PLA_2) and in 1,2-dodecanoyl-sn-glycero-3-phosphocholine (DDPC) monolayer on 10 mM Tris buffer, pH 8.0, containing 20 mM $CaCl_2$ and 100 mM NaCl at 25 °C. (•) Relationship between specific activity of PLA_2 and equilibrium relative humidity in a soy flour solid matrix (Rao and Damodaran 2004)

activity at $\Pi = 17–20$ mN m^{-1} which corresponds to $a_w^s \approx 0.3$. If surface pressure dependency of enzyme activity is really related to water activity, then such a phenomenon ought to be system independent. That is, these enzymes should exhibit maximum specific activity at $a_w \approx 0.3$ in systems other than the lipid-water interface, such as in a solid phase system, as well. The specific activity of phospholipase A_2 and sphingomyelinase in a soy flour matrix at various equilibrium relative humidity (ERH) is shown in Figs. 10.6 and 10.7, respectively, along with the data of the monolayer studies (Rao and Damodaran 2004, 2005). As the ERH was progressively decreased from 0.85 to 0.19, the specific activity of phospholipase A_2 and sphingomyelinase passed through a maximum at ERH (a_w) ≈ 0.3. It should be noted that the water activity – specific activity profiles of these two enzymes in the monolayer and in soy flour matrix are similar, which strongly suggests that the surface pressure-dependency of activity of these enzymes at the lipid-water interface is indeed related to interfacial water activity.

Subtle structural changes in enzymes as a function of the water activity of the environment can be best understood from water sorption isotherms of proteins. As water is progressively stripped from a protein's surface by lowering relative humidity, the multilayer water is removed first, followed by removal of water associated with hydrophobic patches of the protein. A monolayer with a hydration value of about 0.08–0.1 g water/g protein is formed when the relative humidity is

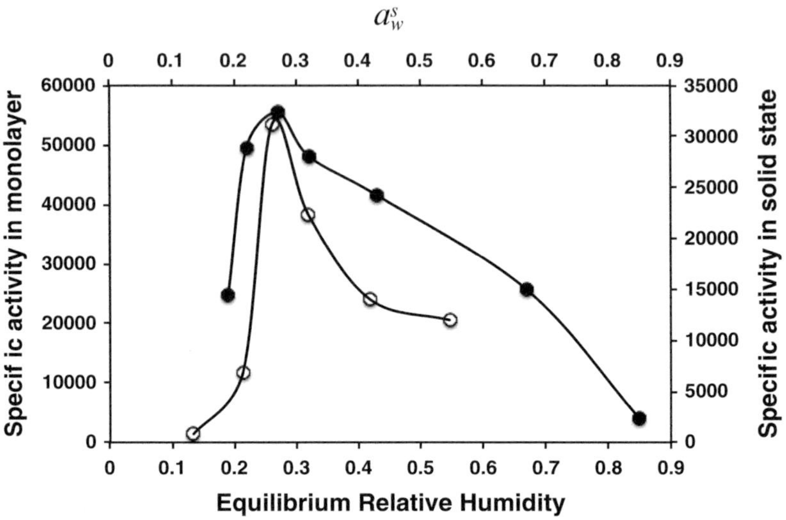

Fig. 10.7 (○) Relationship between specific activity of sphingomyelinase and in sphingomyelin (bovine brain) monolayer on 10 mM Tris buffer, pH 8.0, containing 3 mM $MgCl_2$ and 100 mM NaCl at 25 °C. (•) Relationship between specific activity of sphingomyelinase and equilibrium relative humidity in a soy flour solid matrix (Rao and Damodaran 2005)

further decreased to about 0.3–0.4 (Rupley et al. 1980; Rupley and Careri 1991). At this hydration range, only the ionic and polar regions of the protein's surface are hydrated. Dehydration of non-ionic hydrophilic surfaces commences when the relative humidity is decreased from 0.4 to 0.2. The fact that the specific activities of lipolytic enzymes show a maximum at about $a_w^s = 0.3$ ($\Pi = 18$ mN m^{-1}) suggests that complete dehydration of the hydrophobic surface and partial dehydration of hydrophilic surfaces of the enzyme that contacts the lipid-water interface is critical for transforming the lipolytic enzymes to a catalytically active mode. Reduction in enzyme activity at higher surface pressures, i.e., at $a_w^s < 0.3$, might be related to two factors: Additional dehydration of the protein may alter the conformation of the active site to a sub-optimal substrate binding state. Or, since water is one of the reactants in the hydrolysis reaction, the availability of water at $a_w^s < 0.3$ may become detrimental for the reaction to proceed. Thus, it appears that the thermodynamic activity of water at the lipid-water interface simultaneously controls the active conformation of the enzyme and the availability of water for the hydrolysis reaction.

In the case of protein kinase C (PKC), which catalyzes a phosphorylation (esterification) reaction in cell membrane and exhibits maximum specific activity at 35 mN m^{-1} (Souvignet et al. 1991), the water activity requirement of this enzyme is very different from those of the hydrolytic enzymes. It should be noted that water is one of the products of the protein phosphorylation reaction, and removal of water from the reaction environment is imperative for maintaining the efficiency of the

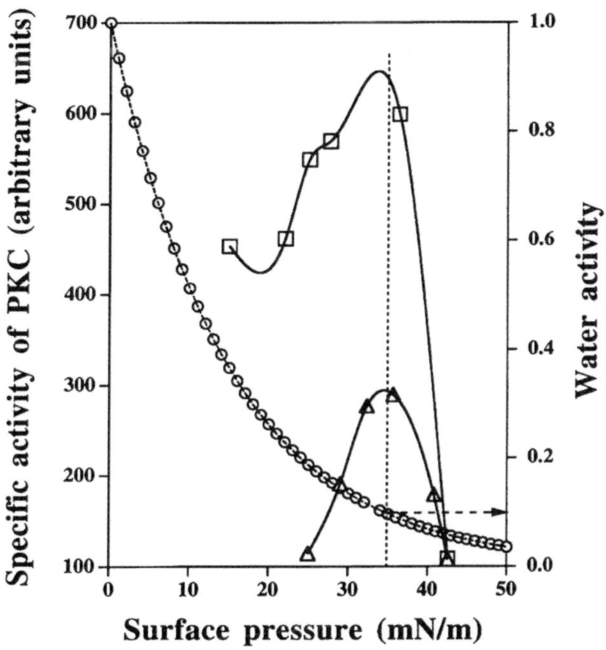

Fig. 10.8 Specific activity of protein kinase C (PKC) in phosphatidylcholine monolayer (Δ) and in phosphatidylserine/diglyceride mixed monolayer (□) as a function of surface pressure of the monolayer (From Souvignet et al. 1991). Superimposed (○) is the relationship between surface pressure and interfacial water activity predicted by Eq. 10.17 (Damodaran 1998)

phosphorylation (esterification) reaction. In the case of PKC this is accomplished by carrying out the reaction at the lipid-water interface at 35 mN m^{-1} at which the interfacial water activity is 0.1 (Fig. 10.8).

These re-interpretations of interfacial activation and surface pressure-dependency of activities of enzymes at the membrane-water interface indicate that, in addition to transforming enzymes from their inactive state to active state at the interface, interfacial water activity might control the availability of water at, or removal of it from, the reaction environment at the membrane-water phase boundary.

The interfacial tension of a spherical lipid bilayer vesicle under resting conditions is about 10^{-3}–10^{-6} mN m^{-1} (de Haas et al. 1997), which translates to an interfacial water activity close to zero. At this interfacial water activity the activity of lipolytic enzymes would be very low, as they would normally require an interfacial water activity of 0.2–0.5 to function. However, it can be visualized that if the activities of these enzymes are under the control of activator/receptor systems, then binding of an activator molecule to membrane-bound receptor would induce conformational changes in the receptor, and this may cause local stretching of the membrane, which will trigger localized increase in the interfacial water activity to a level that is required for the activation of lipolytic enzymes (Damodaran 1998). It has been reported that changes in vesicle morphology caused by lateral phase separation and lipid bilayer heterogeneities modulated the activity of phospholipase A2 (Burack et al. 1997; Burack and Biltonen 1994). These observations also might be fundamentally related to local changes in interfacial water activity resulting from distortion of the spherical shape of the membrane and at the boundary between the lipid expanded and lipid condensed regions of membranes.

10.3 Functional Role of Membrane-Bound Receptors

The above conjecture raises the possibility of an interesting mechanism for the activator-receptor-mediated activation of metabolic pathways in cells. It is known that the inner and outer leaflets of the plasma membrane are impregnated with numerous receptor proteins. The receptors by themselves are not enzymes, but are in many cases a part of a membrane-bound enzyme complex. Their primary function is to act as sensors of chemical signals from outside and inside of the cell and translate that signal into a cellular response. The first step in this signal transduction process is a specific conformational change in the receptor protein when an activator molecule (e.g., hormone) binds to it. Although several theories exist, it is not very well understood how this conformational change in the receptor propagates the signal to the nearby enzyme system, which initiates a cascade of events resulting in activation of a metabolic pathway in the cell.

Intuitively, at the very fundamental level, biology must be using a simple paradigm for operating all the numerous membrane-bound activator-receptor systems in the cell membrane. This universal mechanism might involve localized manipulation of interfacial water activity near the enzyme complex and use of the excess free energy of non-hydrogen-bonded interfacial water molecules to either activate or inactivate the enzyme complex. The essential steps involved in this mechanism are shown schematically in Fig. 10.9. According to this mechanism, binding of the activator to the membrane-bound receptor causes a conformational change in the receptor. This conformational change induces a precise localized perturbation, i.e., localized stretching (or compression) of the membrane. This stretching action, which may involve only a few nm^2 of the membrane surface, will cause a localized increase of interfacial tension and an increase in *local* a_w^s. The high-energy water molecule(s), released as a consequence of this localized stretching action, may activate the target enzyme located in the neighborhood of the receptor by altering its conformation via breaking an intra-molecular hydrogen bond. In essence, according to this mechanism, the receptor/hormone combination may act as a molecular tweezer to cause localized (a few nm^2 area) perturbation of interfacial water activity.

A cell membrane is embedded with hundreds of receptor-activator systems coupled to various membrane-bound enzyme systems. The need for a wide array of receptors to communicate signals from the outside to inside of a cell can be explained as follows: The free energy of interfacial water molecules can be manipulated by varying the extent of local stretching of the membrane. If the activation of a membrane-bound enzyme coupled to a receptor requires, for example, 4 kT of energy to break an intra-molecular hydrogen bond in the membrane-bound enzyme, then the activator-receptor system coupled to this enzyme might be designed in such

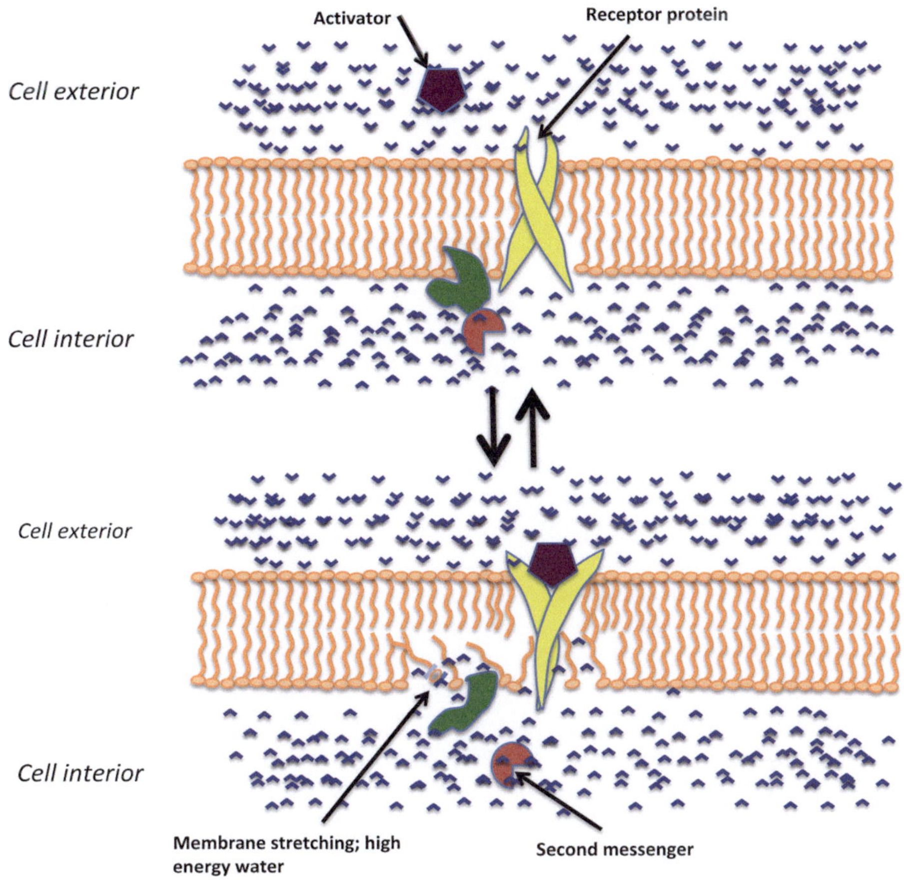

Fig. 10.9 Schematic model for interfacial water activity-dependent modulation of receptor-activator- mediated metabolic pathways in biological systems. See text for details

a way that the extent of conformational change in the receptor would stretch the membrane to an extent that the local water molecules at the interface attain this energy to cleave the hydrogen bond. Likewise, if breaking of an intra-molecular hydrogen bond in another membrane-bound enzyme system requires 6 kT of energy to transform it to the active state, then the activator-receptor system of that enzyme might be designed in such a way that conformational changes in the receptor stretches the membrane to a greater extent so that the interfacial water molecules at that location attain 6 kT of energy to cleave the hydrogen bond. Thus, the functional role of various activator-receptor systems in a cell membrane might be designed to act as mechanical stretching (and compressing) devices at the membrane-water interface. In terms of the thermodynamics of energy transduction, the chemical energy of activator-receptor interaction is transformed into mechanical energy involving a conformational change in the receptor (which stretches the membrane), and this mechanical energy is again converted into chemical energy in the form of release of high energy interfacial water molecule, and this chemical energy is used by the cell to cleave a hydrogen bond and induce desired conformational changes in the enzyme to initiate a cascade of cellular response. In this respect, the free energy of water at the membrane-water interface may act as a latent free energy reservoir for energy and/or signal transduction in cells.

Indirect evidences exist in the literature to support this paradigm. For instance, cell membranes of several species contain many mechanosensitive ion channels (Sukharev et al. 1994; Martinac et al. 1987, 1990). In addition, some of the conventional receptor-activated metabolic pathways also have been found to be mechanosensitive; that is, they become activated even in the absence of activator molecules (such as hormones) when a shear stress is applied (Gudi et al. 1998; Davies 1995). An example of the latter is the G-protein-dependent signaling pathways in cells (Gudi et al. 1998). The fundamental mechanism by which an externally applied mechanical force (pressure) acts in lieu of an activator to instantaneously activate a metabolic pathway or open an ion channel is largely unknown (Sukharev et al. 1994; Davies 1995; Kapoor et al. 2013).

One of the well-studied mechanosensitive ion channels is the large-conductance mechanosensitive channels (MscL) of prokaryotes (Sukharev et al. 1994; Martinac et al. 1987, 1990; Chang et al. 1998; Perozo et al. 2002). MscL is a multi-state channel consisting of four conducting states and a closed state (Sukharev et al. 1999). Using the patch-clamp electrical recording technique, it has been shown that the MscL ion channels in giant spheroplasts of *Escherichia coli* became activated when a negative (suction) or a positive pressure was applied on the spheroplasts (Martinac et al. 1987). The extent of channel gate opening was dependent on the applied suction pressure, however, and the complete opening of the channel typically required about 70–180 mm Hg (1 mm Hg = 133 Pa) suction pressure (Sukharev et al. 1999). Since the pressure gradient across the bilayer is related to interfacial tension according to the Laplace equation, $\Delta P = 2\gamma/R$, where γ is the bilayer tension and R is the radius of curvature, it is generally believed that the mechanosensitivity of the channel is related to an increase in the bilayer tension resulting from the pressure gradient-induced areal deformation of the bilayer (Sukharev et al. 1999). In other words, the current hypothesis contends that the increase in bilayer tension creates a tensile force pulling the subunits of the channel protein to open the gate (Wiggins and Phillips 2005; Sukharev et al. 1999). Experimental results have shown a sigmoidal correlation between the probability of gate opening and the net increase in bilayer tension (predicted from the Laplace equation), with the midpoint of the transition from the closed to the open state occurring at about 11.8 mN m^{-1} (Sukharev et al. 1999). It is believed that the pressure-induced excess interfacial energy, $\Delta\gamma\Delta A$ (where $\Delta\gamma$ is the net increase in bilayer tension and ΔA is the net change in interfacial area), overcomes the activation energy ΔE needed to open the gate (Sukharev et al. 1999). However, detailed analysis of the open channel structure of MscL has raised doubts about this hypothesis (Perozo et al. 2002).

One of intriguing observations made in the above study was that the subconducting states were mostly independent of the pressure-induced bilayer tension and the complete opening of the channel, which occurs at a suction pressure of 70 mm Hg, needed a critical conformational change in the channel protein (Sukharev et al. 1999). This disconnection between the bilayer tension and actual gate opening has not been explained satisfactorily.

The mechanism of MscL channel operation can be explained using the water activity hypothesis. If ΔP is the initial pressure difference between the inside and outside of a spherical membrane, R is its radius of curvature, and γ is the initial interfacial (bilayer) tension, then net change in the bilayer tension upon application of a suction pressure is given by the Laplace equation

$$\Delta\left(\Delta P\right) = \frac{2\left(\Delta\gamma\right)}{R} \tag{10.25}$$

If $\Delta(\Delta P) = 70$ mm Hg and radius R of the vesicle is 3 μm (Sukharev et al. 1999; Martinac et al. 1987), the net increase in bilayer tension Δγ is about 14 mN m^{-1}, which is the value reported by Sukharev et al. (1999). This net increase in bilayer tension is related to an increase in the membrane surface area, which is given by (Chiruvolu and Zasadzinski 1993)

$$\Delta\gamma = K\alpha \tag{10.26}$$

where K is the area expansion modulus of the bilayer and α is the fractional increase in area of the bilayer. Assuming that the average area expansion modulus of a typical cell membrane vesicle is about 150 mN m^{-1} (Miyamoto et al. 1988), Eqs. 10.25 and 10.26 predict that $\alpha = 0.09$ for a suction pressure of 70 mm Hg ($\Delta\gamma = 14$ mN m^{-1}). However, it should be recognized that a typical biological membrane is heterogeneous in nature as it contains phospholipids, sphingomyelins, cholesterol, and proteins. Furthermore, the lipids in a typical cell membrane are phase separated/segregated as cholesterol + sphingomyelin-rich and phospholipid-rich regions impregnated with proteins/enzymes. It has been reported that the value of K varied from 200 mN m^{-1} for a purely phospholipid vesicle to about 1200 mN m^{-1} for a phospholipid + cholesterol (10/90) vesicle (Needham and Nunn 1990). Therefore, it is reasonable to assume that the area expansion modulus of the bilayer would be different at different regions of the bilayer, depending on the local lipid composition and presence of incompressible proteins. Therefore, for a heterogeneous membrane vesicle, Eq. 10.26 can be expressed as

$$\Delta\gamma_{ave} = \alpha\frac{\sum^{n} K_i}{n}. \tag{10.27}$$

In other words, the Δγ determined from Eq. 10.25 represents an average increase in bilayer tension and the K in Eq. 10.26 represents an average value for the bilayer.

If we assume that the value of K_i in the neighborhood of the MscL channel is about 500–700 mN m^{-1}, which is a reasonable value for membranes containing cholesterol and embedded proteins (Needham and Nunn 1990), and $\alpha = 0.09$, then the local change in bilayer tension $\Delta\gamma_i$ will be about 45–63 mN m^{-1}. Since $\Delta\gamma = -\Delta\Pi$, Eq. 10.17 takes the form,

$$\Delta\Pi = RT\Gamma \ln\left(\frac{a^s_{w,2}}{a^s_{w,1}}\right) \tag{10.28}$$

where $a^s_{w,1}$ and $a^s_{w,2}$ are interfacial water activity before and after area expansion of the bilayer. Equation 10.28 predicts that for a local change in $\Delta\gamma$ by 45–63 mN m^{-1}, the pressure-induced stretching of the bilayer would cause a net increase in the free energy (RT ln $(a^s_{w,2}/a^s_{w,1})$) of local interfacial water by about 7.8–10.87 kJ/mol at 25 °C. We hypothesize that conformational changes in the MscL protein and channel opening might occur as a result of breaking of a crucial inter-subunit hydrogen bond in the channel protein by the high-energy local interfacial water molecules. There is indirect evidence in the crystal structure of the MscL protein to support this interpretation. The MscL channel is a homopentamer, where each subunit consists of two transmembrane α-helical segments, TM1 and TM2 (Chang et al. 1998). In the channel structure, each TM1 helix contacts with two TM1 helices of adjacent subunits and two TM2 helices within the same subunit. The radius of the channel pore in the closed state is about 0.2 nm and the pore is lined with a series of four threonine residues of the TM1 helix (residues 25, 28, 32, and 35) and one lysine (residue #33) and one aspartate (residue #36) (Chang et al. 1998). It is quite likely that the contacts between TM1 helices from adjacent subunits within the pore domain might involve hydrogen bonding between the threonine residues. We hypothesize that breaking of one or more of these threonine-threonine inter-subunit hydrogen bonds by the high energy local interfacial water molecules, generated by bilayer deformation, might be the basic mechanism by which channel opening occurs. In terms of energy transformation, the mechanical energy (pressure) is transformed into chemical energy (high energy interfacial water molecules) and this chemical energy is transformed into mechanical energy (conformational change in the channel protein).

10.4 Oscillatory Reactions in Biology

Certain metabolic pathways in cells are known to exhibit oscillatory behavior (Ross and Richter 1984; Schreiber et al. 1996). That is, when the rate of a reaction in a metabolic pathway is measured as a function of time, it exhibits a sinusoidal behavior. Oscillations in metabolic pathways that are controlled primarily by hormone secretions (extracellular activators) can be explained using the scheme shown in Fig. 10.9: In this scheme, first the hormone binds to the receptor and induces a conformational change. This stretches the membrane and increases local water activity; the high-energy water molecules activate the enzyme by breaking a key hydrogen bond, which sends an initial pulse to the metabolic pathway. If the hormone remains bound to the receptor, the metabolic reaction would continue at a constant rate. However, if the affinity of the receptor to hormone were lower in the altered state than in the initial state, then the hormone would dissociate from the hormone-receptor complex. Upon dissociation, the receptor would return back to its original structure. This would relax the membrane, decrease the local water activity, facilitate intra-molecular hydrogen bond formation in the enzyme, and thus return the enzyme to its inactive state. Since the receptor is now in a high affinity mode, a second molecule of the hormone will bind to the receptor, initiating a second cycle of events. During each cycle, the local water activity would rise and fall, and accordingly the enzyme will be repeatedly turned on and off, creating pulses in the metabolic pathway. The periodicity of these pulses can be in the order of microseconds to milliseconds.

10.5 Conclusions

The major thesis of this chapter is that biology uses the water activity-restricted membrane lipid-water interface as the site for catalyzing organic reactions and to control metabolic pathways in cells. The functional role of the membrane and membrane-bound proteins/receptors in this regard is to act as devices to manipulate the free energy (activity) of interfacial water through stretching and compression of the membrane bilayer. The mechanical energy of membrane stretching/compression is converted into chemical energy of interfacial water and this chemical energy is then used to cause conformational changes in second messenger molecules to transmit signals. In other words, the free energy of interfacial water is used as a source of potential energy to propagate metabolic signals in cells. It is shown that interfacial activation of lipolytic enzymes and the functioning of mechanosensitive large pore ion channels can be adequately re-interpreted using the above hypothesis.

References

Auge N, Andrieu N, Negresalvayre A, Thiers JC, Levade T, Salvayre R (1996) The sphingomyelin-ceramide signaling pathway is involved in oxidized low-density lipoprotein-induced cell proliferation. J Biol Chem 271:19251–19255

Bianco ID, Fidelio GD, Yu RK, Maggio B (1992) Concerted modulation by myelin basic-protein and sulfatide of the activity of phospholipase-a2 against phospholipid monolayers. Biochemistry 31:2636–2641

Brady L, Brzozowski AM, Derewenda ZS, Dodson E, Dodson G, Tolley S, Turkenburg JP, Christiansen L, Huge-Jensen B, Norskov L, Thim L, Menge U (1990) A serine protease triad forms the catalytic centre of a triacylglycerol lipase. Nature 343:767–770

Broadskaya EN, Eriksson JC, Laaksonen A, Rusanov AI (1996) Local structure and work of formation of water clusters studied by molecular dynamics simulations. J Colloid Interface Sci 180:86–97

Brockman HL, Law JH, Kezdy FJ (1973) Catalysis by adsorbed enzymes: the hydrolysis of tripropionin by pancreatic lipase adsorbed to siliconized glass. J Biol Chem 248:4965–4970

Burack WR, Biltonen RL (1994) Lipid bilayer heterogeneities and modulation of phospholipase A2 activity. Chem Phys Lipids 73:209–222

Burack WR, Dibble AR, Allietta MM, Biltonen RL (1997) Changes in vesicle morphology induced by lateral phase separation modulate phospholipase A2 activity. Biochemistry 36:10551–10557

Cantor RS (1997a) The lateral pressure profile in membranes: a physical mechanism of general anesthesia. Biochemistry 36:2339–2344

Cantor RS (1997b) Lateral pressures in cell membranes: a mechanism for modulation of protein function. J Phys Chem B 101:1723–1725

Carriere F, Thirstrup K, Hjorth S, Ferrato F, Nielsen PF, Withers-Martinez C, Cambillau C, Boel E, Thim L, Verger R (1997) Pancreatic lipase structure-function relationships by domain exchange. Biochemistry 36:239–248

Cernia E, Battinelli L, Soro S (1996) Biocatalysed hydrolysis of triglycerides in emulsion and as monolayers. Thin Solid Films 284–285:727–730

Chang G, Spencer RH, Lee AT, Barclay MT, Rees DC (1998) Structure of the MscL homolog from Mycobacterium tuberculosis: a gated mechanosensitive ion channel. Science 282:2220–2226

Chen CS, Rosenwald AG, Pagano RE (1995) Ceramide as a modulator of endocytosis. J Biol Chem 270:13291–13297

Chiruvolu S, Zasadzinski AN (1993) Membrane elasticity effects on permeability measurements in vesicles. Am Inst Chem Eng J 39:647–652

Chmura SJ, Nodzenski E, Weichselbaum RR, Quintans J (1996) Protein kinase C inhibition induces apoptosis and ceramide production through activation of a neutral sphingomyelinase. Cancer Res 56:2711–2714

Dahmen-Levison U, Brezesinski G, Moehwald H (1998) Enzymic hydrolysis of monolayers. A polarization modulated-infrared reflection absorption spectroscopy study. Prog Colloid Polym Sci 110:269–275

Damodaran S (1998) Water activity at interfaces and its role in regulation of interfacial enzymes: a hypothesis. Colloids Surf B Biointerfaces 11:231–237

Damodaran S, Song KB (1986) The role of solvent polarity in the free energy of transfer of amino acid side chains from water to organic solvents. J Biol Chem 261:7220–7222

Davies JT (1956) A surface equation of state for charged monolayers. J Colloid Sci 11:377–390

Davies PF (1995) Flow-mediated endothelial mechanotransduction. Physiol Rev 75:519–560

Davies JT, Rideal EK (1963) Interfacial phenomena, 2nd edn. Academic, New York

De Haas KH, Blom C, van den Ende D, Duits MHG, Mellema J (1997) Deformation of giant lipid bilayer vesicles in shear flow. Phys Rev E 56:7132–7137

Derewenda U, Brozozowski AM, Lawson DM, Derewenda ZS (1992) Catalysis at the interface – the anatomy of a conformational change in a triglyceride lipase. Biochemistry 31:1532–1541

Entressangies B, Desnuelle P (1974) Action of pancreatic lipase on monomeric tripropionin in the presence of water-miscible organic compounds. Biochim Biophys Acta 341:437–446

Fendri A, Sayari A, Gargouri Y (2004) Kinetic properties of turkey pancreatic lipase: a comparative study with emulsified tributyrin and monomolecular dicaprin. Chirality 17:57–62

Fowkes FM (1962) Ideal two-dimensional solutions. II. A new isotherm for soluble and gaseous monolayers. J Phys Chem 66:385–389

Gershfeld NL (1970) Intermolecular energies in condensed, lipid monolayers on water. J Colloid Interface Sci 32:167–172

Gomez-Fernandez JC, Aranda FJ, Villalain J, Micol V, Ortiz A, Hernandez T (1990) Modification by diacylglycerols of phospholipid vesicles structure and physical properties. Prog Clin Biol Res 343:53–58

Gronberg L, Slotte JP (1990) Cholesterol oxidase catalyzed oxidation of cholesterol in mixed lipid monolayers: effects of surface pressure and phospholipid composition on catalytic activity. Biochemistry 29:3173–3178

Gudi S, Nolan JP, Frangos JA (1998) Modulation of GTPase activity of G proteins by fluid shear stress and phospholipid composition. Proc Natl Acad Sci U S A 95:2515–2519

Hiemenz PC (1986) Principles of colloid and surface chemistry, 2nd edn. Marcel Dekker, New York

Hirche F, Ulbrich-Hofman R (1999) The interfacial pressure is an important parameter for the rate of phospholipase D catalyzed reactions in emulsions systems. Biochim Biophys Acta 1436:383–389

Honger T, Jorgensen K, Stokes D, Biltonen RL, Mouritsen OG (1997) Phospholipase A_2 activity and physical properties of lipid-bilayer substrates. Methods Enzymol 286:168–190

James SR, Paterson A, Harden TK, Demel RA, Downes CP (1997) Dependence of the activity of phospholipase Cβ on surface pressure and surface composition in phospholipid monolayers and its implications for their regulation. Biochemistry 36:848–855

Jungner M, Ohvo H, Slotte JP (1997) Interfacial regulation of bacterial sphingomyelinase activity. Biochim Biophys Acta 1344:230–240

Kapoor S, Werkmouller A, Goody RS, Waldmann H, Winter R (2013) Pressure modulation of ras-membrane interactions in intervesicle transfer. J Am Chem Soc 135:6149–6156

Koynova RD, Boyanov AL, Tenchov BG (1987) Gel state metastability and nature of the azeotropic points in mixtures of saturated phosphatidylcholines and fatty acids. Biochim Biophys Acta 903:186–196

Lio Y-C, Dennis EA (1998) Interfacial activation, lysophospholipase and transacylase activity of Group VI Ca^{2+}-independent phospholipase A_2. Biochim Biophys Acta 1392:320–332

Marguet F, Douchet I, Cavalier J-F, Buono G, Verger R (1999) Interfacial and/or molecular recognition by lipases of mixed monomolecular films of 1,2-dicaprin and chiral organophosphorus glyceride analogues? Colloids Surf B Biointerfaces 13:37–45

Martinac B, Buechner M, Delcour AH, Adler J, Kung C (1987) Pressure-sensitive ion channel in *Escherichia coli*. Proc Natl Acad Sci U S A 84:2297–2301

Martinac B, Adler J, Kung C (1990) Mechanosensitive ion channels of E. coli activated by amphipaths. Nature 348:261–263

Middleton SR, Pallas NR, Mingins J, Pethica BA (2011) Thermodynamics of ionized monolayers: surface manometry on very low density spread monolayers of sodium octadecyl sulfate at the air/water interface and analysis of ionic double layer contributions to the isotherms. J Phys Chem 115:8056–8063

Mingins J, Stigter D, Dill KA (1992) Phospholipid interactions in model membrane systems. 1. Experiments on monolayers. Biophys J 61:1603–1615

Miyamoto S, Maeda T, Fujime S (1988) Change in membrane elastic modulus on activation of glucose transport system of brush border membrane vesicles studied by osmotic swelling and dynamic light scattering. Biophys J 53:505–512

Moreau H, Pieroni G, Jolivet-Reynaud C, Alouf JR, Verger R (1988) A new kinetic approach for studying phospholipase C (*Clostridium perfringens α toxin*) activity on phospholipid monolayers. Biochemistry 27:2319–2323

Muderhwa JM, Brockman HL (1990) Binding of pancreatic carboxylester lipase to mixed lipid films: implications for surface organization. J Biol Chem 265:19644–19651

Muderhwa JM, Brockman HL (1992) Lateral lipid distribution is a major regulator of lipase activity: implications for lipid-mediated signal transduction. J Biol Chem 267:24184–24192

Myers D (1991) Surfaces, interfaces, and colloids. VCH Publishers, New York

Needham D, Nunn RS (1990) Elastic deformation and failure of lipid bilayer membranes containing cholesterol. Biophys J 58:997–1009

Obeid LM, Hannun YA (1995) Ceramide – a stress signal and mediator of growth suppression and apoptosis. J Cell Biochem 58:191–198

Pallas NR, Pethica BA (2009) Intermolecular forces in lipid monolayers. Two-dimensional virial coefficients for pentadecanoic acid from micromanometry on spread monolayers at the air/water interface. Phys Chem Chem Phys 11:5028–5034
Pattus F, Slotboom AJ, de Hass GH (1979) Regulation of phospholipase A2 activity by the lipid-water interface: a monolayer approach. Biochemistry 18:2691–2697
Perillo M, Yu RK, Maggio B (1994) Modulation of the activity of *Clostridium perfringens* neuraminidase by the molecular organization of gangliosides in monolayers. Biochim Biophys Acta 1193:155–164
Perozo EP, Cortes DM, Sompornpisut P, Klode A, Martinac B (2002) Open channel structure of MscL and the gating mechanism of mechanosensitive channels. Nature 418:942–948
Peters GH, Dahmen-Levison U, de Meijere K, Brezesinski G, Toxvaerd S, Mohwald H, Svendsen A, Kinnunen PKJ (2000) Influence of surface properties of mixed monolayers on lipolytic hydrolysis. Langmuir 16:2779–2788
Pogram LM, Record MT Jr (2007) Hofmeister salt effects on surface tension arise from partitioning of anions and cations between bulk water and the air-water interface. J Phys Chem 111:5411–5417
Prosser AJ, Frances EI (2001) Adsorption and surface tension of ionic surfactants at the air-water interface: review and evaluation of equilibrium models. Colloids Surf A 178:1–40
Ransac S, Ivanova M, Verger R, Panaiotov I (1997) Monolayer techniques for studying lipase kinetics. Methods Enzymol 286:263–292
Rao CS, Damodaran S (2004) Surface pressure dependence of phospholipase A2 activity in lipid monolayers is linked to interfacial water activity. Colloids Surf B Biointerfaces 34:197–204
Rao CS, Damodaran S (2005) Activation of sphingomyelinase in lipid monolayer is related to interfacial water activity > Evidence from two disparate systems. Colloids Surf B Biointerfaces 45:49–55
Rogalska E, Nury S, Douchet I, Verger R (1995) Lipase stereoselectivity and regioselectivity toward three isomers of dicaprin: a kinetic study by the monomolecular film technique. Chirality 7:505–515
Ross J, Richter PH (1984) Dissipation regulation in oscillatory reactions – application to glycolysis – commentary. Adv Chem Phys 55:169–170
Roussel A, Yang Y, Ferrato F, Verger R, Cambillau C, Lowe M (1998) Structure and activity of rat pancreatic lipase-related protein-2. J Biol Chem 273:32121–32128
Rupley JA, Careri G (1991) Protein hydration and function. Adv Protein Chem 41:37–172
Rupley JA, Yang P-H, Tollin G (1980) Thermodynamic and related studies of water interacting with proteins. In: Rowland SP (ed) Water in polymers. American Chemical Society, Washington, DC, pp 111–132
Salah AB, Sayari A, Verger R, Gargouri Y (2001) Kinetic studies of Rhizopus oryzae lipase using monomolecular film technique. Biochimie 83:463–469
Sasaki T, Hazeki K, Hazeki O, Ul M, Katada T (1995) Permissive effect of ceramide on growth factor-induced cell-proliferation. Biochem J 311:829–834
Schmid RD, Verger R (1998) Lipases: interfacial enzymes with attractive applications. Angew Chem Int Ed 37:1608–1633
Schrag JD, Yunge L, Wu S, Cygler M (1991) Ser-His-Glu forms the catalytic triad of lipase from *Geotrichum candidum*. Nature 351:761–764
Schreiber I, Hung YF, Ross J (1996) Categorization of some oscillatory enzymatic reactions. J Phys Chem 100:8556–8566
Scott DL, Otwinowski Z, Gelb MH, Sigle PB (1990) Crystal structure of bee-venom phospholipase A2 in a complex with a transition state analog. Science 250:1563–1566
Smaby JM, Muderhwa JM, Brockman HL (1994) Is lateral phase-separation required for fatty-acid to stimulate lipases in a phosphatidylcholine interface. Biochemistry 33:1915–1922
Souvignet C, Pelosin J-M, Daniel S, Chambaz EM (1991) Activation of protein kinase C in lipid monolayers. J Biol Chem 266:40–44
Sukharev SI, Blount P, Martinac B, Blattner FR, Kung C (1994) A large-conductance mechanosensitive channel in *E. coli* encoded by *mscL* alone. Nature 368:265–268

Sukharev SI, Sigurdson WJ, Kung C, Sachs F (1999) Energetic and spatial parameters for gating of the bacterial large conductance mechanosensitive channel, MscL. J Gen Physiol 113:525–539
Tanaka K, Manning PA, Yu H (2000) Lipase catalysis on monolayers at the air/water interface. 1. Kinetic rate constants on quasi-two-dimension. Langmuir 16:2665–2671
Thuren T, Wilcox RW, Sisson P, Waite M (1991) Hepatic lipase hydrolysis of lipid monolayers. J Biol Chem 266:4853–4861
Tsujita T, Brockman HL (1987) Regulation of carboxylester lipase adsorption to surfaces. 1. Chemical specificity. Biochemistry 26:8423–8429
Tsujita T, Muderhwa JM, Brockman HL (1989) Lipid-lipid interactions as regulators of carboxyl esterase activity. J Biol Chem 264:8612–8618
van Tilbeurgh H, Egloff MP, Martinez C, Rugani N, Verger R, Cambiliau C (1993) Interfacial activation of the lipase – procolipase complex by mixed micelles revealed by X-ray crystallography. Nature 362:814–820
Verger R, De Haas GH (1973) Enzyme reactions in a membrane model. 1. A new technique to study enzyme reactions in monolayers. Chem Phys Lipids 10:127–136
Verger R, de Haas GH (1976) Interfacial enzyme kinetics of lipolysis. Annu Rev Biophys Bioeng 5:77–117
Verger R, Rietsch J, Van Dam-Mieras MCE, de Haas GH (1976) Comparative studies of lipase and phospholipase A_2 acting on substrate monolayers. J Biol Chem 251:3128–3133
Verkleij AJ, Zwaal FRA, Roelofsen B, Comfurius P, Kastelijn D, Van Deenen LLM (1973) The asymmetric distribution of phospholipids in the human red cell membrane. Biochim Biophys Acta 323:178–193
Wiggins P, Phillips R (2005) Membrane-protein interactions in mechanosensitive channels. Biophys J 88:880–902
Wilcox RW, Thuren T, Sisson P, Schmitt JD, Kennedy M, Mosely W (1993) Regulation of rate hepatic lipase by the composition of monomolecular film lipid. Biochemistry 32:5752–5758

Chapter 11
Anhydrobiosis: An Unsolved Problem with Applications in Human Welfare

John H. Crowe

Abstract Anhydrobiosis (Life Without Water) has been known for millennia, but the underlying mechanisms have not been understood until recent decades, and we have achieved only a partial understanding. One of the chief sites of damage from dehydration is membranes, and we and others have provided evidence that this damage may be obviated by the production of certain sugars, particularly trehalose. The sugar stabilizes membranes by preventing fusion and fluidizing the dry bilayers. The mechanism by which this is accomplished has been controversial, and I review that controversy here. In the past decade evidence is accumulating for a role of stress proteins in addition to or as a substitute for trehalose. Genomic studies on anhydrobiotes are yielding rapid progress. Also in the past decade, numerous uses for trehalose in treating human diseases have been proposed, some of which are in clinical testing. I conclude that the mechanisms underlying anhydrobiosis are more complex than we thought 20 years ago, but progress is being made towards elucidating those mechanisms.

Keywords Anhydrobiosis • Trehalose • Microdomains • Water entrapment

11.1 Introduction

Investigations on the phenomenon of anhydrobiosis ("life without water") have a long history, dating back to the time of Leeuwenhoek (see Keilin 1959, for a fascinating, scholarly review of the old history). Nevertheless, the underlying mechanisms by which a wide array of organisms such as seeds of many plants, yeasts, mosses, cysts of crustaceans, and certain microscopic animals such as nematodes, rotifers, and tardigrades all survive loss of essentially all of their water without being killed were completely unknown until the last few decades. The first concrete suggestion concerning the mechanism came from studies on anhydrobiotic nematodes (reviewed in Crowe and Hoekstra 1992). The worms with which these

---

J.H. Crowe (✉)
Department of Molecular and Cellular Biology, University of California, Davis, CA 95618, USA
e-mail: jhcrowe@ucdavis.edu

E.A. Disalvo (ed.), *Membrane Hydration*, Subcellular Biochemistry 71,
DOI 10.1007/978-3-319-19060-0_11

studies were done must be dried slowly, during which time they synthesize large quantities of the disaccharide, trehalose. There is a strong correlation between survival in the dry state and the production of trehalose. This sugar had been known to be present at high concentrations in a large number of anhydrobiotes, but it was thought to be a metabolic storage product, and had not been implicated as having a role in anhydrobiosis. Subsequent investigations established a similar correlation between survival of many organisms in anhydrobiosis and the presence of disaccharides, usually trehalose or, in the case of higher plants, sucrose.

Studies with model systems (isolated biological membranes, liposomes, and proteins) showed that trehalose has a remarkable ability to preserve these biomolecular assemblages in the dry state (reviewed in Crowe 2008). Physical studies on the dry materials indicated a mechanism by which it does so; it depresses the gel to liquid crystalline phase transition in dry phospholipids, often by values approaching 100 °C (reviewed in Crowe 2008). In other words, the presence of the sugar fluidizes dry phospholipid bilayers and mimics the presence of water. We first reported this phenomenon more than 30 years ago (Crowe et al. 1984), but the mechanism by which the sugar depresses Tm is still not entirely clear. Based on spectroscopic evidence presented in the original report, we suggested that the sugar hydrogen bonds to polar residues in dry phospholipids (mainly the phosphate of the headgroup), which changes their lateral spacing, much as water does. This mechanism is known as the water replacement hypothesis (reviewed in Crowe 2008).

Thus, because the sugar has these effects both in vitro and in vivo and because its presence was correlated with survival of anhydrobiosis, we suggested some decades ago that a central role for the sugar in anhydrobiosis seemed to be reasonable and remains so to this day, at least for many, but not all, anhydrobiotes. Nevertheless, the role of trehalose has been based on such correlations or on in vitro data, which in themselves are not convincing that it is important in intact cells and organisms. Furthermore, a key part of this phenomenon is the mechanism by which the sugar depresses the transition temperature, which has come under some considerable discussion. I have summarized more recent findings in this regard in sections following.

Is trehalose special in its ability to preserve biomolecules and assemblages? Yes and no. Other sugars can be just as effective, under special conditions. Trehalose works particularly well for the following reasons: (a) the glycosidic bond linking the two glucose monomers is remarkably stable, far more stable than the bond in sucrose, for example (Schebor et al. 1999); (b) trehalose forms a glass in the dry state, which is essential to its ability to stabilize biomaterials. The temperature at which this glass undergoes its transition to a more fluid state is remarkably high for trehalose—much higher than in comparable sugars. The significance of the glass in vivo has been demonstrated a number of times (see Hengherr et al. 2009 for an interesting example). If conditions are such that the glycosidic bond is protected and the temperature is low enough to maintain the glassy state other sugars can be just as effective as trehalose.

Trehalose has been used successfully to preserve mammalian blood platelets (Wolkers et al. 2001), with excellent recovery and functionality in the rehydrated cells (e.g. Auh et al. 2004). Several groups have tried to use trehalose to preserve nucleated cells in the dry state, and it became apparent that adding trehalose alone was not sufficient, as I will describe below. The cells survived drying initially, but they rapidly died. However, when they were transfected with the gene for a stress protein obtained from anhydrobiotic cysts of a crustacean, Artemia, survival was improved, and the ability of the rehydrated cells to reproduce was elevated dramatically (Ma et al. 2005; Zhu et al. 2006). The stress protein alone was insufficient to protect the cells during drying, but the stress protein and trehalose acted synergistically to do so. Clearly, there are requirements in addition to trehalose to stabilize nucleated cells in the dry state. Recent studies on the apparent synergism between trehalose and stress proteins will be discussed in a following section.

In some anhydrobiotic microorganisms only small amounts of disaccharides have been detected (Tunnacliffe et al. 2005; reviewed in Wharton 2014), suggesting that there must be alternative ways to achieve the same end as those seen with the sugars present. It is not clear yet just what those means might be, but several groups around the world are seeking elucidation of what might be the significant adaptations, using genomics and proteomics (e.g. Leprince and Buitink 2010; Terrasson et al. 2013; Dupont et al. 2014). These powerful methods look well beyond the rather primitive approaches we were using decades ago in which we looked for metabolic products that are downstream from the transcripts and translated products that are now being detected. This approach has been remarkably productive, as I will describe below.

11.2 Lipid Phase Transitions in Membranes and Anhydrobiosis

11.2.1 An Example of Effects of Trehalose on Dry Phospholipids

Most of the following was reviewed in detail in Crowe (2008). Liposomes were prepared from a lipid with low *T*m, palmitoyloleoylphosphatidylcholine (POPC), with a fluorescent marker, carboxyfluorescein trapped in the aqueous interior. When the liposomes were freeze-dried with trehalose and rehydrated, the vesicles were seen to be intact, and nearly 100 % of the carboxyfluorescein was retained. It quickly emerged that stabilization of POPC liposomes, and other vesicles prepared from low-melting-point lipids, had two requirements, inhibition of fusion between the dry vesicles and depression of *T*m in the dry state. In the hydrated state, *T*m for POPC is about −1 °C and rises to about +70 °C when it is dried without trehalose. In the presence of trehalose, *T*m is depressed in the dry state to −20 °C. Thus, the lipid is maintained in the liquid-crystalline phase in the dry state, and phase transitions are not seen during rehydration at room temperature.

The significance of this phase transition during rehydration is that when phospholipids pass through such transitions, the bilayer becomes transiently leaky, which resembles the effects resulting from passage through a main phase transition in fully hydrated membranes during changes in temperature, as discussed earlier. The physical basis for this leakiness has been investigated in some detail by Hays et al. (2001). Thus, the leakage that normally accompanies this transition must be avoided if the contents of membrane vesicles and whole cells are to be retained. During drying, leakage is probably not a problem because *T*m is not affected until all the bulk water has been removed. But during rehydration, it is a serious problem; the membranes are placed in water and will undergo the phase transition in the presence of excess bulk water, thus allowing leakage. In addition to the damage that occurs during passage through the phase transition, and perhaps even more importantly in the present context, phase separation of membrane components can occur in the absence of trehalose during drying, as the membranes undergo the transition into the gel phase, an event that is often irreversible.

Low-temperature-melting lipids such as POPC all seem to behave as described previously; *T*m is depressed to a minimal value immediately in the presence of trehalose after drying, independent of the thermal history. Saturated lipids with high *T*m, such as DPPC, behave quite differently, and effects of trehalose on *T*m depend strongly on the thermal history. When *T*m in DPPC dried without trehalose is measured, it is seen to rise from 41 °C in the hydrated lipid to 110 °C when it is dried. In the presence of trehalose, *T*m is about 60 °C until the acyl chains are melted once, after which *T*m is depressed to 24 °C (Crowe 2008). If the lipid is then incubated at temperatures <24 °C, *T*m rapidly reverts to about 60 °C. Thus, the stable *T*m renders DPPC in the gel phase at physiological temperatures, regardless of whether it is hydrated or dry. This effect may have special relevance for biological membranes, because microdomains contain lipids with elevated *T*m.

11.2.2 Trehalose Maintains Microdomains in Dry Membranes

We have carried out modeling studies in an attempt to discover whether trehalose can maintain the structural integrity of rafts in the dry state and, if so, to elucidate the mechanism. Because the phase behavior of the binary mixture DLPC/DSPC is so well characterized in the hydrated state, we started there, rather than with the more complex mixture seen in a native membrane. Using deuterated DPPC and hydrogenated DLPC, it has been possible to monitor phase behavior of the two lipids in a 1:1 mixture, using FTIR, with the following results: (i) the freshly prepared fully hydrated mixture is completely phase-separated in the gel/liquid-crystalline coexistence temperature regime, as expected; (ii) when this mixture is dried without trehalose, mixing occurs; (iii) if the mixture is dried with trehalose, the lipid phase separation is maintained, although a small fraction of the DLPC is mixed with the DSPC; and (iv) in the mixture dried with trehalose, most of the DLPC fraction has a transition below 0 °C in the dry state, while the DSPC transition is seen at about 80 °C (Leidy et al. 2004).

We propose that trehalose maintains phase separation in this mixture of lipids in the dry state by the following mechanism (Ricker et al. 2003). The DLPC fraction, with its low *T*m in the hydrated state, might be expected to behave like unsaturated lipids described earlier, in that *T*m in the dry state is reduced to a minimal and stable value immediately after drying with trehalose, regardless of the thermal history. That appears to be the case. The DSPC fraction, by contrast, would be expected to be in the gel phase in the hydrated state at room temperature, and it remains in the gel phase when it is dried with trehalose. In other words, we are proposing that by maintaining one of the lipids in the liquid-crystalline phase during drying, while the other remains in the gel phase, trehalose maintains the phase separation. We suggest that this is the fundamental mechanism by which trehalose maintains microdomains in native membranes during drying (Ricker et al. 2003). By maintaining the liquid-ordered/liquid-crystalline microdomain structure that we observed in platelets during drying and rehydration, trehalose could preserve the small-scale phase separation that appears to be important for membrane function and thus prevent macroscopic phase separation.

Moiset et al. (2014) recently presented data that would appear to be in conflict with the mechanism proposed above. They found both by experiment and modeling that trehalose and sucrose destabilize phase separated model membranes and lead to mixing of the components. However, these studies were all done in the presence of excess water. Furthermore, the mixing effect was seen at high sugar concentrations, in the molar range. The effects of trehalose on stabilizing domains were seen at low sugar concentrations, and the studies were done in the dry state, which is a completely different thing.

Thus, it is likely that effects of sugars on thermal transitions in the dry state are a central factor in stabilizing membranes in the absence of water, and the mechanism by which this is accomplished is critical to our understanding of this phenomenon.

11.3 How Does Trehalose Depress *T*m in the Dry State?

The mechanism of depression of *T*m has received a great deal of attention since the discovery of this effect. Three main hypotheses have emerged:

11.3.1 The Water Replacement Hypothesis

The water replacement hypothesis suggests that sugars can replace water molecules by forming hydrogen bonds with polar residues, thereby stabilizing the structure in the absence of water (Crowe 2008). Direct interaction, on the other hand, has been demonstrated by a wide variety of physical techniques, including IR spectroscopy (Crowe et al. 1984; Tsvetkova et al. 1998), NMR (Lee et al. 1986; Tsvetkoa et al. 1998; Wolkers et al. 1998), and X-ray (e.g. Nakagaki et al. 1992). Theoretical analyses have contributed greatly to this field in recent years. Chandrasekhar and Gaber (1988) and Rudolph et al. (1990), in the earliest studies, showed that trehalose can form energetically stable conformations with phospholipids, binding three adjacent phospholipids in the dry state. Sum et al. (2003) showed by molecular simulations that the sugars adapt molecular conformations that permit them to fit onto the surface topology of the bilayer through hydrogen bonds. The sugars interact with up to three adjacent phospholipids. Golovina et al. (2010) confirmed these results with molecular simulations and showed that trehalose increases the area per lipid in the dry state under conditions that seem to be inconsistent with any model that does not require direct interaction between the sugar and polar head group. Modifications of the gel state of hydrated phospholipids by trehalose can only be achieved if a drastic dehydration is performed in the presence of the sugar. The results suggest that trehalose is still intercalated between the phospholipids after restoring water to the dried liposomes either at temperatures below or above the phase transition (Viera et al. 1993).

The water activity in dimyristoylphosphatidylcholine (DMPC) decreases by 60 % when the lipid is dehydrated in the presence of trehalose concentrations higher than 0.02 M. FTIR in these conditions indicated that trehalose binds to the carbonyl groups, replacing 11 of 14 water molecules per lipid molecule. About four are displaced by changes in the water activity of the bulk solution, and seven by specific interactions with the phospholipids. In this last case, at least two of them are linked to the carbonyl group. This appears to be the cause of the decrease in the dipole potential of lipid monolayers spread on an air/water interface from 480 mV in pure water to 425 mV in 0.1 M trehalose (Luzardo et al. 2000).

Molecular dynamics simulations from this same group of investigators showed that trehalose binds to the phospholipid headgroups with its main axis parallel to the membrane normal. It establishes hydrogen bonds with the carbonyl and phosphate groups and replaces water molecules from the lipid headgroup. Notably, the number of hydrogen bonds that the membrane made with its environment was conserved after trehalose binding. The H-bonds between lipid and trehalose have a longer lifetime than those established between lipid and water. The binding of the sugar does not produce changes either in the lipid area or in the lipid order parameter. The effect of trehalose on the dipole potential is in agreement with experimental results. The contribution of the different components to the membrane dipole potential was analyzed. It was observed that the binding of trehalose produces changes in the different components and the sugar itself contributes to the surface potential due to the polarization of its hydroxyl in the interface (Villarreal et al. 2004).

11.3.2 The Water Entrapment Hypothesis

The water entrapment hypothesis suggests that sugars concentrate water near surfaces, thereby preserving its salvation (Belton et al. 1994; Cottone et al. 2002; Lins et al. 2004). While there is some evidence, both experimental (Crowe et al. 1987) and modeling (Golovina et al. 2010) that significant amounts of water are retained in "dry" membranes in the presence of trehalose, there is little evidence that there is sufficient water to depress Tm to the levels seen experimentally. Indeed, since Tm is often well below that of the fully hydrated lipid it is difficult to see how residual water can be responsible for depression of Tm to such a level. That is not to say that a few water molecules around the headgroup are not important, as well they might be for the details of the interaction.

11.3.3 The Hydration Forces Explanation

The hydration forces explanation invokes the mediation of the forces between bilayers by the sugars as membranes come close together, and the subsequent reduction in the induced lateral compression in the plane of the bilayer responsible for deleterious phase transitions (Bryant et al. 1992, 2001; Koster 2001). Lenne et al. (2009) reported that The presence of sugars has no effect on the average spacing between the phospholipid chains in either the fluid or gel phase. Using this finding, they observed that for low sugar concentrations only a small amount of sugar exclusion occurs and that under these conditions, the effects of sugars on the membrane transition temperatures can be explained quantitatively by the reduction in hydration repulsion between bilayers due to the presence of the sugars. They suggested that specific bonding of sugars to lipid headgroups is not required to explain this effect. More recently, Kent et al. (2014) used neutron diffraction to localize deuterated trehalose between opposing bilayers. According to the water replacement hypothesis the trehalose should be found preferentially in association with the polar headgroups. Instead, Kent et al. (2014) reported that the sugar was distributed predominantly in the aqueous phase between adjacent bilayers. This finding would seem to be in disagreement with the proposal for direct interaction with the headgroups. However, since the measurements were done in excess water it is difficult to extend these results to exclude direct interaction between trehalose and dry bilayers. Furthermore, since all these studies were done with multilamellar vesicles, for technical reasons the results cannot easily be compared with those done with unilamellar vesicles. In the former, the solutes are restricted to confined spaces, as opposed to the single bilayer to bilayer seen in unilamellar vesicles used in the studies described above. Furthermore, solutes can be excluded from multilamellar vesicles, as Koster et al. (2003) reported. But such exclusion may be absent in unilamellar vesicles, which has led to some confusion. For example, when unilamellar vesicles are dried with a large polymer fusion during drying was strongly inhibited (Crowe et al. 1998). It is difficult to suppose that the polymers and the vesicles are in different phases under these conditions since fusion is inhibited. Since Tm in the dry lipids is not affected (Crowe et al. 1998), this finding seems clearly inconsistent the hydration forces explanation. Such a polymer is excluded from the interbilayer spaces in multilamellar vesicles and thus would not be expected to have the same effects.

A consensus is slowly emerging that these three mechanisms might not be mutually exclusive. First, in the dry state, vitrification may occur simultaneously with direct interactions between the sugar and polar residues, that is, the sugar and phospholipids form a glass together, with direct interaction between the lipid and sugar, which seems consistent with all three hypotheses. The situation for hydrated systems is perhaps more complex, but data from Andersen et al. (2011) suggest a solution to this apparent disagreement. They reported data using small angle neutron scattering and thermodynamic measurements that seem to show that sugars may be either bound or expelled, depending on the concentration of sugar. At low concentration, small sugars bind quite strongly to a lipid bilayer, and the accumulation of sugar at the interface makes the membrane thinner and laterally expanded, apparently in agreement with the water replacement hypothesis. Above ~0.2 M the sugars gradually become expelled from the membrane surface, and this repulsive mode of interaction counteracts membrane thinning. The dual nature of sugar–membrane interactions, Andersen et al. suggest, offers a reconciliation of conflicting views in earlier reports on sugar-induced modulations of membrane properties.

I suggest that the hydration forces explanation probably does apply under the conditions described by Andersen et al. (2011). Indeed, we have reported data on effects of sugars on Tm in fully hydrated systems that are consistent with this viewpoint (Crowe and Crowe 1991). However, when the bulk water is removed the solute must come in contact with the bilayer surface; there is no water left into which the sugar can partition. At this point I believe the matter is still unresolved, mainly because of technical differences in the way the studies have been done.

11.4 Is Trehalose Necessary and Sufficient for Cellular Stabilization in Anhydrobiotes?

Nearly 40 years ago we reported that a nematode, *Aphelenchus avenae* can be induced to enter an anhydrobiotic state by slowly drying them over a three day period. During the slow dehydration the animals synthesized trehalose, and the ability to survive more extensive water loss was strictly correlated with this synthesis. Similar studies have been reported by Watanabe et al. for (2002) and Mitsumasu et al. (2010). These animals are of special interest because they are the largest anhydrobiotes known In vitro studies by numerous investigators showed the ability of trehalose to stabilize dry biological structures, which led to the widely accepted paradigm that trehalose is a key factor in anhydrobiosis. Nevertheless, that paradigm was based on correlations without convincing evidence for a role *in vivo*. Several groups have addressed that need in recent years, as follows.

11.4.1 Genetic Models

In a remarkable series of recent papers, a group of investigators at research institutes in Dresden have made rapid progress in this regard in the last 3 years. First, Erkut et al. (2011) have shown that trehalose is absolutely required for survival in a nematode and that mutants lacking the ability to synthesize the sugar did not survive even mild dehydration. They used the popular genetic model organism *Caenorhabitidis* as a model organism and reported that the dauer larva is a true anhydrobiote: under defined conditions it can survive even after losing 98 % of its body water. This ability is correlated with a several fold increase in the amount of trehalose. Mutants unable to synthesize trehalose cannot survive even mild dehydration. Light and electron microscopy indicate that one of the major functions of trehalose is the preservation of membrane organization. Fourier-transform infrared spectroscopy of whole worms suggests that this is achieved by preserving the native packing of lipid acyl chains. In the next paper in this series Erkut et al. (2013) used microarray analysis, proteomics, and bioinformatics to identify genes, proteins, and biochemical pathways that are upregulated during the induction process. These pathways were validated by testing the desiccation tolerances of mutants. The data suggest that the desiccation response is activated by sensing the desiccative environment via neurons, leading to elimination of reactive oxygen species and xenobiotics, expression of heat shock and intrinsically disordered proteins, polyamine utilization, and induction of fatty acid desaturation pathways. Erkut et al. (2013) suggest that this response is specific and involves a small number of functional pathways. They make the remarkable suggestion that these pathways represent the genetic toolkit for anhydrobiosis in both animals and plants. In the most recent paper in the series, Abusharkh et al. (2014) showed that, in addition to trehalose accumulation, the dauer larvae reduce their phosphatidylcholine (PC) content. Using Langmuir – Blodgett monolayers, they found that phospholids from preconditioned larvae with reduced PC content exhibit a higher trehalose affinity, a stronger hydration-induced gain in acyl chain free volume, and a wider spread of structural relaxation rates of their transitions and headgroup H-bond interactions. I suggest that this genetic approach, using this well-defined model, will continue to provide powerful evidence concerning the fundamental mechanisms underlying anhydrobiosis.

Using another well-defined genetic model, the yeast using Saccharomyces cerevisiae, Tapia and Koshland (2014) showed that intracellular trehalose is essential for survival to long-term desiccation, in agreement with previous investigations (e.g. Eleutherio et al. 1993). Maintaining long-term desiccation tolerance consists of a balance of trehalose stockpiled prior to desiccation and trehalose degradation by trehalases in the desiccated cells. The activity of trehalases in desiccated cell reveals a surprising enzymatic activity while desiccated. Interestingly, the protein chaperone Hsp104 compensates for loss of trehalose during short-term but not long-term desiccation. Tapia and Koshland (2014) found that desiccation induces protein misfolding/aggregation of cytoplasmic and membrane proteins and demonstrated that trehalose but not Hsp104 mitigates the aggregation of both cytoplasmic and membrane prions. They proposed that cells are initially protected against desiccation by both protein and chemical chaperones, like Hsp104 and trehalose, respectively. As desiccation extends, the activities of the protein chaperones are lost because of their complexity and requirement for energy, leaving trehalose as the major protector against the aggregation of cytoplasmic and membrane proteins. Tapia and Koshland (2014) suggest that trehalose is both a more stable and more versatile protectant than protein chaperones, explaining its important role in desiccation tolerance and emphasizing the translational potential of small chemical chaperones as stress effectors.

11.4.2 The Dessicome

Clearly, it is emerging that multiple adaptations, not just trehalose synthesis, are required for survival of anhydrobiosis. A number of investigators around the world are investigating such adaptations, and among the most fruitful approaches are genetic ones, involving microarray analysis to elucidate the chorus of genes and downstream products that are involved. The large suite of genes, proteins, and metabolites involved in protection against dehydration damage and in repair has been called "the desiccome" (Leprince and Buitink 2010; Terrasson et al. 2013). Out of this array of participants, four categories have been recognized, which appear to act synergistically (for reviews see Hand et al. 2011; Hincha and Thalhammer 2012; Hoekstra et al. 2001; Leprince and Buitink 2010; Gaff and Oliver 2013; Terrasson et al. 2013; Tunnacliffe et al. 2010; Forster et al. 2012). Protection is accomplished by: (1) stabilization of proteins by non-reducing sugars, late embryogenesis abundant (LEA) proteins and heat shock protein (HSP); (2) preventing oxidative damage by a range of antioxidant compounds such as tocopherols, glutathione, together with a coordinated response of metabolism during drying; (3) obviating damage from structural stresses imposed by drying such as cell wall modification, reorganization of intracellular membranes and cytoskeleton, vacuolization and chromatin condensation; (4) altering the regulatory mechanisms and signaling pathways controlling the induction of these protective mechanisms.

11.4.3 LEA Proteins

The adaptation that has seen the most intense investigation in recent years is synthesis of the so-called "late embryogenesis abundant" (LEA) proteins, which were discovered in plant embryos, but more recently have been found in a wide variety of other anhydrobiotic organisms (e.g. Tunnacliffe 2007; Hand 2011). Despite the fact that they were discovered many years ago and have now been found to be widespread among anhydrobiotic organisms, the role of LEA proteins in anhydrobiosis is still not clear, but it appears that they may play a role similar to that ascribed to trehalose. For example, Tolleter et al. (2007) showed that LEAM, a mitochondrial LEA protein expressed in seeds, is a natively unfolded protein, which reversibly folds into a-helices upon desiccation. Structural modeling revealed an analogy with class A amphipathic helices of apolipoproteins that coat low-density lipoprotein particles in mammals. LEAM appears spontaneously modified by deamidation and oxidation of several residues that contribute to its structural features. LEAM interacts with membranes in the dry state and protects liposomes subjected to drying. The overall results suggest that LEAM protects the inner mitochondrial membrane during desiccation. According to sequence analyses of several homologous proteins from various desiccation tolerant organisms, a similar protection mechanism likely acts with other types of cellular membranes. Subsequently, Tolleter et al. (2010) used Fourier transform infrared and fluorescence spectroscopy to gain insight into the molecular details of interactions of LEAM with phospholipid bilayers in the dry state and their effects on liposome stability. LEAM interacted specifically with negatively charged phosphate groups in dry phospholipids, increasing fatty acyl chain mobility. This led to an enhanced stability of liposomes during drying and rehydration, and also upon freezing. Protection depended on phospholipid composition and was strongly enhanced in membranes containing the mitochondrial phospholipid cardiolipin. Collectively, the results provide strong evidence for a function of LEAM as a mitochondrial membrane protectant during desiccation and highlight the role of lipid composition in the interactions between LEA proteins and membranes. Hand (2015) has estimated that the concentration of LEAM required for stabilization of the inner mitochondrial membrane (from Tolleter et al.'s data) closely matches the concentration of LEAM actually observed in mitochondria from another anhydrobiote, *Artemia*, which is reassuring that the suggested role is realistic.

Hand et al. have in the last few years been investigating the role of a number of LEA proteins in *Artemia*, with very instructive results. First, fully hydrated in solution, the CD spectrum of one of the LEAS from *Artemia* exhibited features typical of a disordered, random-coiled protein, but when the protein was dried a shift in α-helix content from 4 % in solution to 46 % was observed. A similar shift in conformation from disorder to order has been seen in a number of LEA proteins from different sources (cf., Tunnacliffe and Wise 2007; Hand et al. 2011b) and certain other stress proteins that have been thought to interact with membranes during chilling or drying (Tomczak et al. 2002; Torok et al. 2003). Such conformational shifts may be key to the ultimate function of these proteins. Hand and Menze (2015) have suggested that the proteins could act as a molecular shield in solution, and the same LEA protein could gain structure as water is removed to further protect the cell in the dry state by interacting with membranes, stabilizing sugar glasses, and forming filamentous networks (Hand et al. 2011; Tunnacliffe and Wise 2007). Hand and Menze (2015) recently studied effects of trehalose and a LEA protein from *Artemia* on stability of a model enzyme, phosphofructokinase, which is very labile; drying with no protection irreversibly reduced enzymatic activity to zero when the enzyme was rehydrated. Addition of trehalose or the LEA protein improved enzymatic activity by a small amount, but when both trehalose and the LEA protein were added nearly 100 % of the original activity was recovered. Clearly, there is a marked synergism between the trehalose and the LEA protein. However, the effects of the protein appear not to be specific; when BSA was added instead of the LEA protein recovery of enzymatic activity was almost as good. Along the same lines, Li et al. (2012) dried human fibroblasts transfected with genes for two LEA proteins and loaded the cells with trehalose using a trehalose transporter. With nothing added the cells showed loss of al membrane integrity, but with both trehalose and LEA proteins membrane integrity approached 100 %, again suggesting a synergism between the LEA proteins and trehalose. Ma et al. (2005) had previously shown that when fibroblasts were transfected with a stress gene from *Artemia* and loaded with trehalose, survival of drying was very poor, but the cells that did survive showed markedly more robust cell divisions following rehydration if the stress gene was present and being expressed—yet another case of a synergism between trehalose and the stress protein.

In summary, there is a growing body of evidence that proteins associated with stress act synergistically with trehalose or other small molecules in stabilizing biological materials, including intact cells. However, very little is understood about the mechanism by which the proteins and sugar interact to produce these effects.

11.5 Some Surprising Uses for Trehalose in Humans

Since there is a large body of evidence that trehalose is an effective stabilizer of biological materials, some considerable attention is being paid to the possibility that it might be useful in humans, particularly as a therapeutic agent in protein folding and protein aggregation diseases, treatment of corneal dysfunctions, preservation of materials of interest in human welfare, and in cosmetics. I will not to attempt to review the large literature in this rapidly growing field and instead refer the reader to an excellent review by Ohtake and Wang (2011). The following is merely a sampling of what is happening in this area.

11.5.1 Cosmetics

Trehalose is being used in a wide variety of cosmetic products, often with little rationale. However, it is also included in some deodorants as a key ingredient in suppressing human body odor, with good evidence concerning its efficacy. Odor in the elderly is caused by the formation of unsaturated aldehydes such as 2-nonenal and 2-octenal, produced by the degradation of unsaturated fatty acid (palmitoleic acid) in the skin. The body of seniors (>55 years) was sprayed with 2 % trehalose after showering, and 20 h later, the amount of unsaturated aldehydes produced was analyzed from the subjects' shirts. Aldehyde production was seen to be reduced by

70 % in subjects with trehalose. Besides the unsaturated aldehydes, free radicals and HPOs are produced during oxidation of fatty acids and may result in more than foul odor. These products can react with proteins and DNA, resulting in cleaving of DNA chain or its irregular production, potentially leading to deleterious conditions, including cancer, as Ohtake and Wang (2011) suggested. This scenario might be avoided with the use of trehalose, which can suppress the breakdown of fatty acids.

11.5.2 Osteoporosis

Osteoporosis results from an imbalance in bone formation and resorption. There are several causes, but the lack of estrogen production after menopause is considered to be the major contributor. Estrogen replacement therapy is effective in preventing bone loss in animals and humans (Gadducci et al. 1997). However, continuous administration of estrogen is accompanied by adverse side effects (Vessey 1984). The effects of trehalose on bone resorption were studied using ovariectomized mice, with trehalose administered orally five times a week for 4 weeks, and the changes in bone weight and calcium/phosphorous contents were analyzed. Bone weight loss was prevented in a dose-dependent manner (Arai et al. 2001). Furthermore, the increase in osteoclast formation was significantly inhibited by trehalose. Trehalose is unlikely to have an estrogen-like function, and the authors of the study speculated that it suppresses osteoclast differentiation.

11.5.3 Huntington's Disease

The cause of the disease is thought to be the aggregation of mutant Huntingtin protein, characterized by long glutamine repeats. When The efficacy of trehalose in reducing aggregation of a model protein (Tanaka et al. 2004) aggregation was decreased by about 50 %. Trehalose was shown to suppress the formation of aggregates in mammalian cells and to improve their viability. Furthermore, oral administration of 2 % trehalose solution to a transgenic model mice resulted in reduction of aggregate formation in the motor cortex and in improved motor function and survival. In addition, trehalose was reported to be effective in reducing the aggregation of amyloid peptides (Liu et al. 2005) which is a key step in the pathogenesis for Alzheimer's disease. Although *in vitro* studies indicate that trehalose suppresses peptide aggregation, their effects *in vivo* are difficult to interpret, as trehalose administered orally should have been degraded to glucose in the intestine. For most of the studies, glucose was examined as a control and was much less effective than trehalose. how trehalose administration resulted in the reduction of aggregate formation in the brain of rats is a mystery. Nevertheless, the efficacy in inhibiting polyglutamine aggregation may place trehalose as one of the leading candidate as a therapeutic compound for the treatment of HD and other protein aggregation diseases. In fact, one clinical trial has already commenced (Couzin 2004). There is even a suggestion that trehalose might emerge as a leading therapeutic agent for treating a wide variety of neurogenerative diseases (Emanuele 2014).

11.5.4 Eye Dysfunctions

Matsuo et al. (2004) have shown that trehalose administered directly on the cornea of patients with dry eye syndrome significantly relieved the discomfort. Part of the discomfort appears to be due to loss of smoothness of the cornea due to desiccation. In more recent studies, the same group showed that administration of the trehalose solutions maintains that smoothness (Izawa et al. 2006). Some recent suggestions concerning how this might work include inhibition of fibroblast differentiation by trehalose (thus, inhibiting formation of scar tissue; Takeuchi et al. 2010), direct protection of the corneal epithelium from desiccation damage (Hovakimyan et al. 2012), and inhibition of cytokine mediated inflammation in the cornea (Cejkova et al. 2011). It is unclear how the sugar imparts any of these effects, but it continues to be of interest in ophthalmic applications (e.g. Luyckx and Baudouin 2011).

11.5.5 Trehalose and Autophagy

Administration of trehalose has been reported to stimulate autophagy in several cells types in vitro. Kruger et al. (2012) reported that autophagy can be activated in neurons in vitro expressing the tau protein, which is thought to be involved in formation of the aggregates in Alzheimer's disease. The cells showed depressed levels of tau aggregation and much less cytotoxicity. In a related study, Lan et al. (2012) showed that production of synuclein (a protein related to Parkinson's disease) was suppressed by trehalose induced autophagy. The authors of both these studies suggested that trehalose might be developed as a therapeutic agent for these diseases, but it is, of course, unclear how such a therapy could be developed for clinical application.

11.6 Conclusions

More than 40 years ago, I wrote a review for The American Naturalist (Crowe 1971) with the same title I've given to this commentary. I've recycled the title because in the not too distant past, at the apex of the excitement about trehalose, we thought for a short time that the problem might have been solved. Now it is clear that we were optimistic in this conclusion, and that what we had thought to be a simple solution is not so simple at all. Furthermore, the outgrowth of the field into far flung applications is at once both gratifying and flabbergasting.

References

Abusharkh SE, Erkut C, Oertel J, Kurzchalia TV, Fahmy K (2014) The role of phospholipid headgroup composition and trehalose in the desiccation tolerance of Caenorhabditis elegans. Langmuir 30:12897–12906

Andersen HD, Wanga C, Arleth L, Peters GH, Westh P (2011) Reconciliation of opposing views on membrane–sugar interactions. Proc Natl Acad Sci U S A 108:1874–1878

Arai C, Kohguchi C, Akamatsu S, Arai N, Yoshizane C, Hasegawa N, Hanaya T, Arai S, Ikeda M, Kuromoto M (2001) Trehalose suppresses lipopolysaccharide-induced osteoclastogenesis bone marrow in mice. Nutr Res 21:993–999

Auh J-H, Wolkers WF, Looper SA, Walker NJ, Crowe JH, Tablin F (2004) Calcium mobilization in freeze-dried human platelets. Cell Preserv Technol 2:180–187

Belton PS, Gil AH (1994) IR and Raman spectroscopic studies of the interaction of trehalose with hen egg lysozyme. Biopolymers 34:957–961

Bryant G, Wolfe J (1992) Interfacial forces in cryobiology and anhydrobiology. Cryo-Letters 13:23–36

Bryant G, Koster KL, Wolfe J (2001) Membrane behaviour in seeds and other systems at low water content: the various effects of solutes. Seed Sci Res 11:17–25

Čejková J, Ardan T, Čejka Č, Luyckx J (2011) Favorable effects of trehalose on the development of UVB-mediated antioxidant/pro-oxidant imbalance in the corneal epithelium, proinflammatory cytokine and matrix metalloproteinase induction, and heat shock protein 70 expression. Graefes Arch Clin Exp Ophthalmol 249:1185–1194

Chandrasekhar I, Gaber BP (1988) Stabilization of the biomembrane by small molecules: interaction of trehalose with the phospholipid bilayer. J Biomol Struct Dyn 5:1163–1171

Cottone G, Cicotti G, Cordone L (2002) Protein-trehalose-water structures in trehalose coated carboxy-myoglobn. J Cell Phys 117:9862–9866

Couzin J (2004) Huntington's disease. Unorthodox clinical trials meld science and care. Science 304:816–817

Crowe JH (1971) Anhydrobiosis: an unsolved problem. Am Nat 105:563–574

Crowe JH (2008) Trehalose as a "chemical chaperone": fact and fantasy. Adv Exp Med Biol 594:143–158

Crowe LM, Crowe JH (1991) Solution effects on the thermotropic phase transition of unilamellar liposomes. Biochim et Biophys Acta – Biomembr 1064:267–274

Crowe JH, Crowe LM, Chapman D (1984) Preservation of membranes in anhydrobiotic organisms: the role of trehalose. Science 1984(223):701–703

Crowe JH, Spargo BJ, Crowe LM (1987) Preservation of dry liposomes does not require retention of residual water. Proc Natl Acad Sci U S A 84:1537–1540

Crowe JH, Hoekstra FA, Crowe LM (1992) Anhydrobiosis. Annu Rev Physiol 54:579–599

Crowe JH, Carpenter JF, Crowe LM (1998) The role of vitrification in anhydrobiosis. Annu Rev Physiol 6:73–103

Dupont S, Rapoport A, Gervais P, Beney L (2014) Survival kit of *Saccharomyces cerevisiae* for anhydrobiosis. Appl Microbiol Biotechnol 98:8821–8834

Eleutherio ECA, Araujo PS, Panek AD (1993) Role of the trehalose carrier in dehydration resistance of Saccharomyces cerevisiae. Biochim Biophys Acta 1156:263–266

Emanuele E (2014) Can trehalose prevent neurodegeneration? Insights from experimental studies. Curr Drug Targets 15:551–557

Erkut C, Penkov S, Khesbak H, Vorkel D, Verbavatz JM, Fahmy K, Kurzchalia TV (2011) Trehalose renders the dauer larva of Caenorhabditis elegans resistant to extreme desiccation. Curr Biol 21:1331–1336

Erkut C, Vasilj A, Boland S, Habermann B, Shevchenko A, Kurzchalia TV (2013) Molecular strategies of the *Caenorhabditis elegans* dauer larva to survive extreme desiccation. PLoS One 8:e82473

Förster F, Beisser D, Grohme MA, Liang C, Mali B, Sieg AM, Engelmann JC, Shkumatov AV, Schokraie E (2012) Transcriptome analysis in tardigrade species reveals specific molecular pathways for stress adaptations. Bioinf Biol Insights 2012:69–95

Gadducci A, Fanucchi A, Cosio S, Genazzani AR (1997) Hormone replacement therapy and gynecological cancer. Anticancer Res 17:3793–3798

Gaff DF, Oliver M (2013) The evolution of desiccation tolerance in angiosperm plants: a rare yet common phenomenon. Funct Plant Biol 40:315–328

Golovina EA, Golovin A, Hoekstra FA, Faller R (2010) Water replacement hypothesis in atomic details: effect of trehalose on the structure of single dehydrated POPC bilayers. Langmuir 26:11118–11126

Hand SC, Menze MA (2015) Molecular approaches for improving desiccation tolerance: insights from the brine shrimp *Artemia franciscana. Planta* (in press)

Hand SC, Menze MA, Toner M, Boswell L, Moore D (2011) LEA proteins during water stress: not just for plants anymore. Annu Rev Physiol 447:115–134

Hays LM, Crowe JH, Wolkers W (2001) Factors affecting leakage of trapped solutes from phospholipid vesicles during thermotropic phase transitions. Cryobiology 42:88–102

Hengherr S, Worland MR, Reuner A, Brümmer F, Schill RO (2009) High-temperature tolerance in anhydrobiotic tardigrades is limited by glass transition. Physiol Biochem Zool 82:749–755

Hincha DK, Thalhammer A (2012) LEA proteins: IDPs with versatile functions in cellular dehydration tolerance. Biochem Soc Trans 40:1000–1003

Hoekstra FA, Golovina EA, Buitink J (2001) Mechanisms of plant desiccation tolerance. Trends Plant Sci 6:431–438

Hovakimyan M, Ramoth T, Löbler M, Schmitz K, Witt M, Guthoff R, Stachs O (2012) Evaluation of protective effects of trehalose on desiccation of epithelial cells in three dimensional reconstructed human corneal epithelium. Curr Eye Res 37(982–989):2012

Izawa YT, Matsuo T, Uchida T (2006) Atomic force microscopic observation of trehalose-treated and dried corneal epithelial Surface. Cell Preserv Tech 4:117–122

Keilin D (1959) The problem of anabiosis or latent life: history and current concept. Proc Roy Soc B 150:149–191

Kent B, Hunt T, Darwish TA, Hauß T, Garvey CJ, Bryant G (2014) Localization of trehalose in partially hydrated DOPC bilayers: insights into cryoprotective echanisms. J R Soc Interface 11:20140069

Koster KL (2001) Effects of sugars on phospholipid phase transitions: relevance to dehydration tolerance. Cryobiol Cryotechnol 47:26–32

Koster KL, Kami AE, Maddocks J, Bryant G (2003) Exclusion of maltodextrins from phosphatidylcholine multilayers during dehydration: effects on membrane phase behavior. Eur Biophys J 32:96–105

Krügera UY, Wanga Y, Kumara S, Mandelkowa E (2012) Autophagic degradation of tau in primary neurons and its enhancement by trehalose. Neurobiol Aging 33:2291–2305

Lan D, Liu F, Zhao J, Chen Y, Wu J, Ding Z, Yue Z, Ren H, Jiang Y, Wang J (2012) Effect of trehalose on PC12 cells overexpressing wild-type or A53T mutant a-synuclein. Neurochem Res 37:2025–2032

Lee CWB, Waugh JS, Griffin RG (1986) Solid-state NMR study of trehalose/1,2-dipalmitoyl-sn-phosphatidylcholine interactions. Biochemistry 25:3737–3742

Leidy C, Gousset K, Ricker JV, Crowe JH (2004) Lipid phase behavior and stabilization of domains in membranes of platelets. Cell Biochem Biophys 40:123–135

Lenné T, Garvey CJ, Koster KL, Bryant G (2009) Effects of sugars on lipid bilayers during dehydration – SAXS/WAXS measurements and quantitative model. J Phys Chem B 113:2486–2491

Leprince O, Buitink J (2010) Desiccation tolerance: from genomics to the field. Plant Sci 179:554–564

Li S, Chakraborty N, Borcara A, Menze MA, Toner M, Hand SC (2012) Late embryogenesis abundant proteins protect human hepatoma cells during acute desiccation. Proc Natl Acad Sci U S A 109:20859–20864

Lins RD, Pereira CS, Hunenberger PH (2004) Trehalose-protein interactions in aqueous solutions. Proteins 55:177–186
Liu R, Barkhordarian H, Emadi S, Park CB, Sierks MR (2005) Trehalose differentially inhibits aggregation and neurotoxicity of beta-amyloid 40 and 42. Neurobiol Dis 20:74–81
Luyckx J, Baudouin C (2011) Trehalose: an intriguing disaccharide with potential for medical application in ophthalmology. Clin Ophthamol 5:577–581
Luzardo MC, Amalfa F, Nuñez AM, Díaz S, Biondi AC, Disalvo EA (2000) Effect of trehalose and sucrose on the hydration and dipole potential of lipid bilayers. Biophys J 78:2452–2458
Ma X, Jamil K, Macrae TH, Clegg JS, Russell JM, Villeneuve TS, Euloth M, Sun Y, Crowe JH, Tablin F, Oliver AE (2005) A small stress protein acts synergistically with trehalose to confer desiccation tolerance on mammalian cells. Cryobiology 51:15–28
Matsuo T (2004) Trehalose versus hyaluronan or cellulose in eye drops for the treatment of dry eye. Jpn J Ophthalmol 48:321–327
Mitsumasu K, Kanamori Y, Fujita M, Iwata K, Tanaka D, Kikuta S, Watanabe M, Cornette R, Okuda T, Kikawada T (2010) Enzymatic control of anhydrobiosis related accumulation of trehalose in the sleeping chironomid, Polypedilum vanderplanki. FEBS J 277:4215–4228
Moiset G, López CA, Bartelds R, Syga L, Rijpkema E, Cukkemane A, Baldus M, Poolman B, Marrink SJ (2014) Disachrides impact the lateral organization of lipid membranes. J Am Chem Soc 136:16167–16175
Nakagaki M, Nagase H, Ueda H (1992) Stabilization of the lamellar structure of phosphatidylcholine by complex-formation with trehalose. J Membr Sci 73:173–180
Ohtake S, Wang J (2011) Trehalose: current use and future applications. J Pharm Sci 100:2020–2053
Ricker JV, Tsvetkova NM, Wolkers WF, Crowe JH (2003) Trehalose maintains phase separation in an air-dried binary lipid mixture. Biophys J 84:3045–3051
Rudolph BR, Chandrasekhar I, Gaber BP (1990) Molecular modeling of saccharide-lipid interactions. Chem Phys Lipids 53:243–261
Schebor C, Burin L, del Pilar Bueras M (1999) Stability to hydrolysis and browning of trehalose, sucrose and raffinose in low-moisture systems in relation to their use as protectants of dry biomaterials. Food Sci Technol 32:481–485
Sum AK, Faller R, de Pablo JJ (2003) Molecular simulation of phospholipid bilayers and insights of the interactions with disaccharides. Biophys J 2003(85):2830–2844
Takeuchi K, Nakazawa M, Ebina Y, Sato K, Metoki T, Miyagawa Y, Ito T (2010) Inhibitory effects of trehalose on fibroblast proliferation and implications for ocular surgery. Exp Eye Res 91:567–577
Tanaka M, Machida Y, Niu S (2004) Trehalose alleviates polyglutamine-mediated pathology in a mouse model of Huntington disease. Nat Med 2004(10):148–154
Tapia H, Koshland DE (2014) Trehalose is a versatile and long-lived chaperone for desiccation tolerance. Curr Biol 24:2758–2766
Terrasson E, Buitink J, Righetti K, Vu BY, Pelletier S, Zinsmeister J, Lalanne D, Leprince O (2013) An emerging picture of the seed desiccome: confirmed regulators and newcomers identified using transcriptome comparison. Front Plant Sci 4:497
Tolleter D, Hincha DK, Macherel D (2010) A mitochondrial late embryogenesis abundant protein stabilizes model membranes in the dry state. Biochim Biophys Acta-Biomembr 1798:1926–1933
Tolleter D, Jaquinode M, Mangavel A, Passirani C, Saulnier P, Manon S, Teyssier E, Payet N, Avelange-Macherel MH, Macherel D (2007) Structure and function of a mitochondrial late embryogenesis abundant protein are revealed by desiccation. Plant Cell 19:1580–1589
Tomczak MM, Hincha DK, Estrada SD, Wolkers WF, Crowe LM, Feeney RE, Tablin F, Crowe JH (2002) A mechanism for stabilization of membranes at low temperatures by an antifreeze protein. Biophys J 82:874–881
Torok Z, Tsvetkova NM, Gabor B, Horváth I, Nagy E, Pénzes Z, Hargitai J, Bensaude O, Csermely P, Crowe JH, Maresca B, Vígh L (2003) Heat shock protein co-inducers specifically modulate the membrane lipid phase. Proc Natl Acad Sci U S A 100:3131–3136

Tsvetkova NM, Phillips BL, Crowe LM, Crowe JH (1998) Effect of sugars on headgroup mobility in freeze-dried dipalmitoylphosphatidylcholine bilayers: solid-state P-31 NMR and FTIR studies. Biophys J 75:2947–2955

Tunnacliffe A, Wise MJ (2007) The continuing conundrum of the LEA proteins. Naturwissenschaften 94:791–812

Tunnacliffe A, Lapinski J, McGee B (2005) A putative LEA protein, but no trehalose, is present in anhydrobiotic bdelloid rotifers. Hydrobiologia 546:315–321

Tunnacliffe A, Hincha DK, Leprince O, Macherel D (2010) LEA proteins: versatility of form and function. In: Lubzens E, Cerda J, Clark M (eds) Sleeping beauties: dormancy and resistance in harsh environments. Springer, Berlin, pp 91–108

Vessey MP (1984) Exogenous hormones in the aetiology of cancer in women. J R Soc Med 77:542–549

Viera LI, Alonso-Romanowski S, Borovyagin V, Feliz MR, Disalvo EA (1993) Properties of gel phase lipid-trehalose bilayers upon rehydration. Biochim Biophys Acta 1145:157–167

Villarreal MA, Díaz SB, Disalvo EA, Montich GG (2004) Molecular dynamics simulation study of the interaction of trehalose with lipid membranes. Langmuir 20:7844–7851

Watanabe M, Kikawada T, Minagawa N, Yukuhiro F, Okuda T (2002) Mechanism allowing an insect to survive complete dehydration and extreme temperatures. J Exp Biol 205:2799–2802

Wharton DA (2014) Anhydrobiosis: the model worm as a model. Curr Biol 21:R578

Wolkers WF, Oldenhof H, Alberda M, Hoekstra FA (1998) A fourier transform infrared study of sugar glasses: application to anhydrobiotic higher plant cells. Biochim Biophys Acta 1379:83–96

Wolkers WF, Walker NJ, Tablin F, Crowe JH (2001) Human platelets loaded with trehalose survive freeze-drying. Cryobiology 42:79–87

Zhu S, Jamil K, Ma X, Crowe JH, Oliver AE (2006) Protection of CANARY cells after drying and rehydration correlates with decrease in apoptotic cell death. Cell Preserv Technol 4:67–77

Concluding Remarks

When water is absorbed in solid, dehydrated phospholipids, it will act as a spacer between lipid molecules. Larger lipid areas require more water. However, the area increase is not linear suggesting different types of water adsorption forming regions of different density and energy of interaction between water molecules themselves and with the lipid chemical groups.

In this book, water has been identified in several states: hydration water, confined water, water exchanged during membrane-membrane and membrane-perturbant interactions, water replaced or displaced by solutes in drying (dessicomes), water as informational units (aquaomics).

As a result of the gradual ingress of the water, the mobility and deformability of the membrane components strongly enhances, which gives to water a role of "external plasticizer".

A fraction of water tightly associates with lipid head groups and others interact with many of the membrane's other functional groups occupying free volume (hydration water). At this point the second hydration shell becomes important.

Presumably, at moderate surface pressure compatible with that in cell membranes, a water population appears, which may be identified as second shell or confined water beyond the hydration of the head groups of the phospholipids. The water organization at this level may be a link between stress states governed by lateral pressure in the membrane as a mechanochemical device and by osmosis that "reacts" to external perturbants, such as amino acid motifs in proteins, peptides or aminoacid residues.

A few number of water molecules, located beyond the hydration shell, is confined in hydrophilic and hydrophobic environments. This number of water molecules associated with each head group is correlated to the packing of acyl chains in the hydrophobic core.

The activity of water in this restricted membrane-water interface modulates the surface pressure and viceversa, surface pressure may modify water activity in these sites for amino acid residues insertion.

E.A. Disalvo (ed.), *Membrane Hydration*, Subcellular Biochemistry 71,
DOI 10.1007/978-3-319-19060-0

In addition, they may provide specific domains for catalyzing organic reactions and to control metabolic pathways in cells.

Complex lipid and sterol compositions, the inclusion of proteins, glycosylation, as well as the formation of lipid domains all result in inhomogeneous local properties of the bilayer in which water molecules can stabilize. The lipid composition can determine a pattern in the interlamellar water structure not only in the bound water population but also farther away from the membrane. Thus, the link between structure and function appears at the water interphase in which changes in the excess free energy in the water arrangements trigger different processes. In this regard, lipidomics meet aquaomics.

Water as antenna for biological signals in lipids may control activities of membrane-bound enzymes through manipulation of the thermodynamic activity of water in the lipid-water interfacial region.

The replacement of water by compounds mimicking water structure in terms of H bonding stereochemistry is not enough. The dynamic properties of water as a liquid are necessary for biological response and function. In this regard, water is far more than the cement to attach membrane structure. It is a component essential for biological activity and hence the understanding of its behavior could be the clue to understand anomalous behavior in human health, plant resistance and preservation of cells and tissues.

Index

E.A. Disalvo (ed.), *Membrane Hydration*, Subcellular Biochemistry 71,
DOI 10.1007/978-3-319-19060-0

MIX
Papier aus verantwortungsvollen Quellen
Paper from responsible sources
FSC® C105338

If you have any concerns about our products,
you can contact us on
ProductSafety@springernature.com

In case Publisher is established outside the EU,
the EU authorized representative is:
Springer Nature Customer Service Center GmbH
Europaplatz 3, 69115 Heidelberg, Germany

Printed by Libri Plureos GmbH
in Hamburg, Germany